Chemical Reactor Design

Chemical Reactor Design

Mathematical Modeling and Applications

Juan A. Conesa

WILEY-VCH

Author

Prof. Juan A. Conesa
Universidad de Alicante
Ingenieria Quimica
Carr. San Vicente del Raspeig
s/n
03690 Alicante
Spain

Cover
© zorazhuang/Getty Images

Library of Congress Card No.:
applied for

British Library Cataloguing-in-Publication Data
A catalogue record for this book is available from the British Library.

Bibliographic information published by the Deutsche Nationalbibliothek
The Deutsche Nationalbibliothek lists this publication in the Deutsche Nationalbibliografie; detailed bibliographic data are available on the Internet at <http://dnb.d-nb.de>.

Print ISBN: 978-3-527-34630-1
ePDF ISBN: 978-3-527-82339-0
ePub ISBN: 978-3-527-82338-3
oBook ISBN: 978-3-527-82337-6

Typesetting SPi Global, Chennai, India
Printing and Binding Markono Print Media Pte Ltd, Singapore

Printed on acid-free paper

10 9 8 7 6 5 4 3 2 1

In loving memory of my parents, Paco y Rosita.

Contents

Preface

The main objective of this book is to introduce the reader to the analysis of complex chemical reactors. In them, reactions with complex kinetics will be carried out and/or they will work in unusual situations, such as unsteady state operation. The mathematical tools necessary for the characterization of flow and kinetic models that can solve complex problems in the design of reactors are given in the book. Reactors operating in a transitory regime are also described, and their design equations are analyzed. Special attention is paid to the description and design of catalytic reaction systems, in which the presence of two or more phases makes their analysis and design complex.

Throughout the book there are many examples of application of the concepts and equations studied, to strengthen the content treated. In the examples, complex reactors are solved in terms of the characterization of their flow, and other situations that are not usually dealt with in textbooks.

For the resolution of many examples, programming in Matlab® (or its equivalent in freeware, GNU Octave) is used. The necessary level of knowledge of this program is very low. The use of big programs that link two or more Matlab scripts is avoided, and simple programs are offered as alternatives, which, although more rudimentary, are capable of performing complex calculations.

The book covers some aspects that are not treated in any textbook dedicated to the reactor design. There are some chapters with material similar to other texts but in many facets the work presents aspects that are not treated at all in any other book, as is the use of numerical methods for solving engineering problems (unsteady state regime included), the use of transfer functions to study residence time distributions, the convolution and deconvolution curves for reactor characterization, forced-unsteady-state-operation, scale-up of chemical reactors, design of multiphasic reactors, and biochemical reactors design (not only Michaelis–Menten nor Monod kinetics). Other aspects considered, mainly in part two, are the design of multiphase gas–liquid–solid reactors, including bubble reactors, agitated and trickle flow reactors.

A special emphasis is done to the numerical solution of differential equations using the finite differences approximation. I know there are available more complex tools for solving such situations, but in my opinion, it is important that students have in mind this simple system for solving PDE. This would give a background for understanding other more complex methods.

At the University of Alicante, the text is the basis for a course of 4.5 credit points in the last year of ChemEng MS. The course is complemented with the study of electrochemical, photochemical, and sonochemical reactors.

Alicante, Spain
May 2019

Juan A. Conesa

Nomenclature

Suggested units are indicated for each variable.

N	Amount of substance (mol)
Q	Volumetric flow rate (m^3/s)
C	Concentration (usually mol/m^3, or kg/m^3)
E	(in the RTD context) → residence time distribution function, RTD (–)
F	Integral form of the residence time distribution function (–)
t_m	First moment of the RTD; mean value of time (s)
$\bar{t}$	Average residence time ($=V/Q$) (s)
V	Volume (m^3)
σ	Standard deviation (s)
s	(in the RTD context) → skewness (s)
s	(in the LT context) → main variable in the Laplace space
δ	Dirac delta function (–)
n_t	Number of tanks in the tanks-in-series model (–)
k	Kinetic constant (units depend on the kinetic law)
D_e	Effective diffusion coefficient (m^2/s)
n_T	Total molar flow (mol/s)
S	Section (m^2)
C_T	Total concentration (mol/m^3)
u	Linear velocity (m/s)
L	Characteristic length (m)
M	Tracer mass (kg or kmol)
V_p	Volume with piston flow regime (m^3)
V_d	Dead volume (m^3)
X_A	Molar conversion of reactant A (–)
r_A	Reaction (or process) rate based on the external catalyst surface ($mol/(s\,m^2)$)
r'_A	Reaction (or process) rate based on the weight of catalyst ($mol/(s\,kg)$)
r''_A	Reaction (or process) rate based on the volume of reacting species ($mol/(s\,m^3)$)
r'''_A	Reaction (or process) rate based on the volume of catalyst particles ($mol/(s\,m^3)$)
C_{As}	Concentration of A in the surface (mol/m^3)
D_{AB}	Diffusion coefficient (diffusivity) of A in B (m^2/s)

q_s	Heat flow in the surface (J/s)
h	Heat transfer coefficient (J/K s)
T	Temperature (K)
k_L	Mass transfer coefficient in the liquid phase $(\text{mol}/(\text{s m}^2)/(\text{mol/m}^3) = \text{m/s})$
Q_c	Flow rate in the cooling zone (m^3/s)
ΔH_r	Enthalpy of a reaction (kJ/mol)
U	Global heat transfer coefficient (J/K s)
E	(in the reaction rates context) → activation energy (J/mol)
$H(s)$	Transfer function (–) (Laplace space)
$X(s)$	Stimulus function (–) (Laplace space)
$Y(s)$	Response to stimulus function (–)(Laplace space)
$h(t)$	Time-dependent transfer function (–)
$y(t)$	Time-dependent response to stimulus function (–)
$x(t)$	Time-dependent stimulus function (–)
$L\{h(t)\}$	Laplace transform of function h(t)
α	Fraction of flow rate (–)
β	Fraction of volume (–)
Cp	Concentration of product "P" (mol/m^3)
c_p	Calorific capacity of the reacting flow (J/K g)
c_{Pc}	Calorific capacity of the cooling fluid (J/K g)
N_c	Dimensionless cooling capacity (–)
N_{ad}	Dimensionless adiabatic heat increment (–)
$\Delta\upsilon$	Dimensionless temperature difference (–)
k_h	Kinetic constant at the temperature of the cooling medium (units depend on the reaction order)
k_0	Pre-exponential factor of the kinetic constant (units depend on the reaction order)
Φ	Thiele modulus (–)
We	Weisz modulus (–)
m	kinetics to diffusion ratio in catalytic reactions (1/m)
K_M	Michaelis constant (g/l)
K_S	Monod constant (g/l)
W	Weight of catalyst (g)
μ	Growth rate per unit of cell (1/s)
ϕ	Tortuosity factor (–)
η	Effectiveness (–)
η_e	Effectiveness (related to external diffusion) (–)
J_0	Bessel function of zero order (–)
J_1	Bessel function of first order (–)
α	Fraction of volume of a subsystem (–)
β	Fraction of flow passing through a subsystem (–)
q	Heat flux (J/s)
q_C	Heat flux in cooling media (J/s)
u	Linear velocity (m/s)
T_C	Temperature of the cooling media (K)
$A_{m\times n}$	Matrix of convolution

g_A	Generation term of a mole balance (mol/s)
n_A	Molar flow of component "A" (mol/s or mol/(s m^2))

Subscripts:

i	Actual position
$i+1$	Position of the following interval
$i-1$	Position of the preceding interval
S	Surface
0	Inlet conditions
mm	Maximum of the maxima curve (critical point)

Superscripts

t	Actual time
$t+1$	Time of the following interval
$t-1$	Time of the previous interval

Part I

Reactor Analysis, Design, and Scale-up

1 Nonideal Flow

1.1 Introduction

Basic chemical reactors (plug flow reactor or PFR, and continuously stirred tank reactor or CSTR) are studied considering their behavior is that of an ideal reactor. Unfortunately, in practice, we often find behaviors that are far from that considered ideal. Consequently, working with them, the chemical engineer must be able to handle and diagnose the behavior of these reactors. At the time of describing the nonideal behavior of a reactor, three concepts are introduced: the residence time distributions (RTDs), the quality of the mixture (not discussed in this book), and the models that can be used to describe the reactor. These three concepts are used to describe the deviations of the mixing assumed in the ideal models and are considered as attributes of the mixture in nonideal reactors.

One way of approaching the study of nonideal reactors is to consider them, in a first approximation, as if the flow model were the one corresponding to a CSTR or a PFR. However, in real reactors, the nonideal flow model implies a minor conversion, so a method that allows for this conversion loss to be considered must be available. Therefore, a higher level of approximation implies the use of information about the RTD.

1.2 Residence Time Distribution (RTD) Function

The idea of introducing the RTD in the analysis of the behavior of reactors occurred thanks to MacMullin and Weber (in 1935), although it was Danckwerts (later, in 1953) who structured this analysis and defined most of the distributions of interest.

In an ideal PFR, all the particles (or units) of material that leave the reactor have remained in it the same time. Analogously, in an ideal (well-mixed) batch reactor, all particles are in the reactor the same period of time. The time that these units have remained in the reactor is what we call the residence time of those particles in that reactor.

The ideal reactor's PFR and batch are the only ones in which all the portions of reactants present in the reactor have the same residence time. In all other reactors, the particles entering the reactor vessel remain inside the reactor for different periods of time; that is, there exists a RTD inside the reactor.

Chemical Reactor Design: Mathematical Modeling and Applications,
First Edition. Juan A. Conesa.

For example, consider an ideal CSTR; the input flow that is introduced to the reactor at a given moment mixes instantaneously and completely with the rest of the material that already exists inside the reactor. In this way, some of the particles that enter the reactor abandon this one almost immediately with the exit current, whereas other atoms remain of almost indefinite form, since all the material is never dragged. Of course, many of the particles leave the reactor after a period close to the average residence time.

The RTD of a reactor is a feature of the mixture that is taking place inside the reactor. Thus, in an ideal PFR there is no axial mixing, and this absence is reflected in the RTD that this type of reactors exhibit. In contrast, in an ideal CSTR there is a great degree of mixing, so the RTD that these reactors exhibit is very different from that of the plug flow. However, not all RTDs are unique to one type of reactor; reactors with marked differences can give identical RTDs. Despite this, the RTD of a certain reactor presents distinctive keys with respect to the type of mixture that is taking place inside it and is one of the ways to characterize the reactor that provides more information.

1.2.1 Measurement of the RTD

The experimental measurement of the RTD is done by injecting a tracer into the reactor at a definite time ($t = 0$). This tracer is a chemical, molecule, or an inert atom. The tracer concentration is then measured at the outlet at different times. The tracer must be inert, easily detectable, with physical properties similar to those of the substances present in the reaction mixture and easily soluble in it. In addition, it should not be adsorbed on walls or other surfaces on the reactor. The objective is to reflect, as best as possible, the behavior of the substances that are flowing through the reactor. The most commonly used tracers are dyes and radioactive material, while the most commonly used injection methods are pulse input ("Dirac delta function") and step input (suddenly increase of the tracer concentration).

1.2.1.1 Pulse Input

This kind of experiment consists of introducing into the current entering the reactor, quickly and at once, an amount N_0 of tracer. The output concentration is subsequently measured as a function of time. Concentration-time curve characteristics for the input and exit of an arbitrary reactor can be observed in Figure 1.1. The concentration-time curve corresponding to the effluent is called curve C in RTD analysis.

Let us consider a system with a single input and a single output, in which a tracer is injected in pulse and in which it is transported, exclusively because of the flow (not because of dispersion), through the system. If a time increment Δt small enough is chosen so that the concentration of tracer, $C(t)$, which leaves the system between the instants "t" and "$t + \Delta t$" is constant, we can express the amount of tracer (in moles or grams), ΔM, that leaves the reactor between "t" and "$t + \Delta t$" as

$$\Delta M = C(t) \cdot Q \cdot \Delta t \tag{1.1}$$

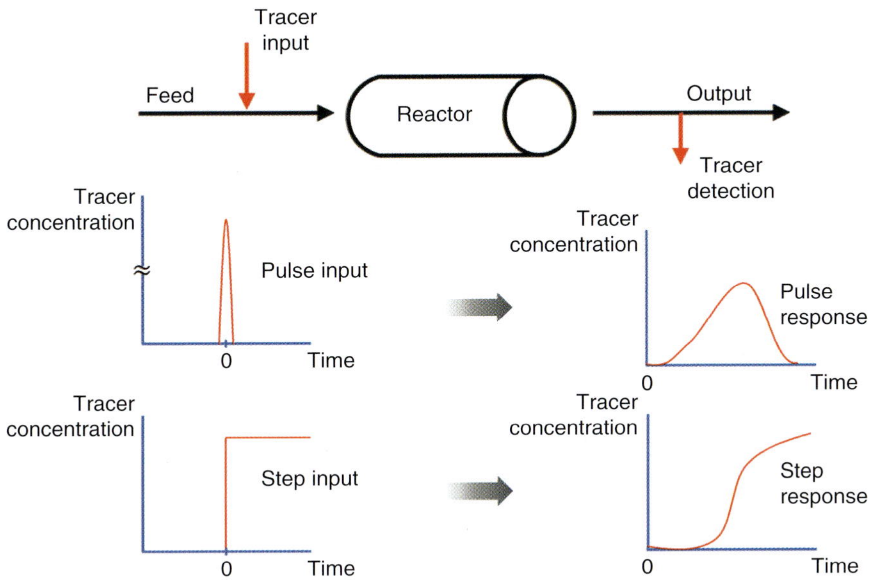

Figure 1.1 Measurement of the RTD.

That is, ΔM is the amount of tracer (in moles or grams, for example) that has remained in the reactor for a time interval comprising between "t" and "$t + \Delta t$." Dividing by the total amount of tracer injected into the reactor (M_0), we obtain the tracer fraction whose residence time in the reactor is between "t" and "$t + \Delta t$":

$$\frac{\Delta M}{M_0} = \frac{C(t) \cdot Q}{M_0} \cdot \Delta t \tag{1.2}$$

For a pulse injection, we define the RTD function, $E(t)$, as

$$E(t) = \frac{Q \cdot C(t)}{M_0} \tag{1.3}$$

This expression describes in a quantitative way how long the different elements of fluid have passed inside the reactor. In consequence:

$$\frac{\Delta M}{M_0} = E(t) \cdot \Delta t \tag{1.4}$$

If M_0 is not directly known, it may be obtained from the output concentrations, adding the different amounts of the tracer that have exited the reactor between 0 and infinity. Expressing in differential form:

$$dM = Q \cdot C(t) \cdot dt$$

and integrating, we obtain

$$M_0 = \int_0^{\infty} Q \cdot C(t) \cdot dt \tag{1.5}$$

Since the volumetric flow (Q) is generally constant, we will define $E(t)$ as

$$E(t) = \frac{C(t)}{\int_0^{\infty} C(t) \cdot dt} \tag{1.6}$$

In that expression, the integral in the denominator is the area under the curve, $C(t)$. In this way, from the tracer concentration, C(t), it is possible to find the curve $E(t)$, as long as that curve is obtained from a perfect pulse of the input tracer.

Another way to interpret the function of residence times is in its integral form:

$$\begin{bmatrix}\text{Amount of tracer exiting the reactor after}\\ \text{passing inside it a time between } t_1 \text{ and } t_2\end{bmatrix} = \int_{t_1}^{t_2} E(t)\cdot dt \tag{1.7}$$

Now, the fraction of tracer that passed inside the reactor a time t, between 0 and infinity is equal to 1; therefore:

$$\int_0^{\infty} E(t)\cdot dt = 1 \tag{1.8}$$

The main drawback of the use of the pulse technique lies in the difficulty in achieving a tracer input to the reactor that is reasonably pulsed (as we explain in Chapter 2, deconvolution of curves). The injection should take place in a very short period compared to the residence times and the tracer dispersion between the injection point and the reactor inlet should be negligible. If these conditions are satisfied, the technique is a simple and direct way to obtain the RTD.

1.2.1.2 Step Input

During a step input experiment, a tracer is added at a steady rate to the reactor feed to give a steady input concentration of C_0 (see Figure 1.1). The concentration of the tracer in the effluent is then monitored from the time of adding the tracer until it reaches a concentration approximating that of C_0.

As stated before, the RTD curve can be easily obtained by injecting a tracer in a pulse input, but now we will formulate a more general relationship between a tracer injection (not necessarily a pulse input) and the corresponding tracer concentration in the effluent (current leaving the system).

Chapter 2 presents a more general equation establishing the relation of the concentration of a tracer leaving a reactor (C_{out}) and the input concentration (C_{in}). This is presented as the convolution integral:

$$C_{out}(t) = \int_0^t C_{in}(t-t')E(t')dt' \tag{1.9}$$

In this equation, $E(t)$ is the corresponding RTD for the reactor. The concentration at the input is evaluated at different times and the resulting signal C_{out} is a convolution of C_{in} and the RTD in the vessel.

In this way, in the following chapters we are able to calculate the $E(t)$ curve given any function of concentration input C_{in} and measuring C_{out}. For now, we continue with more simple techniques, considering that the inlet concentration is introduced in the form of a step input (Figure 1.1).

Let us consider a system of constant volumetric flow (Q) in which a tracer input is introduced in the form of a step, with a rate of addition of tracer to the constant input flow and which begins to occur at time $t = 0$. Before that moment, it is considered that no tracer enters with the inlet current. Under these conditions:

$$C_{in}(t) = 0 \quad \text{for } t < 0$$

$$C_{in}(t) = \text{constant} = C_0 \quad \text{for } t \geq 0$$

The tracer concentration in the feed should be maintained at this value until the concentration of tracer in the effluent is practically C_0, i.e. equal to that of the feed, at which time the test can be interrupted. Figure 1.1 shows a typical output concentration curve for this type of input.

Since the input concentration (C_0) remains constant over time, we can extract it from the integral using Eq. (1.9):

$$C_{\text{out}}(t) = C_0 \int_0^t E(t')dt' \tag{1.10}$$

Obviating that t' is mathematically equal to t, we see that the signal at the output of this experiment is a cumulative function of $E(t)$, as it evaluates the integral of all $E(t)$ from $t = 0$ to the instant "t." This cumulative distribution is called the "F curve" and can be directly determined from a step input.

$$\left[\frac{C(t)}{C_0}\right]_{\text{step input}} = \int_0^t E(\text{t})dt = F(t) \tag{1.11}$$

Differentiating this expression, we obtain the function of RTD, $E(t)$:

$$E(t) = \frac{d}{dt}\left[\frac{C(t)}{C_0}\right]_{\text{step input}}$$

The determination of the $E(t)$ using a step input is, in general, easier to carry out experimentally than the pulse input. It has some other advantages, as in that it is not necessary to know the total amount of tracer introduced during the test period. In return, it has some disadvantages: (i) sometimes, it is not easy to maintain a constant concentration of tracer in the feed; (ii) for calculating the $E(t)$ curve, this procedure implies the differentiation of the data, which can lead, sometimes, to errors; (iii) as the feed should be maintained for a long period of time, the amount of tracer required is usually very large, so if the tracer is expensive, a pulse input is usually used.

Other techniques to introduce the tracer are possible: negative step (dilution of the food), a periodic signal, or a random signal. However, they tend to be more difficult to carry out and are discussed in the following chapters.

1.2.2 RTD Concept in Heterogeneous Systems

Just as in the homogeneous reactors, the concept of RTD is valid in heterogeneous systems, with the particularity that there will exist an RTD related to the flow of each phase involved in the reaction. For example, let us think about the flow produced in a catalytic reactor where a gaseous species is introduced on one side, and, simultaneously, the solid catalyst is introduced on the other. Figure 1.2 presents such a situation.

The experimental techniques to measure these RTD functions are equivalent to the ones used in one-phase reactors. The application of such measurements is extremely important in the design of reactors. For example, let us think of a catalyzed fluidized bed, where gas is flowing at its bottom and circulating up. The flow of the solid is very similar to that of the ideal CSTR, as the fluidized bed is supposed to be perfectly mixed due to the fact that the mass transfer is

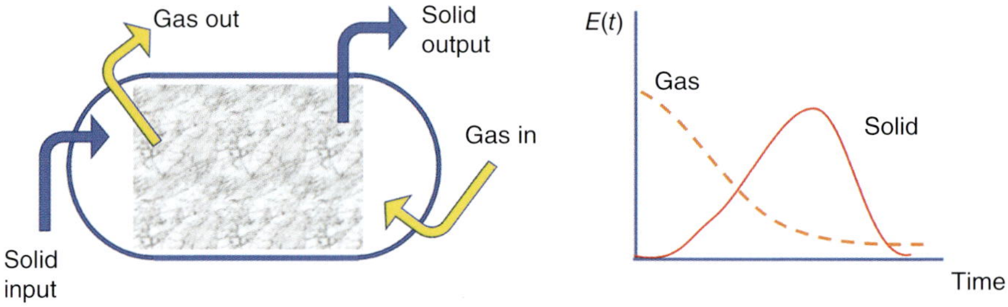

Figure 1.2 Two $E(t)$ functions are needed to characterize a reactor with two moving phases.

extremely fast. Nevertheless, the gas will flow through this bed, and will present a specific and complicated flow pattern, as we know. This pattern will depend on different factors and usually is between the plug flow and the CSTR behavior. The measurement of the precise RTD of both phases will give important information for the design and scale-up of the reactors.

1.2.3 Characteristics of RTD

Going back to the use and measurement of RTD, it is important to know some characteristics. If we consider the age of a reacting particle as the time it has remained under the conditions in which the reaction is carried out, then $E(t)$ is related to the age distribution of the effluent. In fact, the $E(t)$ is also called function of distribution of exit ages.

The fraction of the output current that has remained in the reactor for a period less than a given value, t, equals the sum over all times of the function $E(t)\cdot\Delta t$, or expressed continuously:

$$\int_0^t E(t) = \begin{bmatrix}\text{Fraction of output current that has remained} \\ \text{in the reactor a time less than "}t\text{"}\end{bmatrix} = F(t)$$

which represents the cumulative RTD function, $F(t)$. Analogously, the integral of the $E(t)$ function from time "t" to infinity will represent the fraction of the output current that has remained in the reactor a time higher than "t."

This function $F(t)$ (cumulative distribution function) can be calculated at various "t" values using the area of the curve of the representation of $E(t)$ vs. time. An example of curve $F(t)$ for a tracer step input can be seen in Figure 1.3, where

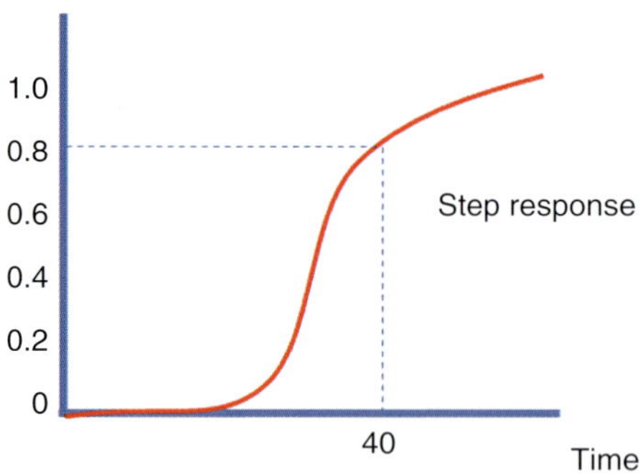

Figure 1.3 Interpretation of the cumulative distribution curve, $F(t)$.

it can be affirmed that 80% of the input tracer spends less than 40 units of time in the reactor.

1.2.3.1 Mean Residence Time

Variables described by distribution functions are characterized by their moments. Correspondingly, function $E(t)$ can also be characterized by its moments. In this sense, the mean value of the variable is equal to the first moment of the RTD function, $E(t)$. So, the mean residence time is defined as

$$t_m = \frac{\int_0^\infty t \cdot E(t)dt}{\int_0^\infty E(t)dt} = \int_0^\infty t \cdot E(t)dt \tag{1.12}$$

When studying the ideal reactors, the spatial time (τ = mass of catalyst/Q) and the average residence time ($\bar{t} = V/Q$) are usually used.

We can demonstrate that, regardless of the RTD that exists in a given reactor, ideal or not ideal, the nominal average residence time, $\bar{t} = V/Q$ is equal to the mean residence time of the RTD distribution, t_m, if the system is of constant density (where Q is constant). For doing this, let us consider a reactor completely filled with a colored fluid (for example, red bromine gas); at an instant $t = 0$, we start to inject yellow gas to replace the bromine filling the reactor. That is, the reactor volume V is equal to the volume occupied by the bromine leaving the reactor. In such a situation, during a certain time dt, the volume of bromine that will leave the reactor is $Q{\cdot}dt$, "Q" being the volumetric flow rate (considered at constant temperature and pressure). In that case, $[1 - F(t)]$ will represent the fraction of the gas that has remained in the reactor for a longer time. Since only this red gas has remained in the reactor more than "t," the volume of bromine, dV, leaving the reactor at time "dt" is

$$dV = (Q \cdot dt)[1 - F(t)] \tag{1.13}$$

If we add all the bromine that has left the reactor in the period between $0 < t < \infty$, we have

$$V = \int_0^\infty Q \cdot [1 - F(t)] \cdot dt \tag{1.14}$$

For constant volumetric flow:

$$V = Q\int_0^\infty [1 - F(t)] \cdot dt$$

$$\frac{V}{Q} = \int_0^\infty [1 - F(t)] \cdot dt \tag{1.15}$$

For the integration of the right-hand side, we will use the integration by parts, indicating that

$$\int u \cdot dv = u \cdot v - \int v \cdot du$$

Applying to Eq. (1.15):

$$\frac{V}{Q} = t[1 - F(t)]_0^\infty + \int_0^1 t \cdot dF \tag{1.16}$$

At $t = 0$, the cumulative function is zero, $F(t) = 0$, and when $t \to \infty$, $[1 - F(t)] = 0$:

$$\frac{V}{Q} = \infty \cdot (1 - 1) - 0 \cdot (1 - 0) + \int_0^1 t \cdot dF$$

In this way, the first term on the right-hand side is zero, and we can write

$$\frac{V}{Q} = \bar{t} = \int_0^1 t \cdot dF \tag{1.17}$$

On the other hand, since $dF = E(t)dt$:

$$\bar{t} = \int_0^\infty E(t) \cdot dt \tag{1.18}$$

Now, the right side of this equality is, precisely, the mean residence time, so we can conclude that spatial time and mean residence time *are equal in systems of constant density, regardless of whether the flow is ideal or not*:

$$\bar{t} = t_m \tag{1.19}$$

This result is only valid for closed systems (that is, without dispersion). The volume of the reactor can now be easily determined:

$$V = Q \cdot t_m \tag{1.20}$$

1.2.3.2 Second and Third Moments of the RTD

When comparing RTDs, it is usual to use the moments associated with the distribution instead of using the entire distribution. To this end, three are the moments that are usually used. The first moment is the average residence time, already defined. The second moment is calculated from the mean and is the variance which is defined as

$$\sigma^2 = \int_0^\infty (t - t_m)^2 \cdot E(t)dt \tag{1.21}$$

This equation corresponds to the square of the standard deviation. The magnitude of this moment gives a measure of the dispersion of the distribution, in such a way that the greater this moment, the greater the dispersion of the distribution (see Figure 1.4).

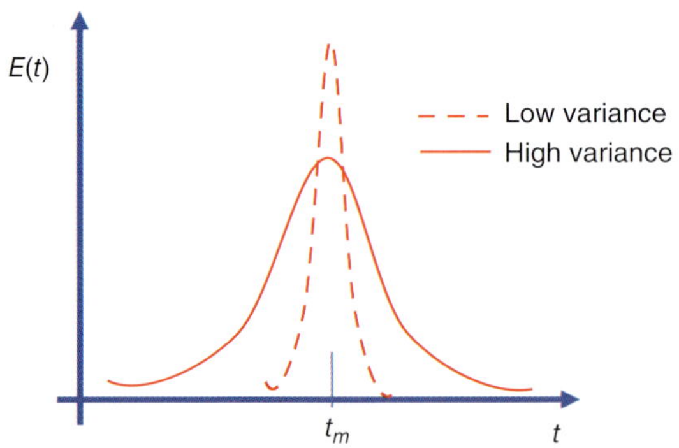

Figure 1.4 Effect of the second moment of the RTD in the curve shape. Both distributions are centered at the same t_m.

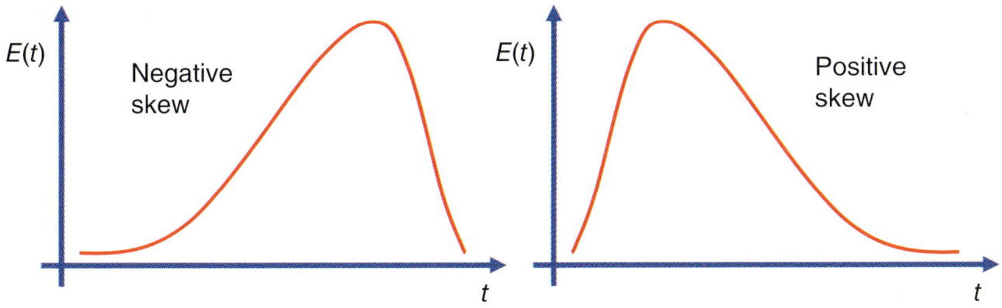

Figure 1.5 Form of the RTD curve for positive and negative skewness.

The third moment is also calculated around the average and is known as skewness. It is defined as

$$s^3 = \frac{1}{\sigma^{3/2}} \int_0^{\infty} (t - t_m)^3 \cdot E(t)dt \tag{1.22}$$

The greatness of this moment measures the extent to which the distribution with respect to the mean is displaced in one direction or another. A positive value of the skewness indicates that a tail to the right is expected in the distribution, and to the left if the skewness is negative. Figure 1.5 shows a scheme.

1.3 RTD in Ideal Reactors

1.3.1 RTD of the Batch and PFR Reactors

The RTD of these two types of reactors is the simplest we can consider, as already mentioned. All the particles (considered very small portions of the fluid, or solid particles if there is a solid moving in the reactor) that leave these reactors have stayed the same time inside the reactor. The output is then infinite at a particular time, and zero otherwise. This will produce an output function similar to that of an arrow of infinite height and zero width but presenting an area equal to one. This arrow appears at $t = V/Q = \bar{t}$. Mathematically, this arrow is represented by the Dirac delta function, in such a way that the $E(t)$ for an PFR is

$$E(t) = \delta(t - \bar{t}) \tag{1.23}$$

The delta function has the following properties:

$$\begin{aligned} &\delta(x) = 0 \quad \text{when } x \neq 0 \\ &\delta(x) = \infty \quad \text{when } x = 0 \end{aligned} \tag{1.24}$$

$$\int_{-\infty}^{\infty} \delta(x)\, dx = 1 \tag{1.25}$$

$$\int_a^b g(x) \cdot \delta(x - \bar{t})\, dx = g(\bar{t}) \quad \text{if } a < \bar{t} < b \tag{1.26}$$

$$\int_a^b g(x) \cdot \delta(x - \bar{t})\, dx = 0 \quad \text{if } \bar{t} \notin [a, b] \tag{1.27}$$

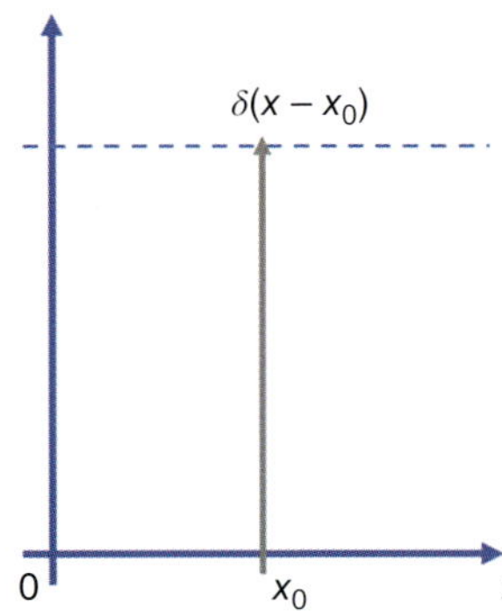

Figure 1.6 Dirac delta function, i.e. response of an ideal PFR to a pulse tracer input when "x" is time and the signal to a tracer is registered. In that case, x_0 would represent the average residence time.

In this way, the variance of this distribution is zero, by the fact that all values of $(t - t_m)$ are zero (all the signal is obtained at $t = t_m$). The function can be seen in Figure 1.6.

1.3.2 RTD of an ideal CSTR

The main characteristic of a CSTR is that the concentration of all substances inside the reactor are identical to those concentrations in the output stream. Imagine that an amount M_0 of inert tracer is injected into a CSTR. Once the tracer in injected, we can do a material balance for inert tracer, and we obtain

$$\begin{aligned} &\text{Input} - \text{Output} = \text{Accumulation} \\ &0 - Q \cdot C = V \cdot \frac{dC}{dt} \end{aligned} \tag{1.28}$$

Note that the input is zero (we consider the balance when the tracer has been already injected, i.e. at $t > 0$). Obviously, the generation term is also zero as the tracer is inert.

Since the reactor is perfectly agitated, C is the tracer concentration at the outlet and also at any point of the reactor. Separating variables and integrating, taking into account that $C = C_0$ for $t = 0$, we obtain

$$C(t) = C_0 \cdot \exp(-t \cdot Q/V) = C_0 \cdot \exp(-t/\bar{t}) \tag{1.29}$$

an expression that allows to obtain the concentration of the tracer in the effluent for any instant "t."

If we take into account the definition of $E(t)$ and introduce the value of $C(t)$, we obtain the RTD of an ideal CSTR:

$$E(t) = \frac{C(t)}{\int_0^\infty C(t) \cdot dt} = \frac{C_0 \cdot \exp(-t/\bar{t})}{\int_0^\infty C_0 \cdot \exp(-t/\bar{t}) \cdot dt} \tag{1.30}$$

Evaluating the integral of the denominator:

$$\begin{aligned} \int_0^\infty C_0 \cdot \exp(-t/\bar{t}) \cdot dt &= C_0 \cdot [\bar{t} \cdot \exp(-t/\bar{t})]_0^\infty \\ &= -C_0 \cdot \bar{t} \cdot [\exp(-\infty) - \exp(0)] = C_0 \cdot \bar{t} \end{aligned}$$

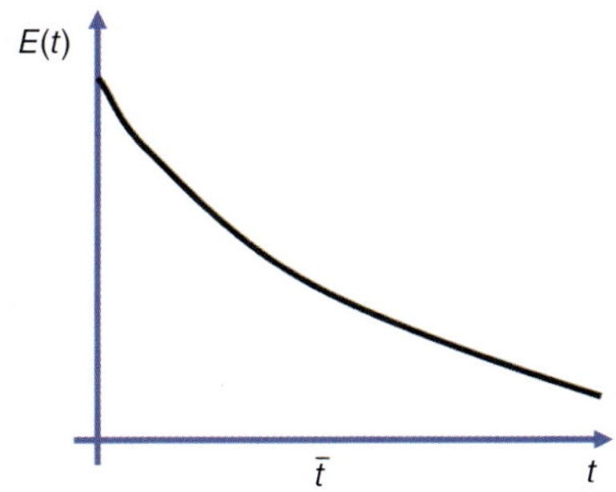

Figure 1.7 Response of a CSTR to a pulse tracer input.

Thus, the RTD is obtained for an ideal CSTR (Figure 1.7):

$$E(t) = \frac{1}{\bar{t}} \cdot \exp(-t/\bar{t}) \tag{1.31}$$

We have already seen that the average residence time in a reactor is given by V/Q or $\bar{t}$. This relationship can now be obtained in a simpler way by applying the definition of the average residence time of an RTD for a CSTR:

$$t_m = \int_0^\infty t \cdot E(t) \cdot dt = \int_0^\infty t \cdot \frac{1}{\bar{t}} \cdot \exp\left(-\frac{t}{\bar{t}}\right) \cdot dt = \bar{t} \tag{1.32}$$

That is to say, the nominal residence time (spatial time) $\bar{t} = V/Q$ coincides with the mean residence time that the material remains in the reactor, t_m.

To know the degree of dispersion in the reactor, we will calculate the second moment, the variance:

$$\sigma^2 = \int_0^\infty \frac{(t-\bar{t})^2}{\bar{t}} \cdot \exp\left(-\frac{t}{\bar{t}}\right) \cdot dt = \bar{t}^2 \tag{1.33}$$

That is to say, $\sigma = \bar{t}$. Or in other words, the standard deviation of the RTD coincides with the average.

1.3.3 RTD of PFR/CSTR in Series

In agitated tank reactors, there are zones, in the vicinity of the agitator, with a high degree of agitation and where an ideal CSTR could be a valid model. Nevertheless, depending on the location of the conductions, the reaction mixture can follow a somewhat tortuous path when entering, or when leaving the area perfectly agitated, or in both cases. This tortuous path can be modeled as if it behaved like a PFR. Thus, the tank-type reactor can be modeled as if it were a CSTR in series with a PFR (Figure 1.8), and the PFR can be before or after the CSTR. Next, we study the RTD for this type of reactor.

Consider, first, the system constituted by a CSTR followed by a PFR (Figure 1.8). The mean residence time in the CSTR will be called $\bar{t}_t$ and the mean residence time in the PFR, $\bar{t}_p$. If we inject a tracer pulse at the entrance of the CSTR, the concentration at the exit of the tank will vary with time according to the expression:

$$C(t) = C_0 \cdot \exp(-t/\bar{t}_t) \quad \text{(from Eq.(1.29))}$$

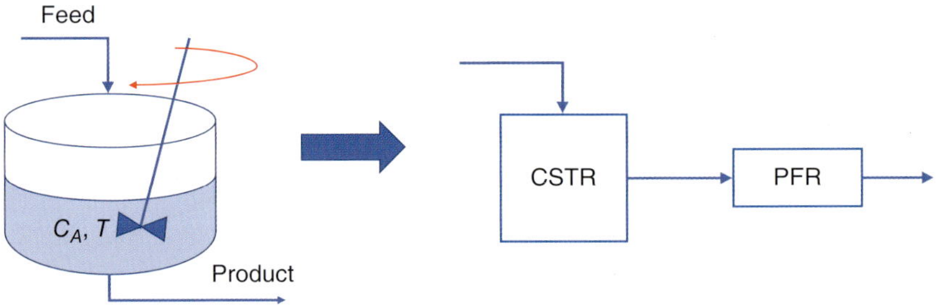

Figure 1.8 Real reactor modeled as a CSTR and a PFR in series.

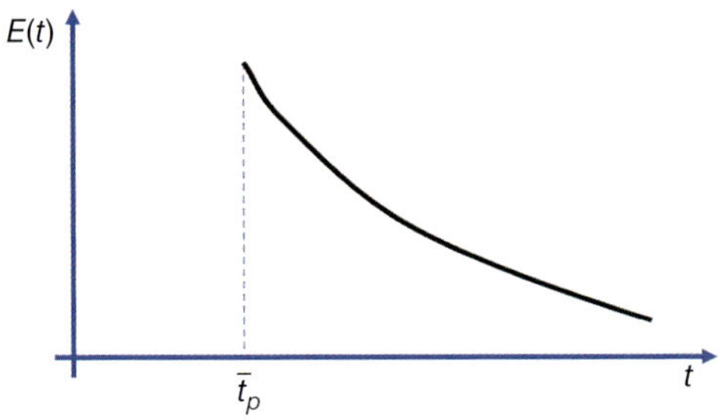

Figure 1.9 RTD for a CSTR and a PFR in series.

This output concentration will exit the PFR in series delayed by a time $\overline{t_p}$. Therefore, the RTD of the series reactor system will be given by (Figure 1.9)

$$E(t) = 0 \quad \text{if } t < \bar{t}_p$$
$$E(t) = \frac{1}{\bar{t}_t} \cdot \exp\left(-\frac{(t - \bar{t}_p)}{\bar{t}_t}\right) \quad \text{if } t \geq \bar{t}_p \tag{1.34}$$

Note that Eq. (1.34) is the same as Eq. (1.31), but the time scaling has been moved $\overline{t_p}$ units.

Let us see now the case of a CSTR preceded by a PFR. If a signal tracer is introduced into the input pulse PFR, the same signal appears at the input of the CSTR, but with a delay of $\overline{t_p}$ seconds, so the system's RTD will be the same as when the CSTR is the first reactor and is followed by a PFR. That is, the order in which both reactors are placed is not important, and the resulting RTD is the same provided that the sum of residence times in the two sections is the same.

However, this is not the only thing that should be considered; in case the reaction that takes place in the system is of the second order, the conversion that would be obtained with both dispositions would be different, as the conversion depends on the concentration. Contrarily, a first-order reaction will produce the same conversion both for PFR + CSTR and CSTR + PFR. This means that the RTD is not a complete description of what happens in a reactor or reactor system. The RTD is unique for a particular reactor; however, the reactor or reactor system is not unique to a particular RTD. In this way, in nonideal reactors, the RTD gives information that is not enough to characterize their behavior, and we need more information.

1.4 Modeling the Reactor with the RTD

On many occasions, the flow inside a reactor is not adjusted either to the complete mixture or to the plug flow, so when trying to use the RTD to predict the conversion that we are going to obtain, we find that the ideal models so far no longer serve us. There will, therefore, be a need to model the real reactor with some type of combination of ideal reactors or introduce new models. In these models, the adjustable parameter is usually evaluated on the basis of the RTD analysis obtained with a tracer test. We classify the models according to the number of adjustable parameters that are extracted from the information provided by the RTD.

1.4.1 Models with One Parameter: Tanks-in-series and Dispersion Models

In these models, we use a single parameter to bear in mind the nonideal behavior of a particular reactor. The parameter is determined by analyzing the measured RTD in a tracer test, as mentioned before.

For modeling the nonideal behavior of CSTRs, usually a dead zone volume (V_d), where the reaction does not take place, is used. Also, it is usual to consider the existence of a part of the fluid that passes through the reactor in short circuit and therefore does not react. In the case of tubular reactors, there are two models that usually represent the flow: the tanks-in-series model and the dispersion model. Both are one-parameter models, being the number of tanks in the first one, n_t, and the dispersion coefficient, D_e, in the last one.

For both types of distributions, once the value of the parameter is known, we will be able to calculate the conversion and or the concentrations at the output of the reactor.

In the case of a nonideal tubular reactor, usually it is assumed that the fluid moves in a plug flow through the reactor, so that each atom passes through the reactor the same time, and that the velocity profile is flat and there is no axial mixing. Both statements are false, to a greater or lesser extent, in all tubular reactors. Two approaches are often used to compensate for failures in the two ideal assumptions. In one case, the real reactor is modeled as a series of CSTRs of the same size. In the other (dispersion model), an axial dispersion is superimposed on the piston flow.

1.4.1.1 Tanks-in-series Model

In this model, we analyze the RTD of a particular tubular reactor to determine the number of CSTRs in series that will present an RTD approximately the same as the actual RTD. Next, we apply the balance equations valid for the ideal CSTR in order to calculate the conversion. We first consider the case of three tanks (Figure 1.10), developing the equations for the expression of the RTD, and then generalize it for "n_t" reactors connected in series. In this way, we obtain an equation that permits to calculate the number of tanks that best correlates the data of the actual RTD.

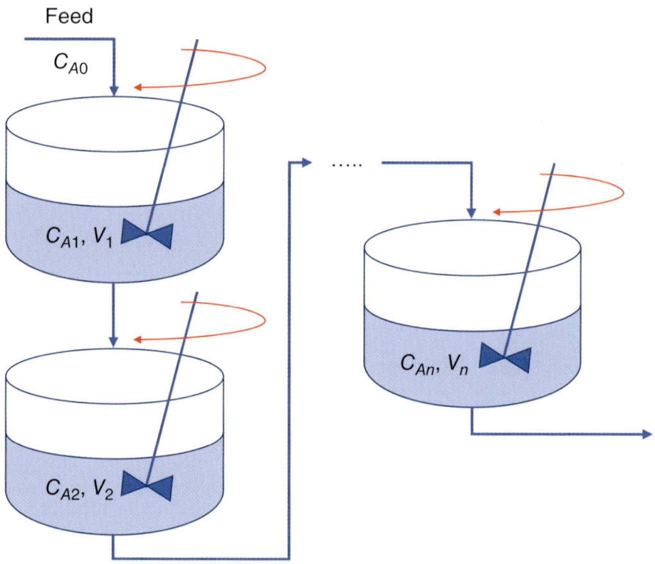

Figure 1.10 Tanks-in-series model.

If an impulse tracer signal is injected into the first tank, the tracer fraction that leaves the third reactor system after remaining in the system for a time between t and $t + dt$ is given by $E(t) \cdot dt$, which can be estimated from the concentration obtained in a pulse tracer experiment:

$$E(t) = \frac{C_{A3}(t)}{\int_0^\infty C_{A3}(t) \cdot dt} \tag{1.35}$$

In this expression, $C_{A3}(t)$ is the concentration of the tracer at the outlet of the third reactor. Now, we must obtain how this concentration varies with time. For a unique CSTR, the mass balance will be

$$\frac{V_1 \cdot dC_{A1}}{dt} = -Q \cdot C_{A1} \tag{1.36}$$

Integrating, we obtain the expression of the tracer concentration at the exit of that reactor:

$$C_{A1} = C_{A0} \exp\left(-Q \cdot \frac{t}{V_1}\right) = C_{A0} \exp\left(-\frac{t}{\overline{t_1}}\right) \tag{1.37}$$

Since the volumetric flow is constant $Q = Q_0$ and that the volume of all the reactors is the same ($V_1 = V_2 = V_3$), the average times will be identical ($\overline{t_1} = \overline{t_2} = \overline{t_3} = \overline{t_i}$), $\overline{t_i}$ being the residence time in each one of the reactors, not in the whole system. Posing a balance of the tracer in the second reactor:

$$\frac{V_2 dC_{A2}}{dt} = Q \cdot C_{A1} - Q \cdot C_{A2} \tag{1.38}$$

Taking into account the expression that we have previously obtained for C_{A1}, we arrive at the differential equation:

$$\frac{dC_{A2}}{dt} + \frac{C_{A2}}{\bar{t}_2} = C_{A0} \exp\left(-\frac{t}{\bar{t}_1}\right) \tag{1.39}$$

which can be solved using an integration factor together with the initial condition $C_{A2} = 0$ for $t = 0$:

$$C_{A2} = \frac{C_{A0}t}{\bar{t}_i} \cdot \exp\left(-\frac{t}{\bar{t}_i}\right) \tag{1.40}$$

Using the same procedure for the third reactor, we obtain the expression for the tracer concentration at the exit of the third tank (and, therefore, of the system):

$$C_{A3} = \frac{C_{A0}t^2}{2\bar{t}_i^2} \cdot \exp\left(-\frac{t}{\bar{t}_i}\right) \tag{1.41}$$

Substituting in the equation for the curve $E(t)$:

$$E(t) = \frac{C_3}{C_0} = \frac{t^2}{2\bar{t}_i^2} \cdot \exp\left(-\frac{t}{\bar{t}_i}\right) \tag{1.42}$$

If we generalize for n_t equal tanks in series:

$$E(t) = \frac{t^{n_t-1}}{(n_t-1)!\bar{t}_i^{n_t}} \cdot \exp\left(-\frac{t}{\bar{t}_i}\right) \tag{1.43}$$

since $\bar{t}_i = \bar{t}/n_t$, where $\bar{t}$ is the quotient of the total volume of the system by the volumetric flow rate Q.

Figure 1.11 shows the RTD for different CSTR numbers in series. As n_t increases, the behavior is closer to piston flow.

The number of reactors in series can be calculated from the dimensionless variance σ^2:

$$\sigma^2 = \int_0^\infty (t - \bar{t})^2 \cdot E(t)dt = \ldots = \frac{\bar{t}^2}{n_t} \tag{1.44}$$

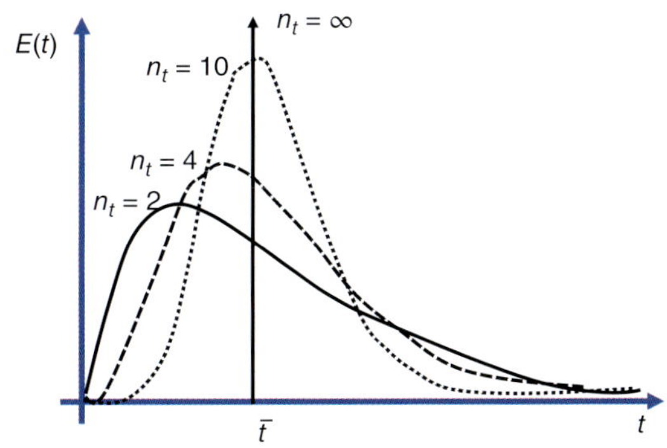

Figure 1.11 Response to a pulse tracer input in function of the number from tanks according to the tanks-in-series model. In the figure, $\bar{t}$ represents the average residence time of the whole system.

That is an expression that gives us the number of tanks needed to model the nonideal reactor as a series of n_t CSTR connected in series.

Let us now calculate the conversion of a reaction in the tanks in series. If the reaction is of the first order:

$$\text{1st reactor}: C_{A1} = \frac{C_{A0}}{1 + \bar{t}_1 k} \tag{1.45}$$

$$\text{2nd reactor}: C_{A2} = \frac{C_{A1}}{1 + \bar{t}_2 k} = \frac{C_{A0}}{(1 + \bar{t}_1 k)(1 + \bar{t}_2 k)} \tag{1.46}$$

As all residence times are equal, $\bar{t}_i$, and the temperature is constant ($k_1 = k_2 = k$), we can generalize the expression as

$$C_{An} = \frac{C_{A0}}{(1 + \bar{t}_i k)^{nt}} \tag{1.47}$$

Therefore, we can express the conversion as

$$X_A = 1 - \frac{1}{(1 + \bar{t}_i \cdot k)^{n_t}} \tag{1.48}$$

In general, the value of n_t obtained from the variance is considered as a non-integer number when calculating the conversion. In this sense, equations of the model can be applied to fractional number of tanks. Nevertheless, if the reaction is not of the first order, sequential molar balances must be made in each reactor (see Example 1.3 in the following sections).

1.4.1.2 The Dispersion Model

In this model, it is considered that there is a dispersion of the material overimposed to the flow, and that this dispersion is governed by an expression analogous to that of Fick for diffusion, which is superimposed on the plug flow. Thus, in addition to the term ($u \cdot S$) due to the flow of the mass of fluid, each component of the mixture will be transported through any section of the reactor with an additional rate [$D_e \cdot S \cdot (dC/dz)$] due to molecular and turbulent diffusion. At first glance, this simple model could only serve the effects of axial mixing. However, it can be seen that it serves to compensate also for the effects of radial mixing and those due to non-flat velocity profiles. These variations in concentration may be due to different speeds and flow paths, as well as to molecular and turbulent diffusion.

Let us imagine a pulse injection of tracer to a tubular reactor of section "S." During the movement of the fluid through the reactor, the pulse widens and becomes more diluted. The molar flow rate of the tracer (n_T), both by dispersion and convection, is

$$n_T = u \cdot S \cdot C_T + \left(-D_e \frac{\partial C_T}{\partial z} \right) \cdot S \tag{1.49}$$

In this expression, "z" is the spatial dimension where the fluid is moving, D_e is the effective dispersion coefficient (m^2/s) and "u" is the superficial velocity.

Note that the term corresponding to the dispersion of the component "*A*" is based on Fick's law for diffusion. If we do an inert tracer balance in a differential volume:

$$\text{In} - \text{Out} = \text{Accumulation}$$

$$-dn_T = dV \cdot \frac{dC_T}{dt} = S \cdot dz \cdot \frac{dC_T}{dt} \tag{1.50}$$

and, using partial derivatives:

$$-\frac{\partial n_T}{\partial z} = S \cdot \frac{\partial C_T}{\partial t} \tag{1.51}$$

Substituting for n_T (Eq. (1.49)) and dividing by the cross-section S:

$$D_e \frac{\partial^2 C_T}{\partial z^2} - \frac{\partial (u \cdot C_T)}{\partial z} = \frac{\partial C_T}{\partial t} \tag{1.52}$$

If we divide by ($u \cdot L$):

$$\frac{D_e}{u \cdot L} \frac{\partial^2 C_T}{\partial z^2} - \frac{\partial C_T}{L \cdot \partial z} = \frac{\partial C_T}{u \cdot L \cdot \partial t} \tag{1.53}$$

The parameter D_e/uL is the so-called ***recipient dispersion module*** or the Peclet–Bodenstein module (Bo = D_e/uL) that measures the degree of axial dispersion. When this module tends to zero, the system is close to piston flow, and when it tends to infinity (large dispersion), we have complete mixing. In the case of a packed bed, the module would be ($\varepsilon \cdot D_e/u \cdot d_p$), where d_p is the particle diameter and ε the porosity of the bed.

Equation (1.52) is only solvable for small values of Bo number, usually Bo < 0.01. In that case, the dispersion modifies the input signal in the reactor, but the tracer widening does not vary in the measuring point with time, in such a way that the boundary conditions are well known. In this situation, the expression for the RTD curve of the reactor is

$$E = \frac{1}{\bar{t} \cdot \sqrt{4 \cdot \pi \cdot \text{Bo}}} \cdot \exp\left[-\frac{\left(1 - \frac{t}{\bar{t}}\right)^2}{4 \cdot \text{Bo}}\right] \tag{1.54}$$

$$t_m = \bar{t} = \frac{V}{Q} \tag{1.55}$$

$$\sigma^2 = 2 \cdot \text{Bo} \tag{1.56}$$

Nevertheless, if the value of the Bodenstein module Bo is higher than 0.01, the response of the tracer to the impulse is wide and passes through the measuring point so slowly that it can change its form during the time of measuring. This produces an asymmetrical *E* curve that, on some occasions, does not have an analytical expression for the *E* curve.

In this case, the *E* curve also depends on what happens in the input and output sections of the reactor vessels. We consider two cases: closed boundary conditions (where there is plug flow behavior outside of the system), and open boundary conditions (where the flow is not affected when passing through the system).

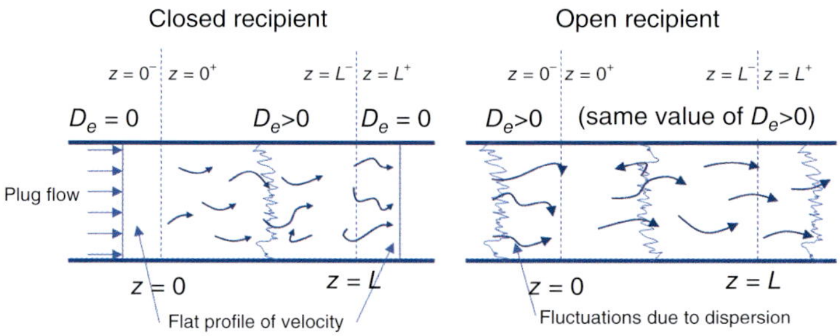

Figure 1.12 Effect of the dispersion on the velocity profile.

For simplifying all possibilities, let us consider only two cases: the closed–closed containers in which there is neither dispersion nor radial variation of the concentration, both upstream and downstream of the reaction zone; and *open–open containers* in which there is dispersion both before and after the reaction zone. Both cases are shown in Figure 1.12, where it is observed that the fluctuations of the concentration due to dispersion overlap the piston flow velocity profile. A closed–open container would have no dispersion at the entrance but only at the exit of the reaction zone.

Boundary Conditions for a Closed–Closed Vessel (Bo > 0.01) In this case, immediately before the reactor entrance zone ($z = 0^-$) and immediately after the exit zone ($z = L^+$), we have piston flow (without dispersion). However, between $z = 0^+$ and $z = L^-$, there is dispersion and convection by the flow movement. The corresponding boundary condition at the input is

$$n_T(0^-, t) = n_T(0^+, t) \tag{1.57}$$

Substituting the value n_T of at each side, we obtain

$$uSC_T(0^-, t) = -SD_e\left[\frac{\partial C_T}{\partial z}\right]_{z=0^+} + uSC_T(0^+, t) \tag{1.58}$$

Taking into account that at the input $C_T(0^-, t) = C_{T0}$ (known concentration):

$$C_{T0} = -\frac{D_e}{u}\left[\frac{\partial C_T}{\partial z}\right]_{z=0^+} + C_T(0^+, t) \tag{1.59}$$

Also, at the exit of the reactor considered, we can write $C_T(L^-, t) = C_T(L^+, t)$, this being later the measured exit concentration. In this way, when $z = L$:

$$C_T(L^-, t) = C_T(L^+, t)$$
$$\left(\frac{\partial C_T}{\partial z}\right)_{z=L} = \frac{C_T(L^-, t) - C_T(L^+, t)}{\partial z} = 0 \tag{1.60}$$

The combination of Eqs. (1.59) and (1.60) are known as the Danckwerts' boundary conditions.

On the other hand, the initial condition of the reactor at $t = 0$ is

$$t = 0, z > 0\ C_T\ (0^+, 0) = 0 \tag{1.61}$$

The injected tracer mass (M_0) is given by

$$M_0 = u \cdot S \cdot \int_0^{\infty} C_T(0^-, t) \cdot dt \tag{1.62}$$

In the closed–closed case, we do not have an analytical expression for the E curve, but the curve can be calculated by numerical methods and, also, we can calculate exactly its mean and variance. Bischoff and Levenspiel (in 1963) found the following relationships for the mean residence time and variance for this case:

$$\bar{t} = t_m$$
$$\left(\frac{\sigma}{t_m}\right)^2 = 2 \cdot \text{Bo} - 2 \cdot \text{Bo}^2 \cdot \left[1 - \exp\left(-\frac{1}{\text{Bo}}\right)\right] \tag{1.63}$$

Experimentally, the dispersion module can be calculated from the values of t_m and σ^2 obtained from the RTD, substituting in the previous expression.

Boundary Conditions for an Open–Open Container (Bo > 0.01) These conditions would be applicable in the case of a packed bed in which the tracer was injected at a point downstream of the inlet, a distance around two to three times the diameter, and whose concentration was measured at a certain distance before the exit. A solution of the differential equation (Eq. (1.52)) could be obtained in the case of a pulse injection.

For an open–open system, the boundary condition at the input is

$$n_T(0^-, t) = n_T(0^+, t) \tag{1.64}$$

Note that the expression is the same as that obtained in the previous case. If the dispersion coefficient is the same at the entrance as in the reaction zone, we will have

$$-D_e \left[\frac{\partial C_T}{\partial z}\right]_{z=0^-} + u \cdot C_T(0^-, t) = -D_e \left[\frac{\partial C_T}{\partial z}\right]_{z=0^+} + u \cdot C_T(0^+, t) \tag{1.65}$$

As we can imagine, the derivatives at $z = 0^+$ and $z = 0^-$ are the same, as no discontinuity is included in the model, so:

$$C_T(0^-, t) = C_T(0^+, t) \tag{1.66}$$

while on the exit:

$$-D_e \left[\frac{\partial C_T}{\partial z}\right]_{z=L^-} + u \cdot C_T(L^-, t) = -D_e \left[\frac{\partial C_T}{\partial z}\right]_{z=L^+} + u \cdot C_T(L^+, t) \tag{1.67}$$

$$C_T(L^-, t) = C_T(L^+, t) \tag{1.68}$$

In addition to these boundary conditions, many other modifications may occur. For example, the dispersion coefficient can have different values in each of the three regions (before the entrance, in the reaction zone, and after the output) and/or the tracer can be injected at a point other than $z = 0$. However, we consider only the case that the dispersion coefficient is the same for any value of z and that the pulse tracer is injected at the point $z = 0$ at time $t = 0$.

In the open–open case, there exists an analytical solution of the differential equation, with an expression of the E curve that is not much complicated:

$$E = \frac{1}{\sqrt{4 \cdot \pi \cdot \text{Bo} \cdot t/\bar{t}}} \cdot \exp\left[-\frac{\left(1 - \frac{t}{\bar{t}}\right)^2}{4 \cdot t \cdot \text{Bo}/\bar{t}}\right] \quad (1.69)$$

The corresponding average residence time is

$$t_m = \bar{t} \cdot (1 + 2 \cdot \text{Bo}) \quad (1.70)$$

where $\bar{t}$ is based on the volume (and flow rate) between $z = 0$ and $z = L$; that is, the reactor volume measured with a calibrated apparatus. Note that, as a result of the previous equations, the average residence time in the open system is longer than the one corresponding to the closed system. The variance will be

$$\left(\frac{\sigma}{t_m}\right)^2 = 2 \cdot \text{Bo} + 8 \cdot \text{Bo}^2 \quad (1.71)$$

In the next two chapters, we discuss the numerical solution to these partial differential equations systems and we present the problem of calculating the conversion in systems with dispersion.

1.4.2 Models with Two Parameters

1.4.2.1 Two CSTR with Exchange of Matter

Consider the case that there is a strongly agitated region near the agitator of an CSTR. Nevertheless, outside this region, the agitation is lower (Figure 1.13). Both regions have an important exchange of material. We also consider that the entrance and exit pipes are connected to the zone of greatest agitation. Each zone will be modeled as a CSTR, both being connected and there is material transfer between them.

If we propose a molar balance on the tracer, with a pulse injected at $t = 0$, for each of the tanks, we obtain

Accumulation = Input − Output (there is no tracer generation)

$$V_1 \frac{dC_{T1}}{dt} = Q_1 C_{T2} - (Q_0 C_{T1} + Q_1 C_{T1}) \text{ (in the first tank)} \quad (1.72)$$

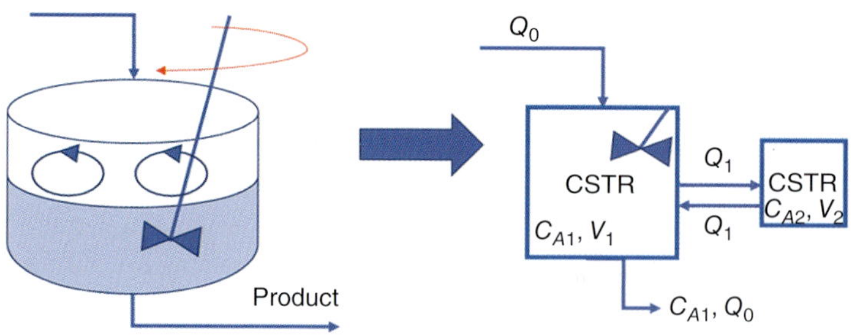

Figure 1.13 Real reactor and modeling using two CSTR.

$$V_2 \frac{dC_{T2}}{dt} = Q_1 C_{T1} - Q_1 C_{T2} \text{ (in the second tank)} \tag{1.73}$$

where C_{T1} and C_{T2} are, respectively, tracer concentrations in both reactors. These two differential equations are coupled and should be solved simultaneously.

In this model, the two adjustable parameters are the flow rate exchanged (Q_1) and the volume of the most agitated region (V_1). Remember that the measured volume (V) is the sum of V_1 and V_2. We will call β the fraction of the total flow that is transferred between both reactors:

$$Q_1 = \beta \cdot Q_0 \tag{1.74}$$

and α the fraction of the total volume that corresponds to the most agitated area:

$$V_1 = \alpha \cdot V \rightarrow V_2 = (1-\alpha) \cdot V \tag{1.75}$$

On the other hand, the average time ($\bar{t}$) is given by the quotient V/Q_0.

The initial conditions ($t = 0$) for this model are (i) $C_{T1} = (C_{T1})_0$, and (ii) $(C_{T2})_0 = 0$

An analytical solution is possible in this case, and is as follows:

$$\left[\frac{C_{T1}}{(C_{T1})_0}\right]_{\text{pulse}} = \frac{(\alpha m_1 + \beta + 1)\exp\left(\frac{m_2 t}{\bar{t}}\right) - (\alpha m_2 + \beta + 1)\exp\left(\frac{m_1 t}{\bar{t}}\right)}{\alpha(m_1 - m_2)} \tag{1.76}$$

being:

$$m_1, m_2 = \left[\frac{1-\alpha+\beta}{2\alpha(1-\alpha)}\right]\left[-1 \pm \sqrt{1 - \frac{4\alpha\beta(1-\alpha)}{(1-\alpha+\beta)^2}}\right] \tag{1.77}$$

However, for more complicated models, an approximate solution would be necessary.

Equation (1.76) shows that, if tank 1 is small compared to 2 (α small) and the transfer speed between both reactors is small (β small), the second exponential term tends to 1 during the first part of the response to an impulse injection. During the second part, the first exponential term tends to 0. If we represent the logarithm of the tracer concentration vs. time, the response curve will tend to a straight line at both ends of the curve and the parameters will be obtained from the slopes (m_1 for $t \rightarrow \infty$ and m_2 for $t \rightarrow 0$) and the cut points of both lines (for $t \rightarrow \infty$, the cut point is $-\{\alpha m_2 + \beta + 1\}/\alpha\{m_1 - m_2\}$).

1.4.2.2 CSTR with Dead Volume and Short Circuit

In this case, the real CSTR is modeled as the combination of an ideal CSTR of volume V_t, a dead zone of volume V_d, and a bypass (short circuit) of volumetric flow rate Q_b (Figure 1.14). Using a tracer injection, we will calculate the parameters of the model V_t and Q_t, since the total volume and the volumetric flow are known.

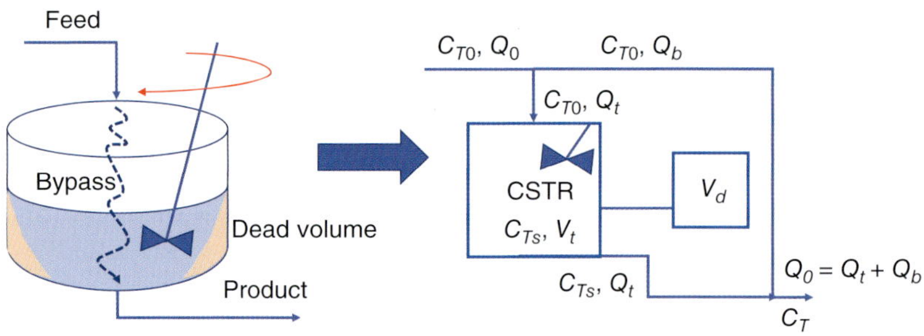

Figure 1.14 Real reactor and modeling using a single CSTR with dead volume and short circuit.

In this case, the derivation of the equations is simpler if we consider the injection of a tracer in positive step. Let us use the scheme in Figure 1.14; the balance in nonstationary regime (at $t > 0$, some tracer is still entering the system) of non-reactive tracer in the volume of reactor V_t, is

$$Q_t \cdot C_{T0} - Q_t \cdot C_{Ts} = \frac{dM_{Ts}}{dt} = V_t \frac{dC_{Ts}}{dt} \tag{1.78}$$

Remembering that for a positive step entry it is fulfilled that:

$$t < 0 \rightarrow C_T = 0$$
$$t \geq 0 \rightarrow C_T = C_{T0}$$

The tracer balance at the point of union of both currents will be

$$C_T = \frac{Q_b \cdot C_{T0} + Q_t \cdot C_{Ts}}{Q_0} \tag{1.79}$$

If we define:

$$V_t = \alpha V \quad \text{and} \quad Q_b = \beta Q_0, \qquad \text{with } \bar{t} = V/Q_0$$

Integrating and replacing in the expression of tracer balance in the reactor, we get

$$\frac{C_{Ts}}{C_{T0}} = 1 - \exp\left[-\frac{1-\beta}{\alpha}\left(\frac{t}{\bar{t}}\right)\right] \tag{1.80}$$

So, the expression of the tracer concentration that leaves the system will be

$$\frac{C_T}{C_{T0}} = 1 - (1-\beta) \cdot \exp\left[-\frac{1-\beta}{\alpha}\left(\frac{t}{\bar{t}}\right)\right] \tag{1.81}$$

If we reorder the equation, we can obtain the parameters of the model (Q_t and V_t or similarly α and β) from a plot of the tracer concentration at the output as a function of time. Representing $\ln[C_{T0}/(C_{T0} - C_T)]$ vs. time, if the model is correct, a straight line of slope $(1-\beta)/\bar{t}$ should be obtained α and an ordinate in the origin of value $\ln[1/(1-\beta)]$.

In order to have in mind the possible RTD curves obtained with this model, Figure 1.15 shows three curves corresponding to three different cases: the ideal

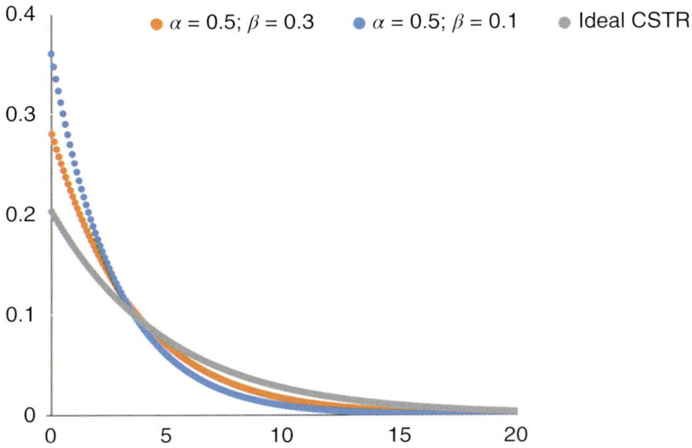

Figure 1.15 $E(t)$ predicted by this model with different values of the parameters, and the one for ideal CSTR for comparison.

CSTR, a reactor where $\alpha = 0.5$ and $\beta = 0.3$ and the other reactor with $\alpha = 0.5$ and $\beta = 0.1$; all of them with the same average residence time of five minutes. As we can see, the differences are quite small and, in practice, it is difficult to affirm without a little more information if the system has or not a dead volume and/or bypass.

1.5 Other Models of Real Reactors Using CSTR and PFR

So far, we have discussed about various reactor models. All of them are based on physical observations, which, in almost all the stirred tanks, are based on the existence of a well-mixed zone in the proximity of the agitator, usually represented by a CSTR. The region out of this well-agitated area can be modeled in various ways. The simplest form is using a model implying a CSTR connected to a dead zone; if it is suspected that some of the feed to the reactor may short out, a bypass current is added.

When the models do not satisfactorily represent the deviations of the ideal flow, we have to try more complicated models. In these models, it is assumed that the real reactor is constituted by a series of regions (flow in piston, flow dispersed in piston, flow in complete mixture, dead volumes) interconnected with each other in different ways (flow in bypass, with recirculation or cross-flow).

The simple types of these models are shown in Figure 1.16, where the form of the tracer response, in terms of E or F curves, for different models can be seen.

Example 1.1 ***Tubular Reactor***
A tubular reactor was designed in order to obtain a conversion of 98% and process $0.03\ m^3/s$. The reaction is a first-order irreversible isomerization. The reactor is 3 m long, with a cross-sectional area of $0.1\ m^2$. In the newly constructed reactor, a tracer pulse test gave the following data: $t_m = 10$ s and $\sigma^2 = 65\ s^2$. What conversion can be expected in the real reactor?

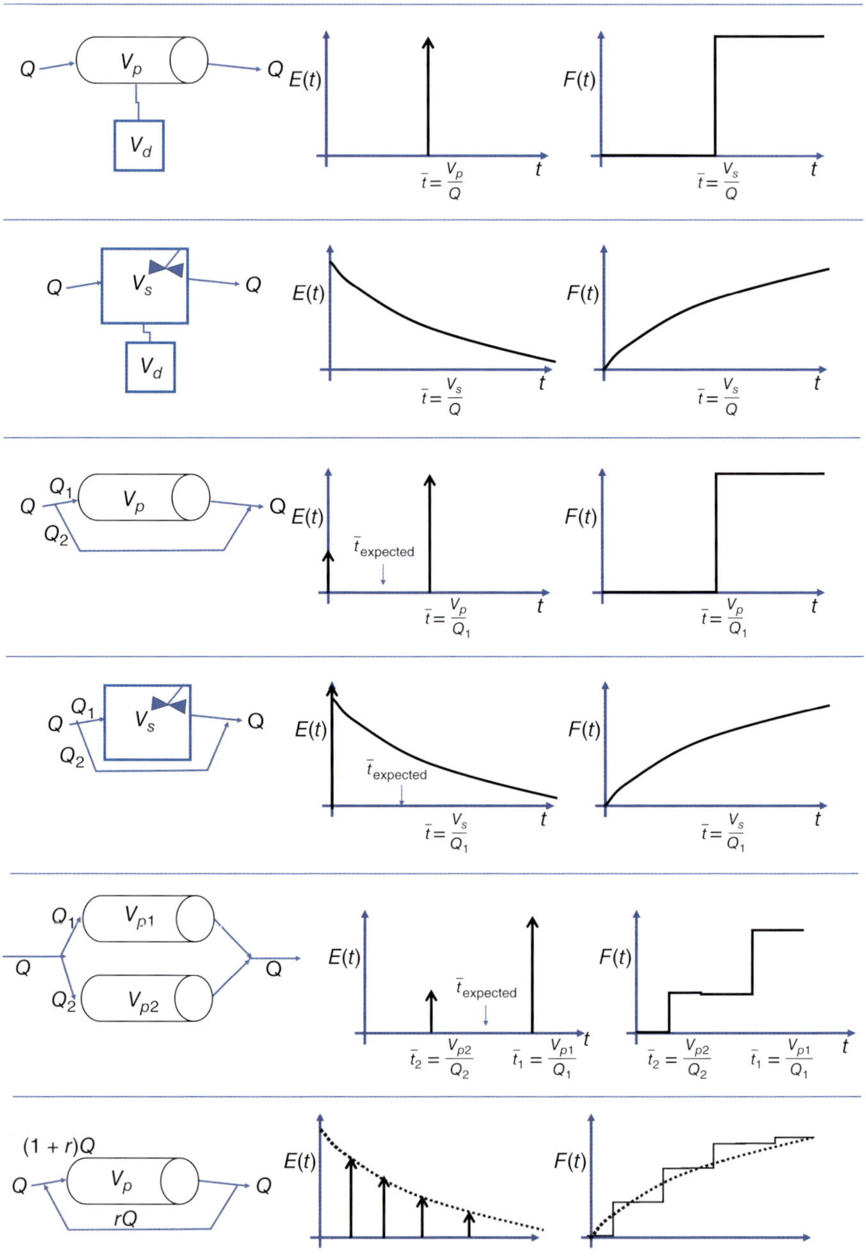

Figure 1.16 Combinations of ideal reactors used to model real reactors.

Solution

During the design of the reactor, a plug flow must be assumed. In this case, the molar balance of the reacting species "A," for a first-order reaction, gives

$$X_A = 1 - \exp(-k\bar{t})$$

In the present case:

$$\bar{t} = \frac{V}{Q} = \frac{0.1 \cdot 3}{0.03} = 10\ \text{s}$$

Coinciding with the value of t_m. From the previous equations: $k = 0.39\ \text{s}^{-1}$

Assuming a dispersion model for the real reactor, Eq. (1.63) is fulfilled, so:

$$\left(\frac{\sigma}{t_m}\right)^2 = 2 \cdot \text{Bo} - 2 \cdot \text{Bo}^2 \cdot \left[1 - \exp\left(-\frac{1}{\text{Bo}}\right)\right]$$

Iterating, we can find the value of the dispersion module: $\text{Bo} = (D_e/u{\cdot}L) = 0.667$

We have not seen in the previous sections the equations for the balances in reacting systems. Let us see this now. If we consider a tubular reactor in which we simultaneously have dispersion and reaction and we can do a molar balance of component A, in a range Δz of the reactor:

$$\text{Input} - \text{Output} + \text{Generation} = \text{Accumulation}$$

$$n_A - (n_A + dn_A) + r_A dV = 0$$

$$-\frac{1}{S}\frac{dn_A}{dz} + r_A = 0 \tag{1.82}$$

$$S \cdot dz = dV$$

Combining this expression with the molar flow of substance A:

$$n_A = -D_e S \frac{\delta C_A}{\delta z} + uSC_A \tag{1.83}$$

we obtain a differential equation of the second order:

$$\frac{D_e}{u}\frac{d^2 C_A}{dz^2} - \frac{dC_A}{dz} + \frac{r_A}{u} = 0 \tag{1.84}$$

which is only linear when the reaction rate is of order 0 or 1.

When the kinetics is of the first order ($r_A = -kC_A$), the following expression is obtained:

$$\frac{D_e}{u}\frac{d^2 C_A}{dz^2} - \frac{dC_A}{dz} - \frac{kC_A}{u} = 0 \tag{1.85}$$

which considers the flow, dispersion, and reaction. The solution of this second-order differential equation can be done analytically. If we consider a closed–closed system, we will apply the Danckwerts' boundary conditions at the input and at the exit of the reactor:

- Taking into account that at the input $C_A(0^-,t) = C_{A0}$(known concentration).

$$C_{A0} = -\frac{D_e}{u} \cdot \left.\frac{\partial C_A}{\partial z}\right|_{z=0^+} + C_A(0^+, t)$$

- At the exit of the reactor considered, we can write $C_A(L^-, t) = C_A(L^+, t)$, this being later the measured exit concentration. In this way, when $z = L$:

$$\left(\frac{\partial C_A}{\partial z}\right)_{z=L} = 0$$

Finally, the solution for the conversion in the reactor can be expressed by

$$1 - X_A = \frac{4a \cdot \exp\left(\frac{1uL}{2D_e}\right)}{(1+a)^2 \cdot \exp\left(\frac{auL}{2D_e}\right) - (1-a)^2 \cdot \exp\left(-\frac{auL}{2D_e}\right)} \tag{1.86}$$

where $a = [1 + 4(\bar{t} \cdot k) \cdot (D_e/uL)]^{1/2}$. This expression allows knowing the conversion that would be obtained for a first-order reaction to be carried out in a tubular reactor or in a bed reactor packed with dispersion.

For the system of the example, the value of "a" results to be 3.376 and the conversion in the real reactor is 0.88

Example 1.2 ***Reaction in a Complex System***

The second-order reaction $2A \rightarrow B$ is going to be carried out in a CSTR showing both a short circuit and a stagnant region. The tracer concentration at the outlet of this reactor is shown in the table, when a step input with initial concentration of tracer equivalent to 10 mg/l was used.

Time (min)	0	2	4	6	7	8	10	12	14	16	18	20
C_T (mol/l)	3.0	5.3	7.2	8.0	8.3	8.6	9.2	9.7	9.7	9.8	9.9	10

The measured volume of the reactor is 1 m^3 and the flow rate to the reactor is 0.1 m^3/min. The reaction rate constant is 150 l/(kmol min). The feeding contains a concentration of A at the input of 2 kmol/m^3. Calculate the conversion that can be expected in this reactor.

Solution

Let us first have a look of the experimental results. The graph C_T-time corresponding o the data is shown in Figure 1.17a.

As we have seen, Eq. (1.81) shows that the tracer concentration for a step input in this model can be expressed by

$$\frac{C_T}{C_{T0}} = 1 - (1-\beta) \cdot \exp\left[-\frac{1-\beta}{\alpha}\left(\frac{t}{\bar{t}}\right)\right] \tag{1.87}$$

From the data in the table, it is easy to calculate the slope and intercept of the representation $\ln(C_T/(C_{T0} - C_T))$ vs. t. The slope would correspond to $\left(-\frac{1-\beta}{\alpha\bar{t}}\right)$ and the intercept to $\ln(1/(1-\beta))$. In Figure 1.17b, we can see the corresponding straight line. Taking into account an intercept of −0.4549 and a slope of 0.2875, as solution of the model we find $\alpha = 0.3$ and $\beta = 0.3$ bearing in mind that the experimental value of average residence time $\bar{t} = \frac{V}{Q} = \frac{1}{0.1} = 10$ minutes should be used as it was defined for the whole system.

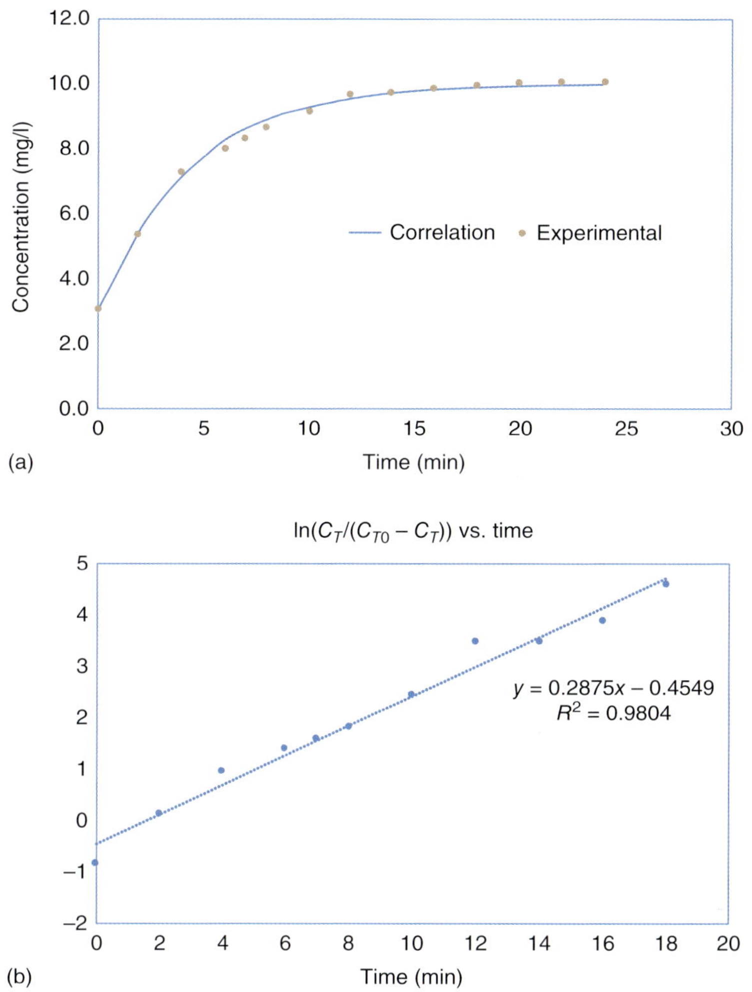

Figure 1.17 (a) Concentration of tracer obtained in a step input experiment and (b) calculation of the parameters.

For the system with reaction, let us see how to obtain the conversion with this model in the case of the first-order reaction (not in this case, but will be useful). For a first-order reaction, the molar balance of A in the reactor where the reaction takes place (V_t) gives

$$Q_t C_{A0} - Q_t C_{As} - k C_{As} V_t = 0$$

$$C_{As} = \frac{C_{A0}(1-\beta)Q_0}{(1-\beta)Q_0 + \alpha V k}$$

If we do a reagent balance A in the point where the short circuit current and the output current of the reactor are mixed, we will obtain

$$C_{A0} Q_b + C_{As} Q_t = C_A (Q_b + Q_t)$$

Clearing C_A, the concentration at the exit of the system:

$$C_A = \frac{Q_b C_{A0} + C_{As} Q_t}{Q_0} \tag{1.88}$$

So, finally:

$$\frac{C_A}{C_{A0}} = 1 - X_A = \beta + \frac{(1-\beta)^2}{(1-\beta) + \alpha \bar{t} k} \tag{1.89}$$

the expression that allows to calculate the conversion based on the parameters of the model.

As in Example 1.2, the reaction is second order, we can write:

$$Q_t C_{A0} - Q_t C_{As} - k C_{As}^2 V_t = 0 \tag{1.90}$$

Rearranging:

$$\frac{V_t \cdot k}{Q_t} \cdot C_{As}^2 + C_{As} - C_{A0} = 0 \tag{1.91}$$

Let us be aware that the flow rate and volume in the previous equations are those in the well-stirred part of the reactor, in such a way that:

$$\frac{V_t}{Q_t} = \frac{\alpha V}{(1-\beta) Q_0} = \frac{\alpha}{(1-\beta)} \bar{t}$$

and clearing, the value of C_{As} results to be

$$C_{As} = \frac{-1 + \sqrt{1 + 4\frac{\alpha}{1-\beta} \bar{t} k C_{A0}}}{2\frac{\alpha}{1-\beta} \bar{t} k} \tag{1.92}$$

Using the data in the example, the concentration at the exit of the CSTR is 0.927 kmol/m^3. The value of C_A at the exit of the system can be calculated from Eq. (1.80). Finally, in the example, $C_A = 1.249$ kmol/m^3

Example 1.3 ***Second-order Reaction in a Series of CSTRs***

An irreversible second-order reaction occurs in an isothermal, but not ideal, CSTR. The reactor volume is 1000 l and the flow velocity of the reagent stream is 1 l/s. At reactor temperature, $k = 0.005$ l/(mol s). The concentration of A in the feed stream is 1 mol/l. The DTR is obtained by a tracer test in this reactor at the desired feed rate and reaction temperature. Calculate the conversion that can be obtained with the tanks-in-series model.

RTD data obtained:

t (s)	0	5	10	25	40	70	100	175	250	325
$E(t)$ (s^{-1})	$9.712\ 18\cdot10^{-6}$	$2.206\ 86\cdot10^{-5}$	$6.411\ 68\cdot10^{-5}$	0.000 1090 4	0.000 199 787	0.000 286 936	0.000 476 181	0.000 619 435	0.000 719 564	$9.712\ 18\cdot10^{-6}$
t (s)	400	700	1000	2 500	4 000	7 000	10 000	15 000	20 000	
$E(t)$ (s^{-1})	0.000 782 749	0.000 791 834	0.000 627 888	$6.954\ 06\cdot10^{-5}$	$4.508\ 28\cdot10^{-6}$	$1.200\ 36\cdot10^{-8}$	$2.505\ 25\cdot10^{-11}$	$6.80\ 646\cdot10^{-16}$	$1.60\ 549\cdot10^{-20}$	

Solution
First of all, we can plot the data obtained with the tracer.

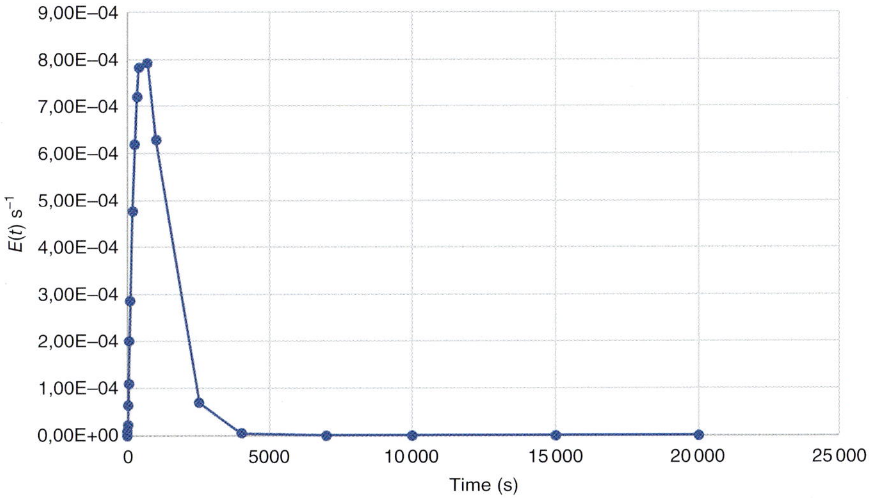

As we can see, the curve is similar to a CSTR, but small differences can be accounted with the use of the tanks-in-series model. From the data, it is easy to calculate:

$$t_m = \int_0^{\infty} t \cdot E(t) \cdot dt = 1129 \text{ s}$$

$$\sigma^2 = \int_0^{\infty} (t - t_m)^2 \cdot E(t) \cdot dt = 5.420 \cdot 10^5 \text{ s}^2$$

We can calculate the number of tanks in the model, that is, $n_t = t_m{}^2/\sigma^2 = 2.221$ tanks.

We have seen in the previous sections the equations for the balances in reacting systems or first order. In that case, Eq. (1.48) with $n_t = 2.221$ would give us the expected conversion. Let us see what occurs if the system is nth order. If we consider a single tank reactor (of volume V_1) we can do a molar balance of component A:

$$\frac{V_1}{n_{A0}} = \frac{C_{A0} - C_A}{C_{A0}(r_A)} = \frac{C_{A0} - C_{A1}}{C_{A0}(k \cdot C_{A1}^n)} \tag{1.93}$$

If the system has a constant density:

$$\bar{t}_1 = \frac{V}{Q} = \frac{C_{A0} - C_{A1}}{(k \cdot C_{A1}^n)} \tag{1.94}$$

$$C_{A1} = \frac{C_{A0}}{(1 + k\bar{t}_1 C_{A1}^{n-1})} \tag{1.95}$$

As we can check, the concentration at the exit can be calculated from this non-linear equation, and an approximation can be used.

In a second reactor connected in series with this, we would have

$$C_{A2} = \frac{C_{A0}}{(1 + k\bar{t}_1 C_{A1}^{n-1})(1 + k\bar{t}_2 C_{A2}^{n-1})} \tag{1.96}$$

For n_t tanks in series, the product in the denominator would have n_t factors and the intermediate concentrations $C_{A1}, C_{A2}, \ldots$ are needed.

In the present example, a value of $n_t = 2.221$ tanks would represent the system, but it is not possible to do the balance in a part of a reactor. In such situations, the best solution is to take the nearest integer to n_t or take both limits, in the present case $n_t = 2$ and $n_t = 3$.

Assuming two tanks $\bar{t}_i = \bar{t}/2$, i.e. all tanks of the same volume, we have $\bar{t}_i = 500$ s

We can now calculate C_{A1} using (Eq. (1.95)) with the corresponding values of k and n. Easily, the equation gives $C_{A1} = 0.461$ mol/l. If a second reactor is considered, (Eq. (1.96)) would give us $C_{A2} = 0.274$ mol/l, that is, a conversion of 0.723.

On the other hand, if three tanks are considered, the residence time in each tank is 333.3 s and the solution $C_{A1} = 0.533$ mol/l, $C_{A2} = 0.326$ mol/l, and $C_{A3} = 0.245$ mol/l. The final conversion would be 0.754.

Example 1.4 ***Nonideal CSTR***
In an isothermal nonideal CSTR, an irreversible second-order reaction is carried out. The values of flow rate, volume, and rate constant are known, and the RTDs has been measured with a tracer. Indicate the equations necessary to determine the concentration at the output of the reactor as a function of the input flow (or average residence time), if the reactor is modeled: (a) as three equal CSTRs in series, or (b) as three equal CSTRs in parallel.

Solution

(a) If the three reactors are connected in series, the equations just discussed in previous example are valid, so the concentration at the exit of each reactor will be given by Eqs. (1.95) and (1.96). The one corresponding to the third reactor is equivalent.
(b) On the contrary, if the reactors are connected in parallel, each of them will have a flow rate equal to 1/3 of the total. The concentration at the exit of the reactors will be the same in all three. And the mix of three currents of the same concentration in the same proportion, will give a current with exactly the same concentration, calculated by Eq. (1.95) bearing in mind the volume and the flow rate at each reactor.

Bibliography

Aris, R. (1969). *Elementary Chemical Reactor Analysis*. Englewood Cliffs: Prentice Hall.

Bourne, J. (1999). *Turbulent mixing and Chemical Reactions*. New York: Wiley.

Carberry, J.J. and Varma, J. (eds.) (1987). *Chemical Reaction and Reactor Engineering*. New York: Marcel Dekker.
Coulson, J.M. and Richardson, J.F. (1979). *Chemical Engineering III. Chemical Reactor Design*, 2ª ed. Oxford: Pergamon Press.
Fogler, H.S. (1998). *Elements of Chemical Reaction Engineering*, 3ª ed. New Jersey: Prentice Hall.
Holland, C.D. and Anthony, R.G. (1992). *Fundamentals of Chemical Reaction Engineering*, 2ª ed. Englewood Cliffs: Prentice-Hall.
Kayode Coker, A. (2001). *Modelling of Chemical Kinetics and Reactor Design*. Elsevier.
Levenspiel, O. (1999). *Chemical Reaction Engineering*, 3rde. New York: Wiley.
Levenspiel, O. (1962). *Chemical Reaction Engineering: An Introduction to the Design of Chemical Reactors*. New York: Wiley.
Levenspiel, O. (1979). *The Chemical Reactor Omnibook*. OSU. Bookstores, Corvalis.
Westerterp, K.R., Van Swaaij, W.P.M., and Beenackers, A. (1984). *Chemical Reaction Design and Operation*, 2nde. Chichester: Wiley.

2

Convolution and Deconvolution of Residence Time Distribution Curves in Reactors

2.1 Introduction

As discussed in Chapter 1, both pulse and step tracer input can be used to characterize the residence time distribution (RTD) in chemical reactors. One of the problems of these techniques is that the pulse or the step is usually not perfect, in the sense that deviations from the ideal pulse or step are usually present. In this chapter, the evolution of the output signal in a reactor is studied for an arbitrary input of tracer.

Also important are the systems where more than one reactor is combined in series. In this case, the signal entering the first reactor is modified according to its RTD, and the modified signal enters the second reactor, where it is modified, and so on.

In this sense, the convolution of the signals enables us to calculate the response to an arbitrary stimulus function, as is explained later. It is also possible to calculate the impulse response $E(t)$ by deconvolution from the measured stimulus and response.

2.2 Convolution

First of all, let us focus on the idea of the convolution operation. This is very important (and maybe difficult) to understand the behavior of signals (tracer curve response) throughout a series of reactors.

During the convolution of two functions "f" and "g," we will calculate a third function that expresses how the sum of one of the functions is modified by the other. Mathematically, a convolution is defined as the integral over all space of one function "$f(x)$" at "x" times displacing another function $g(x)$ given at "$u - x$." The integration is taken over the variable x, typically from minus infinity to infinity over all the dimensions (in case the variable is more than one dimension). In this way, the convolution is a new function of a new variable u, as shown in the following equations:

$$\begin{aligned} C(u) &= f(x) \otimes g(x) = \int_{\text{space}} f(x) \cdot g(u-x) \cdot dx \\ &= g(x) \otimes f(x) = \int_{\text{space}} g(x) \cdot f(u-x) \cdot dx \end{aligned} \tag{2.1}$$

Chemical Reactor Design: Mathematical Modeling and Applications,
First Edition. Juan A. Conesa.

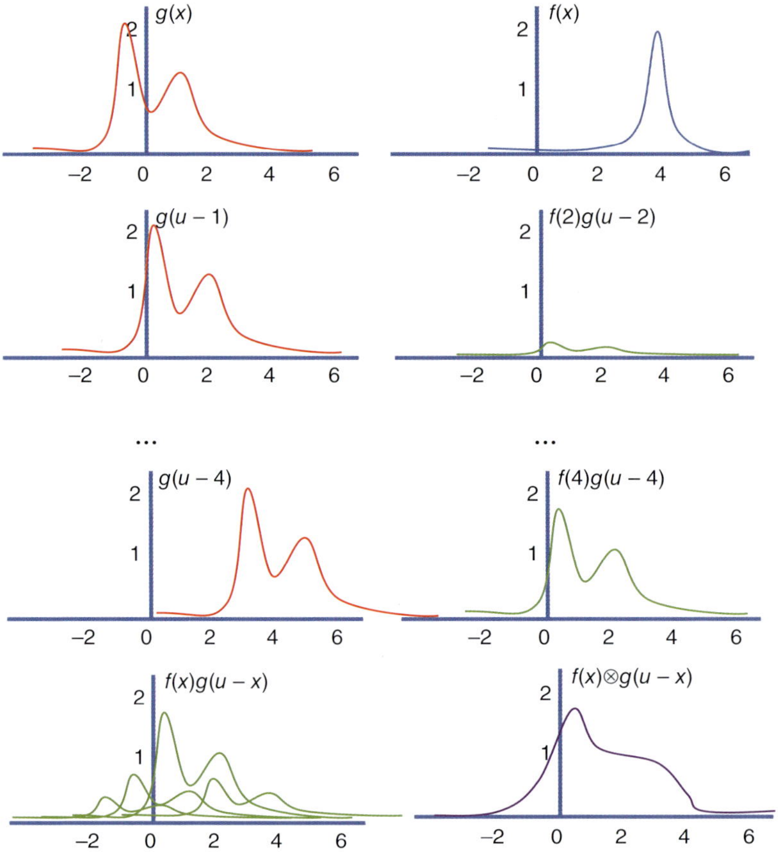

Figure 2.1 Visual example of the convolution of two functions.

The cross in the circle is used to indicate the convolution operation. Another way of figuring out the concept of convolution is to see it as an integral stating the amount of overlap of one function "g" as it is moved over another function "f." In some manner, the convolution "blends" one function with another.

Figure 2.1 shows a graphical example of the convolution of two given functions, $f(x)$ and $g(x)$. At the top of the figure, we have the original functions. Basically, the figure shows how we can consider the convolution, as a weighted sum of copies of one function displaced in the ordinate axis. In the sum, the fractions are given by the function value of the second function at the shift vector. The next three pairs of graphs show (on the left) the function "g" shifted by various values of x and, on the right, that shifted function "g" multiplied by "f" at the value of x.

The last two graphs show the superposition of all weighted and displaced copies of "g", and, on the right, the integral (i.e. the sum of all the weighted, shifted copies of "g"). We can see that the biggest contribution comes from the copy shifted by 4, the position of the peak of "f."

If one of the functions is a Dirac delta centered at $\overline{x}_1$, the other function will be shifted by a vector equivalent to the position of the peak, i.e. it will be centered at

$(\overline{x}_1 + \overline{x}_2)$, $\overline{x}_2$ being the original position of the second function. This is discussed later when treating the convolution of RTD signals in chemical reactors.

Other interesting properties of the convolution of curves is that the moments of the convolution are the sum of the moments of the individual curves, that is, the mean residence time of the convoluted curve is the sum of the mean residence times, and this is also true for the variances.

2.2.1 Convolution Properties

Convolution operation has the mathematical properties of commutativity, associativity, and distributivity, i.e.:

$$f(x) \otimes g(x) = g(x) \otimes f(x) \tag{2.2}$$

$$f(x) \otimes [g(x) \otimes h(x)] = [f(x) \otimes g(x)] \otimes h(x) \tag{2.3}$$

$$f(x) \otimes [g(x) + h(x)] = [f(x) \otimes g(x)] + [f(x) \otimes h(x)] \tag{2.4}$$

It is easy to see a demonstration of the commutative property. First, we start with the convolution integral written one way:

$$C(u) = f(x) \otimes g(x) = \int_{-\infty}^{+\infty} f(x) \cdot g(u - x) \cdot dx \tag{2.5}$$

Now we can substitute the variables, replacing $(u - x)$ with a new variable x^*:

$$x^* = u - x, \quad dx^* = -dx \tag{2.6}$$

$$C(u) = -\int_{+\infty}^{-\infty} f(u - x^*) \cdot g(x^*) \cdot dx^* \tag{2.7}$$

In Eq. (2.7), the sign of the variable of integration has changed, and so we have to change the signs of the limits of integration. Because these limits are infinite, the shift of the origin does not modify the magnitude of the limits.

Now we can reverse the order of the limits, which will change the sign of the equation, and interchange the order of the functions "g" and "f."

$$C(u) = \int_{-\infty}^{+\infty} f(u - x^*) \cdot g(x^*) \cdot dx^* \tag{2.8}$$

It does not matter if we change the name of the integration variables, so we can call it "x" again and we have the result we wanted to show.

In this way, the function $C(u)$ will contain the convoluted values of functions $f(x)$ and $g(x)$ in the whole space. Nevertheless, in chemical reactor characterization, the independent variable will be time, so surely the lower limit is fixed at $t = 0$. In the same way, sometimes we are required to calculate the convolution until a given time "t," so the upper limit of the integration will be this time, in such a way that, usually:

$$C(u) = \int_{0}^{t} f(u - t') \cdot g(t') \cdot dt' \tag{2.9}$$

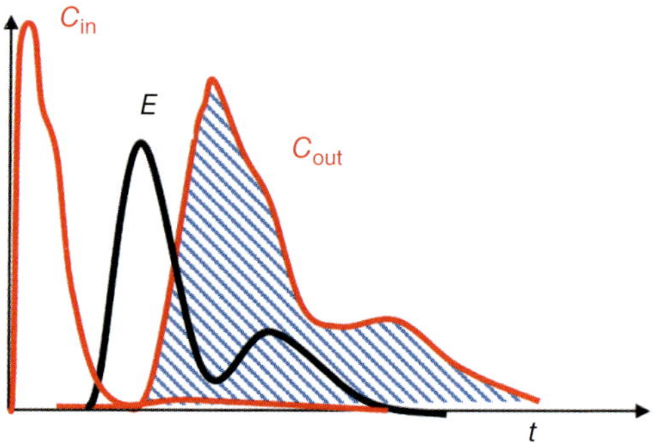

Figure 2.2 Input signal to a reactor, RTD, and output convoluted signal.

2.2.2 Application to a Reactor RTD

Let $C_{in}(t)$ and $C_{out}(t)$ be the concentration of tracer (in mg/l, for example) at the inlet and outlet of a continuous system (Figure 2.2). When the convolution theorem is applied to the RTD in a reactor or a combination of reactors, for example, to calculate the signal exit of a reactor with a given RTD when the input signal is known, the solution of the system also implies the calculation of the associated times to the input and output signals (see Figure CDC-2.2). In the case shown in the figure, the reactor distribution is E and the signal input is C_{in}, and we need to convolute both signals to get C_{out}:

$$C_{out} = E \otimes C_{in} \tag{2.10}$$

2.2.3 Calculating Convolution Functions

For functions of a discrete variable "x" (i.e. arrays of numbers), the definition of the convolution operation is

$$C(u) = \sum_{u=-\infty}^{\infty} f(x) \cdot g(u - x) \tag{2.11}$$

As pointed out later, it is not possible to calculate the convolution of two given functions if both are given at points with different Δx.

Assume that $f(x)$ is a vector of "n" values, i.e. $f(x) = [f_1, f_2, f_3, \ldots, f_n]$ and $g(x)$ has "m" values so that $g(x) = [g_1, g_2, g_3, \ldots, g_m]$.

As an example, if $n = 5$ and $m = 4$, for the convolution operation the desired result is

$$\begin{aligned}
C_1 &= f_1 \cdot g_1 \\
C_2 &= f_2 \cdot g_1 + f_1 \cdot g_2 \\
C_3 &= f_3 \cdot g_1 + f_2 \cdot g_2 + f_1 \cdot g_3 \\
C_4 &= f_4 \cdot g_1 + f_3 \cdot g_2 + f_2 \cdot g_3 + f_1 \cdot g_4 \\
C_5 &= f_5 \cdot g_1 + f_4 \cdot g_2 + f_3 \cdot g_3 + f_2 \cdot g_4 \\
C_6 &= f_5 \cdot g_2 + f_4 \cdot g_3 + f_3 \cdot g_4 \\
C_7 &= f_5 \cdot g_3 + f_4 \cdot g_4 \\
C_8 &= f_5 \cdot g_4
\end{aligned}$$

We can write this in matrix form as

$$\begin{pmatrix} C_1 \\ C_2 \\ C_3 \\ C_4 \\ C_5 \\ C_6 \\ C_7 \\ C_8 \end{pmatrix} = \begin{bmatrix} f_1 & 0 & 0 & 0 \\ f_2 & f_1 & 0 & 0 \\ f_3 & f_2 & f_1 & 0 \\ f_4 & f_3 & f_2 & f_1 \\ f_5 & f_4 & f_3 & f_2 \\ 0 & f_5 & f_4 & f_3 \\ 0 & 0 & f_5 & f_4 \\ 0 & 0 & 0 & f_5 \end{bmatrix} \cdot \begin{pmatrix} g_1 \\ g_2 \\ g_3 \\ g_4 \end{pmatrix} \tag{2.12}$$

or

$$C = \mathbf{A} \cdot g \tag{2.13}$$

with **A** being the "matrix convolution," with values of $f(x)$ in the form of columns. For a general case, when vectors of n and m values are combined, the resulting matrix convolution is an $(n + \mathrm{m} - 1) \times m$ matrix and the resulting C is a vector of $(n + m - 1)$ values:

$$C_{(n+m-1) \times 1} = [\boldsymbol{A}_{(n+m-1) \times m)} \cdot g_{(m \times 1)}] \tag{2.14}$$

Let us call "v" to the amount of rows of the C vector, i.e. $v = (n + m - 1)$. Let us apply these equations to the convolution of an RTD curve, E, and the signal input to a reactor, C_{in}. As mentioned before, we will need to convolute both signals, in such a way that:

$$C_{\text{out}} = E \otimes C_{\text{in}} \tag{2.15}$$

If C_{in} has "m" values and E has "n" values, the convolution is given by

$$\begin{pmatrix} C_{\text{out},1} \\ C_{\text{out},2} \\ C_{\text{out},3} \\ C_{\text{out},4} \\ \vdots \\ C_{\text{out},v} \end{pmatrix} = \begin{bmatrix} E_1 & 0 & 0 & \cdots & 0 \\ E_2 & E_1 & 0 & \cdots & 0 \\ E_3 & E_2 & E_1 & \cdots & 0 \\ \vdots & \vdots & \vdots & \vdots & \vdots \\ 0 & E_n & E_{n-1} & \cdots & \cdots \\ & & E_n & & \\ & & \vdots & & \\ 0 & 0 & 0 & 0 & E_n \end{bmatrix} \cdot \begin{pmatrix} C_{\text{in},1} \\ C_{\text{in},2} \\ \vdots \\ C_{\text{in},m} \end{pmatrix} \tag{2.16}$$

This is

$$C_{\text{out}\,(v\times 1)} = C_{\text{out}(n+m-1)\times 1} = [\boldsymbol{A}_{(n+m-1)\times m} \cdot C_{\text{in}(m\times 1)}] \tag{2.17}$$

In order to calculate the *time vector* associated to C_{out}, we would need the corresponding times associated to C_{in} and E. In general, let us assume that the first and last element of C_{in} corresponds to times $t_{0,C_{\text{in}}}$ and $t_{m,C_{\text{in}}}$ while the first and last element of the E vector are $t_{0,E}$ and $t_{n,E}$. For the convolution to be possible (as mentioned before), it is necessary that the time increments in both vectors (Δt) are equal, i.e.:

$$\Delta t = \frac{t_{m,C_{in}} - t_{0,C_{in}}}{m-1} = \frac{t_{n,E} - t_{0,E}}{n-1} \tag{2.18}$$

where "n" and "m" are, respectively, the number of points of vectors E and C_{in}.

Let us note that the convoluted signals C_{out} would contain ($\nu = n + m - 1$) values in a vector, and the corresponding time vector could be calculated by

$$t_{0,\text{convolution}} = t_{0,C_{in}} + t_{0,E} \quad \text{(initial time is the } \textit{sum} \text{ of the initial times)}$$

$$t_{\nu,\text{convolution}} = t_{m,C_{in}} + t_{n,E} \quad \text{(final time is the } \textit{sum} \text{ of the final times)}$$

$$\Delta t_{\text{convolution}} = \frac{t_{\nu,\text{convolution}} - t_{0,\text{convolution}}}{m+n-2} \tag{2.19}$$

Applying Eq. (2.15), we can also see that the following is true:

$$\Delta t_{\text{convolution}} = \frac{(m-1)\cdot\Delta t + (n-1)\cdot\Delta t}{m+n-2} = \left(\frac{m+n-2}{m+n-2}\right)\cdot\Delta t = \Delta t \tag{2.20}$$

That is, the time increment of the convoluted curve is the same as those of the curves (remember that this time increment should be the same for both curves).

Example 2.1 ***Manual Calculation of Convolution***

In an experimental reactor, a tracer is injected with the following distribution:

t (s)	0	1	2	3	4
C (mol/l)	0	8	4	6	0

In the reactor, the RTD is given by

t (s)	2	3	4	5	6
E	0	0.05	0.50	0.35	0

Calculate the evolution of the output signal with time.

Solution

We must calculate both the concentration and the time vectors. The time vector is easy to calculate. We have $m = 5$ points and $n = 5$ points, so the resulting vector will have 9 points ($m + n - 1$). The first value of time is $(0 + 2) = 2$ and the last one is $(6 + 4) = 10$. The corresponding $\Delta t_{\text{convolution}}$ is $(10 - 2)/(\text{m} + \text{n} - 2) = 8/8 = 1$; therefore:

$t_{\text{convolution}}$ (s)	2	3	4	5	6	7	8	9	10

To calculate the concentration:

$$
\begin{aligned}
C_{\text{convolution1}} &= C_{\text{in1}} \cdot E_1 \\
C_{\text{convolution2}} &= C_{\text{in2}} \cdot E_1 + C_{\text{in1}} \cdot E_2 \\
C_{\text{convolution3}} &= C_{\text{in3}} \cdot E_1 + C_{\text{in2}} \cdot E_2 + C_{\text{in1}} \cdot E_3 \\
C_{\text{convolution4}} &= C_{\text{in4}} \cdot E_1 + C_{\text{in3}} \cdot E_2 + C_{\text{in2}} \cdot E_3 + C_{\text{in1}} \cdot E_4 \\
C_{\text{convolution5}} &= C_{\text{in5}} \cdot E_1 + C_{\text{in4}} \cdot E_2 + C_{\text{in3}} \cdot E_3 + C_{\text{in2}} \cdot E_4 + C_{\text{in1}} \cdot E_5 \\
C_{\text{convolution6}} &= C_{\text{in5}} \cdot E_2 + C_{\text{in4}} \cdot E_3 + C_{\text{in3}} \cdot E_4 + C_{\text{in2}} \cdot E_5 \\
C_{\text{convolution7}} &= C_{\text{in5}} \cdot E_3 + C_{\text{in4}} \cdot E_4 + C_{\text{in3}} \cdot E_5 \\
C_{\text{convolution8}} &= C_{\text{in5}} \cdot E_4 + C_{\text{in4}} \cdot E_5 \\
C_{\text{convolution9}} &= C_{\text{in5}} \cdot E_5
\end{aligned}
$$

Tanking the values of C_{in} = tracer input and E = RTD of the reactor, we can easily calculate:

$$
\begin{aligned}
C_{\text{convolution1}} &= 0 \cdot 0 \\
C_{\text{convolution2}} &= 8 \cdot 0 + 0 \cdot 0.05 \\
C_{\text{convolution3}} &= 4 \cdot 0 + 8 \cdot 0.05 + 0 \cdot 0.5 \\
C_{\text{convolution4}} &= 6 \cdot 0 + 4 \cdot 0.05 + 8 \cdot 0.5 + 0 \cdot 0.35 \\
C_{\text{convolution5}} &= 0 \cdot 0 + 6 \cdot 0.05 + 4 \cdot 0.5 + 8 \cdot 0.35 + 0 \cdot 0 \\
C_{\text{convolution6}} &= 0 \cdot 0.05 + 6 \cdot 0.5 + 4 \cdot 0.35 + 8 \cdot 0 \\
C_{\text{convolution7}} &= 0 \cdot 0.5 + 6 \cdot 0.35 + 4 \cdot 0 \\
C_{\text{convolution8}} &= 0 \cdot 0.35 + 6 \cdot 0 \\
C_{\text{convolution9}} &= 0 \cdot 0
\end{aligned}
$$

And, finally:

$C_{\text{convolution}}$ (mol/l)	0	0	0.4	4.2	5.2	4.4	2.1	0	0

2.3 Deconvolution

Once we have outlined the convolution operation, let us propose the inverse operation, i.e. the deconvolution of curves. This is a more complicated issue and is rarely treated in specialized books and papers. Let us assume that we want to know the E curve of a reactor, where we feed a given amount of tracer in the form C_{in} (associated to a time vector) and where we are measuring the exit concentration C_{out}. This is a very common problem because the calculation of E from an impulse input signal (C_{in} = Dirac delta function) or a step function is not always possible or easy. Assuming the dimension of vectors is "m" for C_{in} and "v" for C_{out} (as used earlier), we have that $C_{\text{out}} = E \otimes C_{\text{in}}$ and

$$C_{\text{out}(v\times 1)} = [\boldsymbol{A}_{(v\times m)} \cdot C_{in(m\times 1)}] \tag{2.21}$$

Note that the dimensions of the convolution matrix $\mathbf{A}$ are $(v \times m)$ and from that we will calculate $(n = v - m + 1)$ values of the vector E curve (according to Eq. (2.16)) (check Example 2.2 for details).

From this equation, it is not possible to calculate directly matrix **A**. If we have a look at Eq. (2.16), we can check that, for the calculation of n values of vector E, we have $(n + m - 1)$ equations, so it is an overdetermined system of equations. This means that there are many solutions for the system and we need to choose in any way the best solution. In this way, for the resolution of the deconvolution of two given signals (C_{in} and C_{out}), it is necessary to pose an optimization by least squared residuals of the components of matrix **A**.

For calculating the E curve, the easiest way to do the optimization is to minimize an objective function (O.F.) in the form of the sum of squared residuals between the C_{out} curve (known) and the product [**A**·C_{in}]:

$$\text{O.F.} = [C_{out(\nu\times 1)} - \boldsymbol{A}_{(\nu\times m)} \cdot C_{in(m\times 1)}]^2 \tag{2.22}$$

For that purpose, an initial value of the vector E is taken, and the use of an optimization method is necessary.

For the calculation of the corresponding time vector for E distribution (deconvoluted signal), we have

$$t_{0,E} = t_{0,C_{out}} - t_{0,C_{in}}$$

(initial time of the convolution is the sum of the initial times)

$$t_{\nu,E} = t_{\nu,C_{out}} - t_{m,C_{in}}$$

(final time of the convolution is the sum of the final times)

$$\Delta t_{deconvolution} = \frac{t_{\nu,E} - t_{0,E}}{\nu - 1} = \Delta t \tag{2.23}$$

Since $(\nu - 1) = (m + n - 2)$, then Eq. (2.23) is the same as Eq. (2.20).

Example 2.2 ***Manual Calculation of Deconvolution***

In an experiment to determine the E(t) of a certain reactor, the tracer is introduced in the reactor following the equation $C_{in} = t \cdot \exp(-t/2)$. The response of the nonideal tank reactor is given in the following table:

t_{out} (min)	4	5	6	7	8	9	10	11	12	13
C_{out} (mg/l)	0.9	1.8	2.1	5.2	3.6	4.5	1.7	0.8	0.7	0.5

Indicate how to calculate the $E(t)$ of the reactor, together with the time vector associated to the $E(t)$.

Solution

For doing this deconvolution, the easiest method is to calculate discrete values of C_{in} and then deconvolute. We can choose the number of points we use for the calculation (with limits), and also at the time values, but the time increment should be the same so that the C_{out} signal (i.e. one minutes) and the C_{in} cannot be zero in all points. Another restriction is that we will need enough data for C_{in}. We have $\nu = 10$ points of the convoluted curve (C_{out}). For example, if we use $m = 10$ points to calculate C_{in}, as we have $\nu = 10$ values of C_{out}, we can only calculate $n = \nu - m + 1 = 1$ point of the RTD. If we use $m = 9$ points in C_{in}, we will be able to calculate two values of $E(t)$. Let us choose to use $m = 5$ points in C_{in}, so we will calculate six points in $E(t)$. Choosing $t_{0,in} = 0$, then:

t_{in} (min)	t	0	1	2	3	4
C_{in} (mg/l)	t*exp(−t/2)	0	0.61	0.74	0.67	0.54

Following the reasoning presented before, the deconvoluted signal will have $(10 - 5 + 1) = 6$ points. The time vector of these points is easy to calculate, as $t_{E,0} = t_{out,0} - t_{in,0}$ and the time increment is one minutes; therefore:

t_E (min)	4	5	6	7	8	9

For calculating the $E(t)$, we should solve the following equations system:

$$\begin{pmatrix} 0.9 \\ 1.8 \\ 2.1 \\ 5.2 \\ 3.6 \\ 4.5 \\ 1.7 \\ 0.8 \\ 0.7 \\ 0.5 \end{pmatrix} = \begin{bmatrix} E_1 & 0 & 0 & 0 & 0 \\ E_2 & E_1 & 0 & 0 & 0 \\ E_3 & E_2 & E_1 & 0 & 0 \\ E_4 & E_3 & E_2 & E_1 & 0 \\ E_5 & E_4 & E_3 & E_2 & E_1 \\ E_6 & E_5 & E_4 & E_3 & E_2 \\ 0 & E_6 & E_5 & E_4 & E_3 \\ 0 & 0 & E_6 & E_5 & E_4 \\ 0 & 0 & 0 & E_6 & E_5 \\ 0 & 0 & 0 & 0 & E_6 \end{bmatrix} \cdot \begin{pmatrix} 0 \\ 0.61 \\ 0.74 \\ 0.67 \\ 0.54 \end{pmatrix}$$

As we can see, we have 6 unknowns and 10 equations, so the system is overdetermined. An optimization in needed, giving the desired values of $E(t)$:

t_E (min)	4	5	6	7	8	9
$E(t)$	1.79	2.33	2.58	0.00	1.24	0.00

The following figure presents graphically the results.

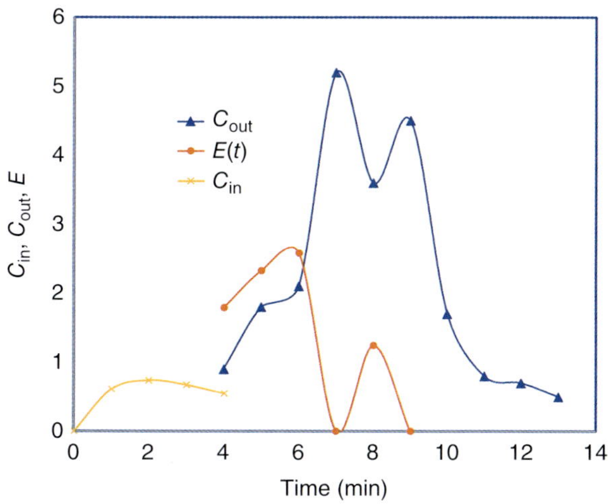

2.4 Computer Program Using Matlab® (Convolution)

Numerical computing tools such as Matlab can facilitate the development of different aspects in engineering. Compared to other computational tools and programming languages, such as C++ and Java, Matlab is easier for students to learn, and it has been widely incorporated in advanced courses in certain engineering disciplines. Even through Matlab is a commercial software, a freeware package is available (the software is called "GNU Octave"); and it can be used with small modifications of the programs, as almost all functions are compatible.

Although Matlab (and GNU Octave) contain specific functions such as "conv" and "deconv" for convolution and deconvolution, respectively, a simple use of these is not valid for the calculation of RTD curves (especially the deconvolution function). Furthermore, those programs are able to correctly calculate the convolution of two signals, but the corresponding time vectors are obviated.

For the *convolution* to be calculated using Matlab, let us assume a system where C_{in} (n values) and E (m values) and the corresponding time vectors ($t_{0,C_{\text{in}}} \dots t_{n,C_{\text{in}}}$) and ($t_{0,E} \dots t_{m,E}$) are known. Note that the time increment in both vectors should be the same.

It is easy to arrange matrix **A** from the given values in a for-end loop in Matlab, and then calculate the convolution simply using relation Eq. (2.16). The use of the "conv" function would give exactly the same result in the major part of the cases.

Calculating the corresponding time vector for C_{out} is simple taking into account that $\Delta t_{\text{convolution}} = \Delta t$, as shown in Eq. (2.20). The time vector goes from the sum of the initial times ($t_{0,C_{\text{in}}} + t_{0,E}$) to the sum of the final times ($t_{n,C_{\text{in}}} + t_{m,E}$). Let us remember that neither Matlab nor GNU Octave calculates this time vector.

The following examples are simple applications of the previous equations. More complicated input–output signals could be used, but maybe it is easier to understand the method if we use simple examples.

Example 2.3 ***Convolution of Two Given Signals***
Let us have the following data for the input signal to a reactor:

t (s)	0	1	2	3	4	5	6	7	8	9	10	11
C_{in}	0	0	8	6	4	5	6	3	1	0	0	0

And the E distribution is given by

t (s)	0	1	2	3	4	5
E	0	0.05	0.5	0.35	0.1	0

Calculate the convolution vector and the corresponding time vector to the convoluted signal.

Solution
The GNU code for the resolution of this example would be the following:

Program Listing 2.1

```
Cin=[0 0 8 6 4 5 6 3 1 0 0]; % Input signal to the reactor
tin=[0 1 2 3 4 5 6 7 8 9 10]; % Time vector associated to Cin
Er=[0 0.05 0.5 0.35 0.1 0]; % RTD distribution in the reactor
tr=[0 1 2 3 4 5]; % Time vector associated to E
inct=tr(2)-tr(1); % Time increment for Cin and E
m=length(Er); % Dimension of vector E
n=length(Cin); % Dimension of vector Cin

A=zeros(m,n); % Initializing the matrix

for i=1:n
    A(i:(m-1)+i,i)=Er; % Construction of the matrix
     convolution
end

Cout=(A*Cin')'; % Calculation of the convoluted curve


% Calculation of time vector for the convoluted curve:

t0convolution=(tin(1)+tr(1)); % Initial time for the
convoluted curve
tendconvolution=(tin(n)+tr(m)); % Final time for the
convoluted curve
tconvolution=t0convolution:inct:tendconvolution;
% Time vector for the convoluted curve

plot(tin,Cin,tr,Er,tconvolution,Cout)
legend('Cin','RTD','Cout')
xlabel('time(s)')
```

In the Matlab program, first of all, there are the inputs of the experimental curves. The time increment is then calculated and following that the matrix **A** is constructed. Finally, the convoluted curve is calculated, and the time vector corresponding to the convoluted curve and the graphs are constructed. The result is shown in Figure 2.3.

Example 2.4 ***Convolution of a Given Vector with an Ideal Reactor. Continuously Stirred Tank Reactor***
Let us have the following data for the input signal to a continuously stirred tank reactor (CSTR):

t (s)	0	1	2	3	4	5	6	7	8	9	10	11
C_{in}	0	0	8	6	4	5	6	3	1	0	0	0

In the CSTR, the residence time is known to be $\bar{t} = 5$ s. What is the signal at the exit of the reactor?

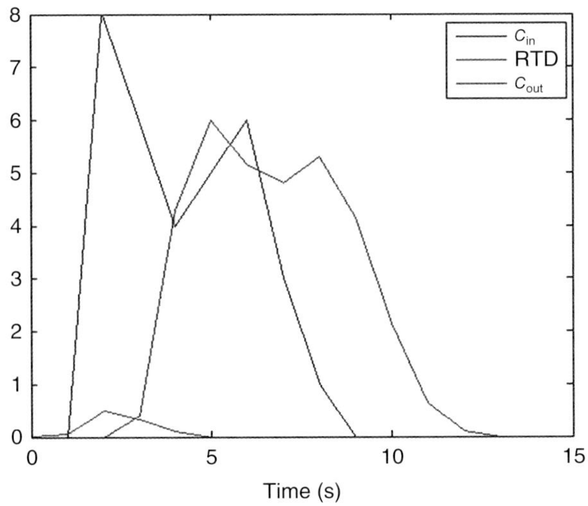

Figure 2.3 Solution to Example 2.3. Convolution of two given signals.

Solution

In this case, the E (RTD) curve should be calculated for each value of the time vector associated to C_{in}, according to the equation:

$$E = \frac{1}{\bar{t}} \exp\left(-\frac{t}{\bar{t}}\right)$$

In the program, the only necessary change is to introduce the following instead of the experimental RTD curve:

```
tr=tin; % Time vector associated to E
Er=(1/5)*exp(-tr/5); % RTD distribution in the reactor
```

The result of this simple calculation is given in Figure 2.4. This is very illustrative and we can "play" with the given parameters to see the effect in the convoluted curve.

Example 2.5 ***Convolution of a Given Vector with an Ideal Reactor. Plug Flow Reactor***

Let us have the following data for the input signal to a plug flow reactor (PFR):

t (s)	0	1	2	3	4	5	6	7	8	9	10	11
C_{in}	0	0	8	6	4	5	6	3	1	0	0	0

In the PFR, the residence time is known to be $\bar{t}$= 5 s. What is the signal at the exit of the reactor?

Solution

In this case, the E (RTD) curve is a Dirac delta centered at five seconds. This is very easy to simulate in GNU Octave (and Matlab®) by just using a vector with zeros at any position, but 1 at the corresponding position to five seconds (= $\bar{t}$). We will use logical variables for that purpose.

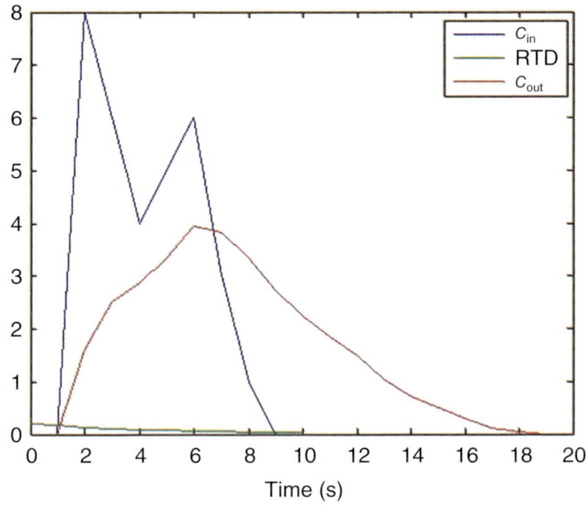

Figure 2.4 Solution to Example 2.4. Convolution of a given signal with the RTD of a CSTR.

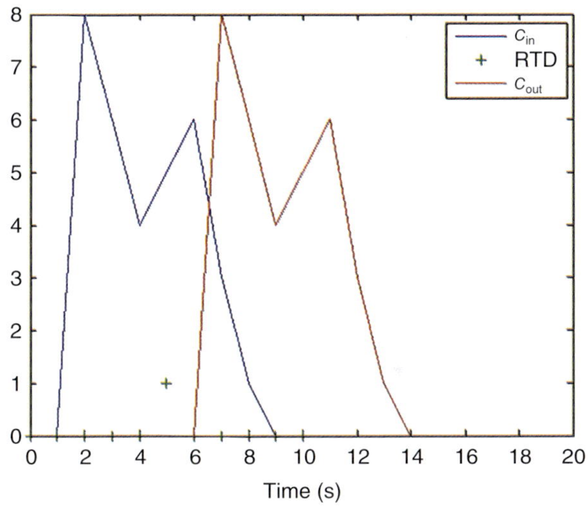

Figure 2.5 Solution to Example 2.5. Convolution of a given signal with the RTD of a PFR.

As in the former example, in the GNU Octave code, the only necessary change is to change the RTD distribution:

```
tr=tin; % Time vector associated to E
Er=(tr==5); % RTD distribution in the reactor
```

In this case, the double equal (==) is a logical operator answering 1 if the condition is met, and zero in the other case. The result of this calculation is given in Figure 2.5.

2.5 Computer Program Using MATLAB (Deconvolution)

It is more complicated to calculate the *deconvolution* of the curves C_{in} and C_{out} to obtain the E function. In this case, the use of the "deconv" function is not able to give a good result. For the calculation, we need to solve the system of linear equations given by Eq. (2.13) in such a way that the first step is to obtain the convolution matrix **A**. As mentioned before, the system is overdetermined, and the minimization of a function is necessary. The O.F. to minimize can be programmed in the following way:

Program Listing 2.2

```
function OF=minimal(f0) % f0 is the E curve being optimised
global Cin Cout
n=length(f0); % Dimension of the resulting E vector
v=length(Cout); % Dimension of vector Cout
m=length(Cin); % Dimension of vector Cin
for i=1:m
    A(i:i+n-1,i)=f0;% Constructing the convolution matrix
end
OF=sum((Cin'-A'*Cout').^2);
```

In that function, the first step is to share the variables C_{in} and C_{out} with other scripts, in order to be able to get the corresponding values of the vectors. Then, the function calculates the corresponding convolution matrix to give the calculated differences between C_{in} and the product ($\mathbf{A}^*C_{out}$). Finally, O.F. is obtained for a given vector f_0 that will be optimized.

On the other hand, the main script for calculation has the following structure:

Program Listing 2.3

```
global Cin Cout

Cin=[0 0 8 6 4 5 6 3 1 0 0]; % Input signal to the reactor
tin=[0 1 2 3 4 5 6 7 8 9 10]; % Time vector associated to Cin

Cout = [0 0 0 0.4 4.3 6 5.1 4.8 5.3 4.1 2.1 0.6 0.1 0 0 0];
tout= [0 1 2 3 4 5 6 7 8 9 10 11 12 13 14 15];

inct=tin(2)-tin(1); % Time increment for Cin and Cout

v=length(Cout); % Dimension of vector Cout
m=length(Cin); % Dimension of vector Cin

n=v-m+1; % Dimension of the resulting E vector

f0=ones(1,n)/2; % First assumption

Erd=fminsearch(@minimal,f0);

t0deconvolution=(tout(1)-tin(1));
tenddeconvolution=(tout(end)-tin(end));
inctdeconvolution=(tenddeconvolution-t0deconvolution)/(n-1);
tdeconvolution=t0deconvolution:inctdeconvolution:tenddeconvolution;
figure
plot(tin,Cin,tout,Cout,tdeconvolution,Erd)
legend('Cin','Cout','Er')
xlabel('time (s)')
```

In the script, after giving values to C_{in} and C_{out}, the calculation of the time increment and dimension of the vectors is done. The initial value for f_0 is taken to be a vector with all values equal to 0.5. Finally, the optimization is done by minimizing the O.F. using "fminsearch" and the graphs are constructed.

Example 2.6 ***Deconvolution of Two Given Signals***
Let us have the following data for the input signal to a reactor:

t (s)	0	1	2	3	4	5	6	7	8	9	10	11
C_{in}	0	0	8	6	4	5	6	3	1	0	0	0

And at the exit of the reactor, the concentration follows the following data:

t (s)	0	1	2	3	4	5	6	7	8	9	10	11	12	13	14	15
C_{out}	0	0	0	0.4	4.3	6	5.1	4.8	5.63	4.1	2.1	0.6	0.1	0	0	0

Calculate the RTD at the reactor by deconvolution of the vectors and the corresponding time vectors.

Solution
This is an unusual deconvolution case. The GNU code in this example would be the following:

Program Listing 2.4

```
Cin=[0 0 8 6 4 5 6 3 1 0 0]; % Input signal to the reactor
tin=[0 1 2 3 4 5 6 7 8 9 10]; % Time vector associated to Cin

Cout = [0 0 0 0.4 4.3 6 5.1 4.8 5.3 4.1 2.1 0.6 0.1 0 0 0];
tout= [0 1 2 3 4 5 6 7 8 9 10 11 12 13 14 15];

inct=tin(2)-tin(1); % Time increment for Cin and Cout

v=length(Cout); % Dimension of vector Cout
m=length(Cin); % Dimension of vector Cin

n=v-m+1; % Dimension of the resulting E vector

f0=ones(1,n)/2; % First assumption

Erd=fminsearch(@minimal,f0); % Minimization of the O.F.

t0deconvolution=(tout(1)-tin(1));
tenddeconvolution=(tout(end)-tin(end));
inctdeconvolution=(tenddeconvolution-t0deconvolution)/(v-1);
tdeconvolution=t0deconvolution:inctdeconvolution:tenddeconvolution;
figure
plot(tin,Cin,tout,Cout,tdeconvolution,Erd)
legend('Cin','Cout','Er')
xlabel('time (s)')
```

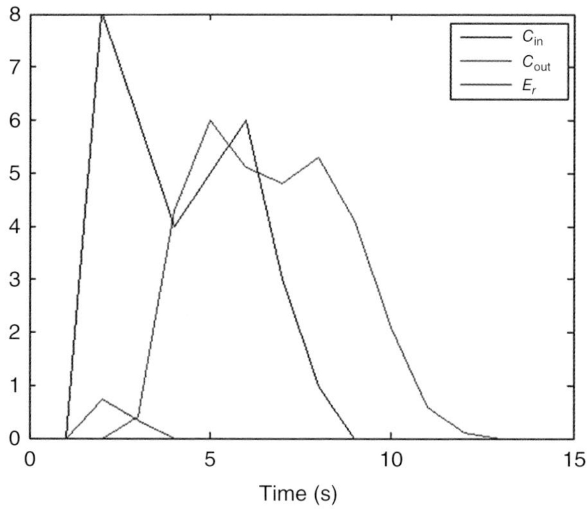

Figure 2.6 Solution to Example 2.6. Deconvolution.

As in the previous cases, first of all, there is the input of the experimental curves. Then the time increment is calculated from C_{in} or C_{out}. The dimensions of the vectors are read and an arbitrary first value of $E(t)$ is given. The optimization is done and the time vector corresponding to the deconvoluted curve, according to the equations shown before, is calculated. The result is shown in Figure 2.6.

The practice with these techniques could be completed by calculating the convolution of the calculated (deconvoluted) E curve with the input signal C_{in}, just to check everything is correct and that we obtain again the C_{out} vector.

2.6 Convolution of Signals in Reactors Connected in Series

The convolution and deconvolution of signals is also important in systems where more than one reactor is combined in series. In such cases, the signal entering the first reactor is modified according to its RTD, and the modified signal enters the second reactor, where it is modified, and so on. In the next chapters, we treat more complex systems in detail, but let us now treat systems with two or three reactors.

Figure 2.7 shows a system of three reactors connected in series. The convolution of the tracer curve with the RTD of each reactor follows the equations:

$$C_{out1} = E_1 \otimes C_{in}$$

$$C_{out2} = E_2 \otimes C_{out1} = E_2 \otimes E_1 \otimes C_{in}$$

$$C_{out3} = E_3 \otimes C_{out2} = E_3 \otimes E_2 \otimes E_1 \otimes C_{in}$$

In this case, we must do three convolution operations to get the final signal, with the difficulties that this would have. If C_{in} were a pulse input, it would

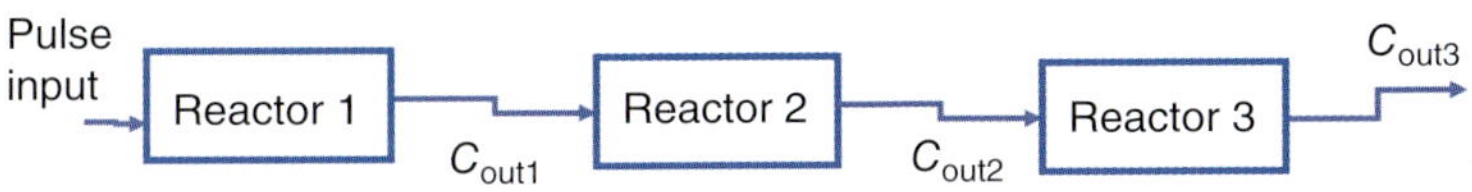

Figure 2.7 Set of three reactors connected in series.

be eliminated in the convolution. Assuming that the input is a perfect pulse, we will have

$$C_{out1}(t) = \frac{M}{Q} E_1(t) \quad \text{(see Chapter 1)}$$

The first convolution, in reactor 2 will be, from Eq. (2.8):

$$C_{out2}(t) = \int_0^t C_{out1}(t')\ E_2(t-t')\,dt' = \sum_{n=1}^{t/\Delta t'} C_{out1}(t - n\Delta t')E_2(n\Delta t')\,\Delta t'$$

Note that the integral is taken over time from 0 to time "t." In the sum, the number of summands will depend on the number of increments used in the calculation. $C_{out1}(t - n\Delta t')$ represents the value of C_{out1} in "n" time intervals with an increment of $\Delta t'$. Similarly, $E_2(n\Delta t')$ is the RTD function of the second reactor at the different times. Note that the number of increments is $t/\Delta t'$ for the integration of the curve until this time "t."

In the same way, for the signal obtained after the third reactor:

$$C_{out3}(t) = \int_0^t C_{out2}(t')\,E_3(t-t')\,dt' = \sum_{n=1}^{t/\Delta t'} C_{out2}(t - n\Delta t')\,E_2(n\Delta t')\,\Delta t'$$

As we see, the combination of several reactors involves the calculation of the convolutions of each curve, which makes it very tedious. There are, as we see in the next chapter, math techniques that allow us to approach the calculation of more complex systems, and involve the use of integral transforms.

Example 2.7 ***Pulse Input Tracer***

The response to a pulse input tracer is given by the following equations:

$$C_f = t - 2 \quad \text{when } 2 \le t \le 5$$

$$C_f = 3 \quad \text{when } 5 \le t \le 8$$

$$C_f = 11 - t \quad \text{when } 8 \le t \le 11$$

$$C_f = 0 \quad \text{elsewhere}$$

Calculate the average residence time and the function $E(t)$ in this reactor. If this signal is fed to a CSTR of $\bar{t} = 1$, calculate the output signal. Note that units of time are not specified.

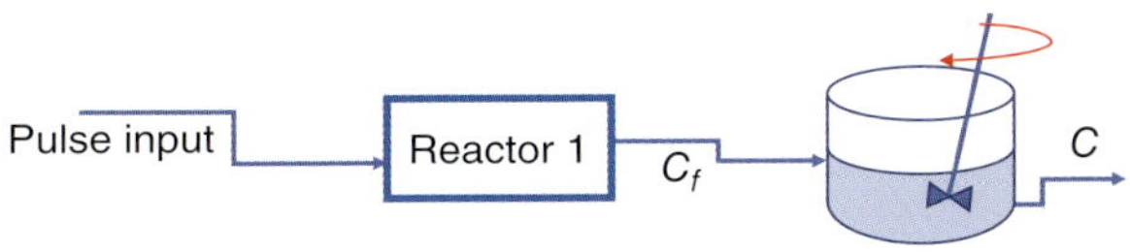

Solution

We can calculate the area over the $C_f(t)$ curve:

$$\int_0^\infty C_f(t)dt = \int_2^5 (t-2)dt + \int_5^8 3dt + \int_8^{11} (11-t)dt = 18$$

In this way:

$$E = (t-2)/18 \quad \text{when } 2 \le t \le 5$$
$$E = 3/18 \quad \text{when } 5 \le t \le 8$$
$$\mathrm{E} = (11-t)/18 \quad \text{when } 8 \le t \le 11$$
$$E = 0 \quad \text{elsewhere}$$

Also, we can calculate the following integral:

$$\int_0^\infty t \cdot C(t)dt = \int_2^5 t(t-2)dt + \int_5^8 3t\,dt + \int_8^{11} (11-t)t\,dt = 117$$

And then:

$$t_m = \frac{117}{18} = 6.5 \text{ (units of time)}$$

If this signal is fed to a CSTR, the balance in the reactor would be

$$\bar{t}\frac{dC}{dt} + C = Cf = \begin{Bmatrix} (t-2) & t = [2,5] \\ 3 & t = (5,8) \\ (11-t) & t = [8,11] \end{Bmatrix}$$

We can do the numerical calculation of the output signal replacing the increments by differences, as we study later. For example, in the first interval (superscripts "t" and "$t+1$" refer to the actual position in time and the next one):

$$\bar{t}\frac{C^{t+1} - C^t}{\Delta t} + C^t = (t-2), \qquad \text{with } C = 0 \text{ for } t = 2$$

i.e.

$$C^{t+1} = C^t + \frac{(t - 2 - C^t) \cdot \Delta t}{\bar{t}}$$

Choosing a correct value of Δt, we can calculate the evolution of the concentration. Doing a similar reasoning for the other two time intervals, the solution is given in the following plot:

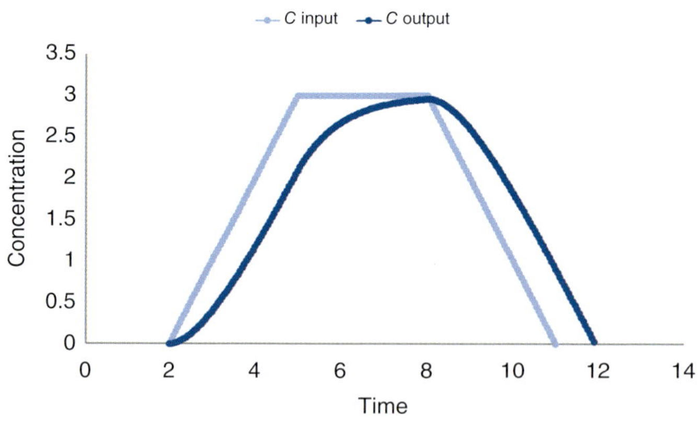

Example 2.8 ***Experimental Solid Flow Reactor***
An experimental flow reactor, initially designed as a CSTR, is fed with small particles of solids, and the response to an impulse signal (solid impregnated in a colored compound) has the function: $C(t) = a(10 - t)$ (with time in minutes).

(a) Represent approximately the function $C(t)$ and calculate $E(t)$ of the solid.
(b) This reactor is followed by a fluidized bed reactor of mean residence time (of the solid) equal to three minutes. Determine the concentration of tracer at the output of the system.

Solution

(a) The first reactor of this system was designed to be a completely mixed flow reactor. Let us see in a graph the form of the $C(t)$ function obtained:

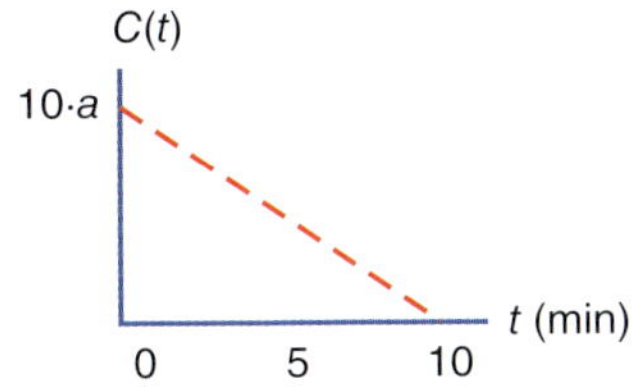

Note that the function for $t > 10$ is zero. As we can see, the response is somewhat similar to a CSTR, so the reactor was quite well scaled-up and constructed, at least in the aspect of the flow of solids. To calculate the $E(t)$ function, we need to calculate the area under the $C(t)$ curve. This can be done in several ways (for example, considering the area is that of a triangle), but the general way is

$$\int_0^\infty C(t)dt = \int_0^{10} a\cdot(10-t)\cdot dt = a\cdot\left(10t - \frac{t^2}{2}\right)\Bigg]_0^{10}$$
$$= a\cdot\left(10\cdot 10 - \frac{10^2}{2}\right) = 50a$$

So, for the RTD we have

$$E_1(t) = \frac{C(t)}{\int_0^\infty C(t)dt} = \frac{a\cdot(10-t)}{50\cdot a} = 0.02\cdot(10-t)$$

The mean residence time for this reactor is

$$t_m = \int_0^\infty t\cdot E_1(t)\cdot dt = \int_0^{10} t\cdot 0.02\cdot(10-t)\cdot dt$$
$$= \left(\frac{0.2t^2}{2} - 0.02\frac{t^3}{3}\right)\Bigg]_0^{10} = \frac{1}{3} = 0.\widehat{3}$$

(b) The $E(t)$ function of the solid being fed to a fluidized bed is that of a CSTR (actually, the reactor closest to the perfectly mixed flow is the fluidized bed), so we can write:

$$E_2 = \frac{1}{t_2}\exp\left(-\frac{t}{t_2}\right)$$

where we know the value of $\overline{t_2} = 3$ min. The convolution of both signals is obtained at the exit of this second reactor, and then:

$$C_{\text{out}} = \int_0^{\infty} E_1(t) \cdot E_2(t - t') \cdot dt'$$

In the next chapter, we see a way to solve this equation analytically, but for now let us use the method mentioned earlier, by taking values of time with the same increment for both functions. Using a value of $\Delta t = 1$ min, we can construct the following table of data:

t (min)	E_1	E_2
0	0.200	0.333
1	0.180	0.239
2	0.160	0.171
3	0.140	0.123
4	0.120	0.088
5	0.100	0.063
6	0.080	0.045
7	0.060	0.032
8	0.040	0.023
9	0.020	0.017
10	0.000	0.012
11	0.000	0.009
12	0.000	0.006

Let us use the Matlab program shown before, and now the code is given.

Program Listing 2.5

```
t=0:12; % Values of time (inct=1);
inct=1;

E1=0.02*(10-t) % Values of E for the first reactor (see Table)
E2=1/3*exp(-t/3); % Values of E for the second reactor (see Table)

m=length(E2); % Dimension of vector E2
n=length(E1); % Dimension of vector E1 (in this case equals to m but
not in general

A=zeros(m,n); % Initializing the matrix

for i=1:n
A(i:(m-1)+i,i)=E2; % Construction of the matrix convolution
end

Cout=(A*E1')'; % Calculation of the convoluted curve
```

```
% Calculation of time vector for the convoluted curve:

t0convolution=(2*t(1)); % Initial time for the convoluted curve
tendconvolution=(2*t(n)); % Final time for the convoluted curve
tconvolution=t0convolution:inct:tendconvolution; % Time vector for
the convoluted curve

plot(t,E1,t,E2,tconvolution,Cout)
legend('E1','E2','Cout')
xlabel('time(s)')
```

The following figure is obtained in the calculation.

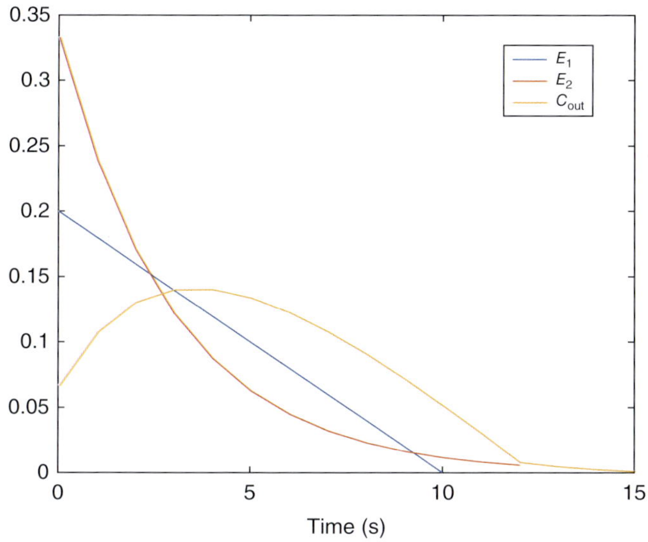

Bibliography

Anderssen, A.S. and White, E.T. (1969). The analysis of residence time distribution measurements using Laguerre functions. *Canadian Journal of Chemical Engineering* 47: 288–295.

Levenspiel, O. (1999). *Chemical Reaction Engineering*, 3e. New York, NY: Wiley.

Mann, U. (2009). *Principles of Chemical Reactor Analysis and Design*. New York, NY: Wiley.

Mecklenburg, J.C. and Hartland, S. (1975). *The Theory of Backmixing*. London: Wiley.

3

Use of Transfer Function for Convolution and Deconvolution of Complex Reactor Systems

3.1 Introduction

Transfer function is a very useful concept for studying nonideal chemical reactors. This function indicates how the changes in the input to a system affect the exit stream. It is valid for several systems (computers, PIPs, control systems, …) and is based on the following definition:

$$H(s) = \frac{\text{Response to stimulus}}{\text{Forcing function}} = \frac{Y(s)}{X(s)}$$

This definition is given using the variable "s," which is related to the Laplace transform of the time-dependent functions in the reactor system.

Using a block diagram, we can say that:

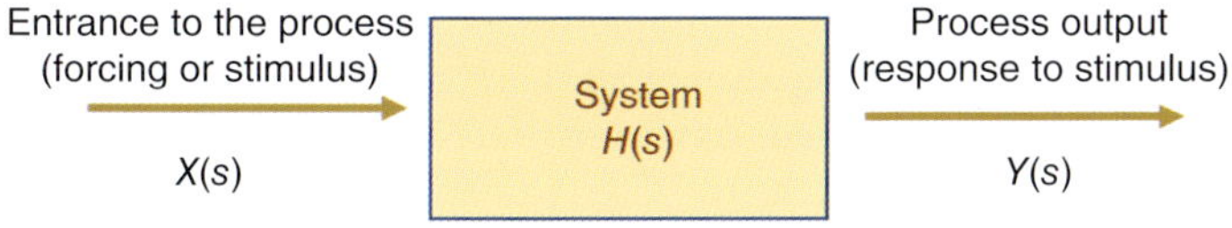

3.2 Definition and Properties of the Transfer Function

The transfer function is defined as the quotient between the Laplace transform of the output function and the Laplace transform of the input function, under the assumption that the initial conditions are zero (equal to zero). This is valid for a linear and time-invariant system.

This condition can be seen as a convolution process, formed by the input excitation convolved with the considered system, resulting in the response within a time interval. Now, in that sense (that of the convolution), it must be observed that the transfer function is formed by the deconvolution between the input signal and the system.

By definition, a transfer function can be determined by the expression:

$$H(s) = \frac{\text{Response to stimulus}}{\text{Forcing function}} = \frac{Y(s)}{X(s)} \tag{3.1}$$

where $H(s)$ is the transfer function, $Y(s)$ is the Laplace transform of the response signal, and $X(s)$ is the Laplace transform of the input signal.

Chemical Reactor Design: Mathematical Modeling and Applications,
First Edition. Juan A. Conesa.

The transfer function also can be considered as the response of a system initially inert to a pulse input signal (as we show later, the Laplace transform of a pulse is the unity):

$$H(s) = L\{h(t)\} = \int_0^\infty e^{-st} \cdot h(t)dt \tag{3.2}$$

The exit is then calculated as

$$Y(s) = H(s) \cdot X(s) \tag{3.3}$$

and the response as a function of time can be calculated through the inverse of the Laplace transform:

$$y(t) = L^{-1}\{Y(s)\} \tag{3.4}$$

3.3 Laplace Transform

The Laplace transform is a type of integral transform frequently used for the resolution of ordinary differential equations (ODEs). The Laplace transform of a defined function $f(t)$ (in differential equations, in mathematical analysis or in functional analysis ...) for all positive numbers $t \geq 0$, is the function $F(s)$, defined by

$$F(s) = L\{f(t)\} = \int_0^\infty e^{-st} \cdot f(t)dt \tag{3.5}$$

The Laplace transform converts a time domain function (where time is an independent variable) to the Laplace domain (where "s" is the independent variable). This transform can also be used, obviously, for other systems where "t" would represent a real variable. Contrarily, "s" is, in general, a complex variable of the form $s_1 + i \cdot s_2$. Nevertheless, in this book, all the Laplace transforms are in the real domain.

Let us discuss and explain some of the properties of the Laplace transform. Table 3.1 shows the most important properties.

One of the most important properties of the Laplace transform is related to the calculation of the derivatives, very common in reactor design. Note that to calculate $df(t)/dt$ in the Laplace space, we just need to evaluate "$s \cdot f(s)$," and then the Laplace transform of an integral can be calculated by "$f(s)/s$."

3.3.1 Laplace Transform of Some Important Functions for Reactor Characterization

Here are some important functions in reactor design and characterization.

Table 3.1 Properties of the Laplace transform.

Linearity:

$L\{af(t) + bg(t)\} = aL\{f(t)\} + bL\{g(t)\}$

Product by scalar:

$L\{Kf(t)\} = KL\{f(t)\} = Kf(s)$

Derivatives:

$L\{f'(t)\} = sL\{f(t)\} - f(0)$

$L\{f''(t)\} = s^2L\{f(t)\} - sf(0) - f'(0)$

$L\{f^n(t)\} = s^nL\{f(t)\} - s^{n-1}f(0) - \cdots - f^{n-1}(0)$

The values at $t = 0$ are usually taken as zero; therefore:

$L\{f'(t)\} = sL\{f(t)\} = s \cdot f(s)$

$L\{f''(t)\} = s^2L\{f(t)\} = s^2 \cdot f(s)$

$L\{f^n(t)\} = s^nL\{f(t)\} = s^n \cdot f(s)$

Integral of a function:

$L\left\{\int_0^t f(t)dt\right\} = \frac{f(s)}{s}$

Pure delay theorem. The Laplace transform of a function $f(t)$ delayed "a" units of time is

$L\{f(t - a)\} = e^{-s \cdot a} \cdot f(s)$

And then:

$L\{e^{at}f(t)\} = f(s - a)$

Final value theorem. If the temporary limit exists, the following fulfills:

$\lim_{t\to\infty} f(t) = \lim_{s\to 0} sf(s)$

Therefore:

$f(\infty) = \lim_{s\to 0}[sf(s)]$

3.3.1.1 Ramp Function

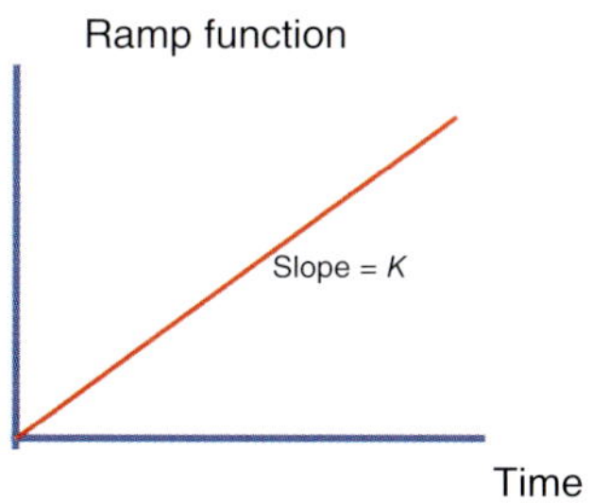

Suppose $f(t) = Kt$, applying the Laplace transform:

$$L\{Kt\} = \int_0^\infty Kte^{-st}dt = K\int_0^\infty te^{-st}dt$$

and integrating by parts, we obtain

$$L\{Kt\} = \frac{K}{s^2}$$

3.3.1.2 Sinusoidal Function

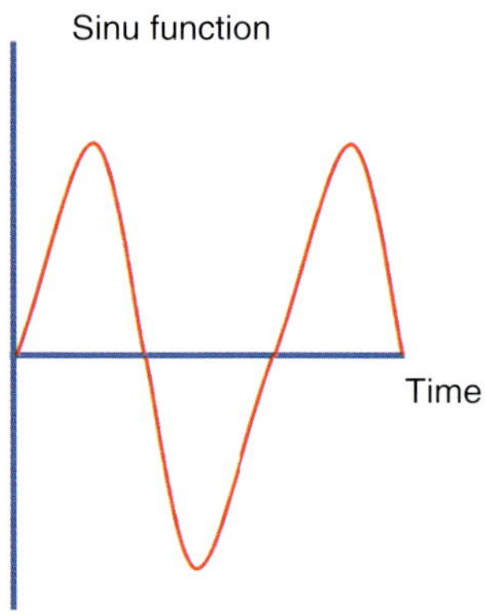

Suppose $f(t) = \sin(wt)$, where "w" is the frequency (rad/time) of the sinusoidal wave. Applying the Laplace transform:

$$L\{\sin(wt)\} = \int_0^\infty e^{-st}\sin(wt)dt$$

and integrating:

$$L\{\sin(wt)\} = \frac{w}{s^2 + w^2}$$

3.3.1.3 Pulse Function

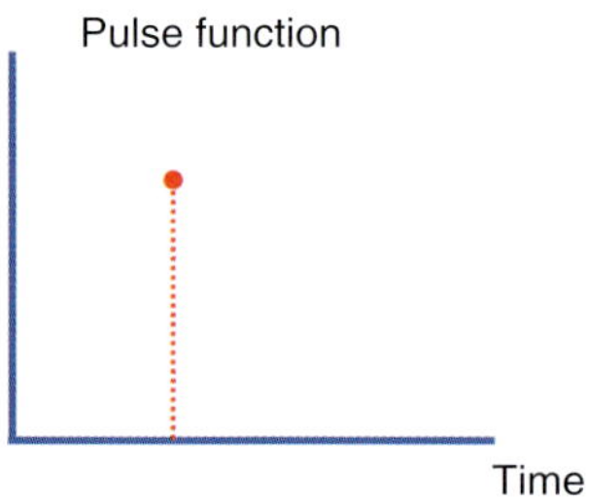

This function is particularly useful in reactor characterization, as it is generally used to determine the residence time distribution (RTD). The function is a Dirac delta, which can be represented by the derivative of a unitary step function:

$$\delta(t) = \frac{d}{dt}U(t)$$

This unitary step can be written as

$$U(t) = \lim_{a\to\infty}(1 - e^{-at})$$

and using these relations, we can write

$$L\{\delta(t)\} = L\left\{\frac{d}{dt}\lim_{a\to\infty}(1 - e^{-at})\right\} = \lim_{a\to\infty} L(ae^{-at})$$

Taking the Laplace transform:

$$\lim_{a\to\infty} L(ae^{-at}) = \lim_{a\to\infty}\frac{a}{s+a} = 1$$

showing that the Laplace transform of a pulse is the unity, as stated before.

3.3.1.4 Other Functions

Table 3.2 shows the most important Laplace transform function generally used.

Table 3.2 Laplace transform of basic functions.

$f(t)$	$F(s) = L\{f(t)\}$
Pulse function, Dirac delta	1
c	$\frac{c}{s}$
$U(t)$, unitary step	$\frac{1}{s}$
$f(t) = \begin{Bmatrix} f(t-a) & t \geq a \\ 0 & t < a \end{Bmatrix}$	$e^{-as}f(s)$
t^n	$\frac{n!}{t^{n+1}}$
e^{at}	$\frac{1}{s-a} \quad s > a$
$t^n \cdot e^{at}$	$\frac{n!}{(s-a)^{n+1}} \quad s > a$
$\sin(at)$	$\frac{a}{s^2+a^2}$
$\cos(at)$	$\frac{s}{s^2+a^2}$
$\sinh(at)$	$\frac{a}{s^2-a^2}$
$\cosh(at)$	$\frac{s}{s^2-a^2}$
$e^{-at}\cos(wt)$	$\frac{s+a}{(s+a)^2+w^2}$
$e^{-at}\mathrm{sen}(wt)$	$\frac{w}{(s+a)^2+w^2}$
$t^{-1/2}$	$\sqrt{\frac{\pi}{s}}$

3.4 Use of Laplace Transform in Chemical Reactor Characterization

If we introduce in a reactor a pulse input, as we have seen, $X(s) = 1$, and then the transfer function of the reactor is the Laplace transform of the signal at the exit, $Y(s)$, that equals the RTD in the reactor:

$$H(s) = \frac{Y(s)}{X(s)}; \quad \text{if } X(s) = 1; \qquad Y(s) = H(s) \tag{3.6}$$

If we introduce in a reactor a step input:

$$X(s) = \frac{1}{s} \tag{3.7}$$

and then,

$$H(s) = Y(s) \cdot s \tag{3.8}$$

Having in mind that:

$$L\left[\frac{df(t)}{dt}\right] = s \cdot f(s) \tag{3.9}$$

it is fulfilled:

$$L^{-1}[s \cdot Y(s)] = \frac{dY(t)}{dt} \tag{3.10}$$

and then,

$$y(t) = \int_0^t H(t)dt \tag{3.11}$$

This means that the exit $Y(t)$ is the integral function of the RTD, as expected when working with a step input.

3.4.1 Study of the RTD in the CSTR

We will study the different reactors using several input signals and using the concept of transfer function or direct solution with ODEs. This will allow us to later solve more complex systems.

(1) Step input, direct solution of the ODE:
If we consider the input and system shown in Figure 3.1.
A mass balance in nonsteady state has the following expression (see Chapter 4 for details):

Input + generation = Output + Accumulation

$$Q \cdot C_{\text{in}} + 0 = Q \cdot C_{\text{out}} + \left(\frac{dC_{\text{out}}}{dt}\right) \cdot V \tag{3.12}$$

Dividing by the flow rate, and rearranging:

$$C_{\text{in}} = C_{\text{out}} + \bar{t}\left(\frac{dC_{\text{out}}}{dt}\right) \tag{3.13}$$

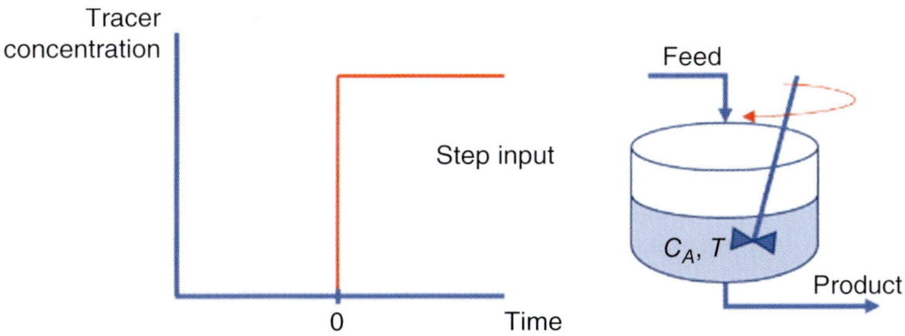

Figure 3.1 Step input of tracer to a CSTR.

where, as usual, $\bar{t} = \frac{V}{Q}$. The former equation has as initial condition: $C_{out} = 0$ at $t = 0$. Then we can write

$$\frac{dC_{out}}{dt} = (C_{in} - C_{out})\frac{1}{\bar{t}}$$

$$\ln(C_{in} - C_{out}) + \frac{t}{\bar{t}} = \ln(C_{in}) \tag{3.14}$$

and, finally:

$$F(t) = \frac{C_{out}}{C_{in}} = 1 - \exp\left(-{}^{t}/_{\bar{t}}\right) \tag{3.15}$$

This is the "F" curve (integration of the RTD) mentioned in many textbooks for the continuous stirred tank reactor (CSTR).

(2) Step input, solution using Laplace transform:
If $u(t)$ is the step input function, the mass balance can be written as

$$\text{In} + \text{Generation} = \text{Out} + \text{Accumulation}$$

$$u(t) = C_{out} + \bar{t}\left(\frac{dC_{out}}{dt}\right) \tag{3.16}$$

As for the derivative, the following is fulfilled:

$$L\left[\frac{dC_{out}(t)}{dt}\right] = s \cdot C_{out}(s) \tag{3.17}$$

and for the step input, the Laplace transform is

$$u(s) = \frac{\Delta u}{s} \tag{3.18}$$

In the previous equation, Δu is the value of the change in the step input. We can write

$$\frac{\Delta u}{s} = C_{out}(s) + \bar{t}sC_{out}(s) \tag{3.19}$$

Solving:

$$C_{out}(s) = \frac{\Delta u}{s(1 + \bar{t}s)} = \frac{\Delta u}{s} - \frac{\Delta u}{(1 + \bar{t}s)} \tag{3.20}$$

Looking for the inverse of the Laplace transform:

$$C_{\text{out}}(t) = \Delta u \left[1 - \exp\left(-\frac{t}{\bar{t}}\right)\right] \tag{3.21}$$

and, therefore:

$$F(t) = \frac{C_{\text{out}}(t)}{\Delta u} = 1 - \exp\left(-{}^{t}\!/_{\bar{t}}\right) \tag{3.22}$$

As we would expect, Eq. (3.22) and (3.15) are the same.

(3) Pulse input, direct solution with the differential equation:
Let us now think about a pulse input injected into a CSTR. In that case, the mass balance is

$$\text{Input} + \text{Generation} = \text{Output} + \text{Accumulation}$$

$$0 = Q \cdot C_{\text{out}} + V\left(\frac{dC_{\text{out}}}{dt}\right) \tag{3.23}$$

But with the initial condition of this ODE as

$$t = 0 \rightarrow C_{\text{in}}(0) = \frac{M}{V} \tag{3.24}$$

Using Eq. (3.23) and the definition of $\bar{t}$, then:

$$\left(-\frac{dC_{\text{out}}}{C_{\text{out}}}\right) = \frac{dt}{\bar{t}} \quad \rightarrow \quad \ln\frac{C_{\text{out}}}{C_{\text{in}}(0)} = -\frac{t}{\bar{t}} \rightarrow \quad \frac{C_{\text{out}}}{C_{\text{in}}(0)} = \exp\left(-\frac{t}{\bar{t}}\right)$$

For calculating the $E(t)$ function, we must calculate the integral of this output signal; therefore:

$$\int_0^{\infty} \frac{C_{\text{out}}}{C_{\text{in}}(0)} dt = \int_0^{\infty} \exp\left(-\frac{t}{\bar{t}}\right) dt = \bar{t} \cdot \left[\exp\left(-\frac{t}{\bar{t}}\right)\right]_0^{\infty} = \bar{t}$$

In this way, we can calculate:

$$E(t) = \frac{\frac{C_{\text{out}}}{C_{\text{in}}(0)}}{\int_0^{\infty} \frac{C_{\text{out}}}{C_{\text{in}}(0)} dt} = \frac{1}{\bar{t}} \exp\left(-\frac{t}{\bar{t}}\right) \tag{3.25}$$

The result is, as we expected, the RTD of the CSTR, as a step input was used as input signal.

(4) Pulse input, using Laplace transform:
This later case can also be treated using integral transform. Following Eq. (3.23):

$$\text{Input} + \text{Generation} = \text{Output} + \text{Accumulation}$$

$$M \cdot \delta(t) + 0 = V\left(\frac{dC_{\text{out}}}{dt}\right) + Q \cdot C_{\text{out}} = (C_{\text{in}}(0) \cdot V) \cdot \delta(t) \tag{3.26}$$

Taking Laplace transform:

$$V \cdot s \cdot C_{\text{out}}(s) + Q \cdot C_{\text{out}}(s) = (C_{\text{in}}(0) \cdot V) \cdot 1 \tag{3.27}$$

Rearranging:

$$C_{\text{out}}(s) = \frac{C_{\text{in}}(0) \cdot V}{Q} \cdot \frac{1}{(1 + \bar{t} \cdot s)} = C_{\text{in}}(0) \cdot \bar{t} \cdot \frac{1}{(1 + \bar{t} \cdot s)} \tag{3.28}$$

Doing the inverse:

$$L^{-1}\left[\frac{1}{(1 + \bar{t} \cdot s)}\right] = \frac{1}{\bar{t}} \exp\left(-\frac{t}{\bar{t}}\right) \tag{3.29}$$

$$C_{\text{out}}(t) = \frac{C_{\text{in}}(0)}{\bar{t}} \exp\left(-\frac{t}{\bar{t}}\right) \tag{3.30}$$

and, finally:

$$E(t) = \frac{\frac{C_{\text{out}}}{C_{\text{in}}(0)}}{\int_0^\infty \frac{C_{\text{out}}}{C_{\text{in}}(0)} dt} = \frac{1}{\bar{t}} \exp\left(-\frac{t}{\bar{t}}\right) \tag{3.31}$$

Eq. (3.31) obviously represents the RTD in the mixed flow reactor. At this point, we have to ask for the transfer function in the CSTR itself, i.e. what is the function $H(s)$ transforming the input signals $X(s)$ into $Y(s)$? If we take a look at the previous equation, the transfer function in a CSTR can be derived from the $E(t)$:

$$E(s) = L\left[\frac{1}{\bar{t}} \exp\left(-\frac{t}{\bar{t}}\right)\right] = \frac{1}{1 + \bar{t}s} \tag{3.32}$$

3.4.2 Study of the RTD in the PFR

Let us now discuss the mass balance of a tracer in different situations in the plug flow reactor (PFR). The mass balance in a differential volume of a PFR in non-steady state is (see Chapter 5):

Input + Generation = Output + Accumulation

$$Q \cdot C + 0 = Q \cdot (C + dC) + \left(\frac{dC}{dt}\right) \cdot dV$$

$$-Q \cdot dC = \left(\frac{dC}{dt}\right) \cdot dV \tag{3.33}$$

Using partial derivatives:

$$Q \cdot \left(\frac{\delta C}{\delta V}\right) = -\left(\frac{\delta C}{\delta t}\right) \tag{3.34}$$

If the input has the form of a step, in the time–space coordinates, the corresponding initial values are

$$\left.\begin{array}{l} C(0, t) = u(t) \\ C(V, 0) = 0 \end{array}\right\} \tag{3.35}$$

In this equation, $u(t)$ represents the step input function. That, taking Laplace transform, is

$$\left.\begin{array}{l} C(0,s) = \frac{\Delta u}{s} \\ C(V,0) = 0 \end{array}\right\} \tag{3.36}$$

Eq. (3.34) should also be transformed with respect to "t," and we get

$$Q \cdot \left(\frac{\delta C}{\delta V}\right) + s \cdot C(s) = 0$$

or, more commonly:

$$Q \cdot dC = -s \cdot C(s) \cdot dV$$

The solution of this linear differential equation is

$$C\,(s) = \frac{\Delta u}{s} \exp\left(-s\frac{V}{Q}\right) \tag{3.37}$$

and doing the inverse transform:

$$F(t) = \frac{C}{\Delta u} = L\left[\frac{1}{s}\exp\left(-s\frac{V}{Q}\right)\right] = u\left(t - \frac{V}{Q}\right) = u(t - \bar{t}) \tag{3.38}$$

This function implies that the step input delays a time equivalent to $\bar{t} = V/Q$ in the reactor, i.e. the residence time in the plug flow.

As we have done for the back mixed flow reactor, we will need a function $H(s)$ characteristic of the reactor. From the previous discussion, we see that, in the PFR, the transfer function is

$$E(s) = L(\delta(t - \bar{t})) = \exp(-\bar{t}s) \tag{3.39}$$

3.5 Complex Network of Ideal Reactors

Combinations of ideal reactor elements connected in series and parallel are often used to model the results of tracer tests performed on real reactors. Such systems are easily analyzed with the Laplace transform (L.T.) technique. Also, the L.T. technique provides a very convenient means of obtaining the moments of RTD functions (Chapter 1) as well as aids considerably in obtaining the solution of partial differential equations (Chapter 4).

We already know how the RTDs and how the transfer functions of the two ideal reactors (PFR and CSTR) are. Using these RTDs, we can derive the functions corresponding to a compartmentalized general model with delays in time. This means that given a real flow system or reactor that can be represented by a set of CSTRs and PFRs in series or in parallel, we will be able to deduce the curve corresponding to the combination. We will use for this purpose the Laplace transforms.

In this sense, any network of CSTRs and PFRs consists of the following:

- Elements in parallel
- Series elements
- Mix points
- Points of separation (split)

Before presenting the general mathematical approach, let us see a simple example that will help us understand more complex systems.

Example 3.1 ***Combination of Two Reactors in Series***
During the analysis of the behavior of a reactor, the curve of the response to a tracer is determined. In view of the results, it is verified that the reactor behaves like a PFR connected with a CSTR in series, each one with a different volume, as shown in the following figure. Determine the transfer function of the entire system.

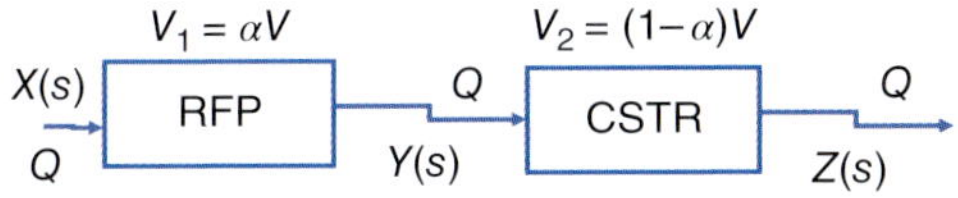

Solution
From the definition of the transfer function, we can write

$$H_1(s) = \frac{Y(s)}{X(s)}$$
$$H_2(s) = \frac{Z(s)}{Y(s)}$$

where $H_1(s)$ and $H_2(s)$ are the transfer functions of the reactors. Easily we can get

$$Z(s) = H_1(s) \cdot H_2(s) \cdot X(s) = X(s) \cdot \exp\left(-\frac{t}{\overline{t_1}}\right) \cdot \frac{1}{1 + \overline{t_2} \cdot s}$$

Note that we have substituted the transfer functions by the ones corresponding to PFR (reactor 1) and CSTR (reactor 2). In this equation, the residence times $\overline{t_1}$ and $\overline{t_2}$ refer to the flow and volume of each reactor, i.e.:

$$\overline{t_1} = \frac{\alpha V}{Q} = \alpha\bar{t}$$
$$\overline{t_2} = \frac{(1-\alpha)V}{Q} = (1-\alpha)\bar{t}$$

In this simple way, we see that a convolution of signals is transformed into a product in the Laplace space. Let us continue with more complex systems.

3.5.1 Systems in Series

For a system of two subsystems in series, such as the one shown in Figure 3.2, we can write

$$Y_1(s) = H_1(s) \cdot X_1(s)$$
$$Y_2(s) = H_2(s) \cdot Y_1(s) \tag{3.40}$$

and then:

$$Y_2(s) = H_1(s) \cdot H_2(s) \cdot X_1(s) \tag{3.41}$$

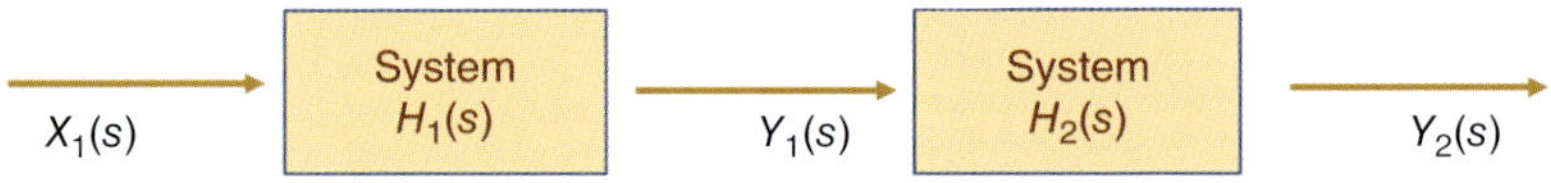

Figure 3.2 Series of coupled systems and signals.

This would be equivalent to the convolution operation discussed in Chapter 2. Nevertheless, as we have seen in the previous example, each transfer function should be evaluated with the corresponding change in volume (or residence time). In this way, it is more correct to write

$$Y_2(s) = H_1(\alpha_1 s) \cdot H_2(\alpha_2 s) \cdot X_1(s) \tag{3.42}$$

where α_1 is the ratio of volume in reactor 1, and α_2 corresponding to reactor 2. If we look for the transfer function of the whole system, i.e. $Y_2(s)/X_1(s)$, we get

$$H(s) = E(s) = Y_2(s)/X_1(s) = H_1(\alpha_1 s) \cdot H_2(\alpha_2 s) \tag{3.43}$$

That corresponds to the Laplace transform of the residence time distribution of the whole system.

In general, for a system of N-subsystems in series:

$$E(s) = \prod_{j=1}^{N} E_j(\alpha_j s) \tag{3.44}$$

where $E_j(s)$ is the L.T. of a pulse introduced in the individual system having a volume V.

Example 3.2 ***N-CSTRs in Series***
If we want to find the E-curve of N equal-sized CSTRs in series, like the system shown in Figure 3.3, we will use the RTD of a single stirred tank of volume V and flow rate Q:

$$E(t) = \frac{1}{\bar{t}} \exp\left(-\frac{t}{\bar{t}}\right) \tag{3.45}$$

So that the transfer function is

$$E(s) = \frac{1}{1 + \bar{t}s}$$

where $\bar{t} = \frac{V}{Q}$. For N equal-sized tanks in series, each tank contains $\alpha_j = \frac{1}{N}$ fraction of the volume. Hence:

$$E_j(\alpha\, s) = E_j\left(\frac{s}{N}\right) = \frac{1}{1 + \frac{\bar{t}}{N}s} \tag{3.46}$$

The overall transfer function is obtained by

$$E(s) = \prod_{j=1}^{N} E_j(\alpha_j s) = \frac{1}{\left(1 + \frac{\bar{t}}{N}s\right)^N} \tag{3.47}$$

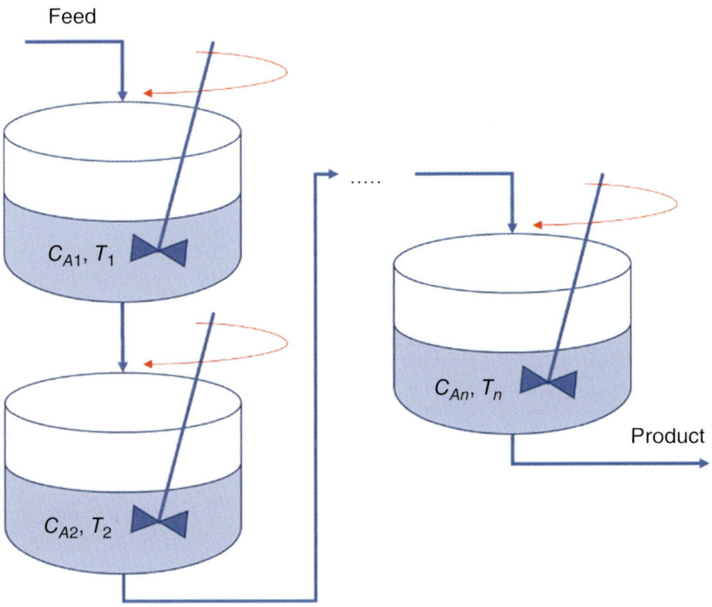

Figure 3.3 System of CSTRs connected in series.

The overall impulse response then is obtained by inversion of the Laplace transform:

$$E(t) = L^{-1}\left\{\frac{1}{\left(1+\frac{\bar{t}}{N}s\right)^N}\right\} = \left(\frac{N}{\bar{t}}\right)^N L^{-1}\left\{\frac{1}{\left(s+\frac{N}{\bar{t}}\right)^N}\right\}$$
$$= \left(\frac{N}{\bar{t}}\right)^N \frac{\bar{t}^{N-1}}{(N-1)!}e^{-Nt/\bar{t}} \tag{3.48}$$

This equation is equivalent to that found in Chapter 1 for the tanks-in-series model, as it could not be otherwise.

3.5.2 Systems in Parallel

Let us consider a system with two branches in parallel, like the one shown in Figure 3.4.

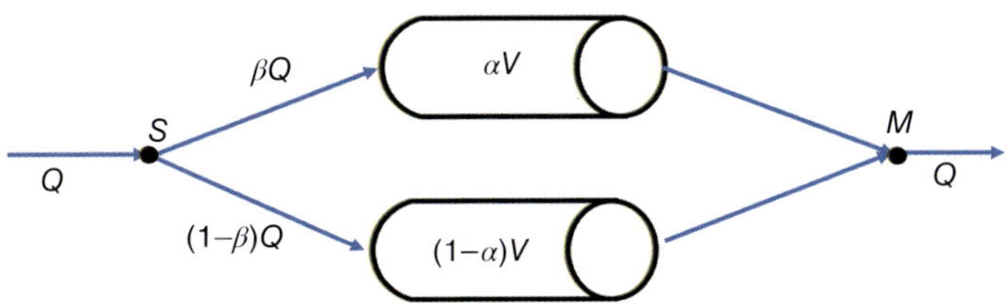

Figure 3.4 Branches in parallel with the split "S" and mix "M" points.

In this system, the upper branch contains a fraction α of the total volume, and the lower one $(1-\alpha)$. Through the upper branch passes a flow βQ and the lower one $(1-\beta)Q$. We will assume that the volume of the split point (S) and the mix point (M) are nil. Let $E_1(s)$ and $E_2(s)$ be the transfer functions (Laplace transform of an impulse and therefore Laplace transform of the reactor RTD) of the upper and lower branches, in the case that this branch had all the volume and all the flow would pass through it. Since only volume αV is in the upper branch and passes βQ through it, the Laplace transform of the response is

$$E_1\left(\frac{\alpha}{\beta}s\right) \tag{3.49}$$

Similarly, for the other branch:

$$E_2\left(\frac{1-\alpha}{1-\beta}s\right) \tag{3.50}$$

The overall response is obtained by the weighted average of the two, since at the point M there is a mixture proportional to the flow:

$$E(s) = \beta\, E\left(\frac{\alpha}{\beta}s\right) + (1-\beta)\overline{E_2}\left(\frac{1-\alpha}{1-\beta}s\right) \tag{3.51}$$

This can be generalized to M-branches in parallel, so that:

$$E(s) = \sum_{j=1}^{M} \beta\, E_j\left(\frac{\alpha_j}{\beta_j}s\right) \tag{3.52}$$

where β_j is the fraction of the total flow passing through branch "j" and α_j is the volume fraction that this branch has.

This implies that if the dimensionless response of a system with volume V and flow Q produces a certain flow pattern $E(t)$, then the response to an impulse when the same system when it is a subsystem (part of a network) containing a fraction α of the volume and a passing fraction β of the flow through it, is

$$E_{\text{sub}}\left(\frac{\alpha}{\beta}t\right) \tag{3.53}$$

Using the L.T.s:

$$L\{E(t)\} = \int_0^\infty e^{-st}E(t)dt = \frac{1}{\bar{t}}\int_0^\infty e^{-st}E_\theta(\theta)dt = E(s) \tag{3.54}$$

$$L\{E_{\text{sub}}(t)\} = \frac{\alpha}{\beta}\frac{1}{\bar{t}}\int_0^\infty e^{-st}E\left(\frac{\alpha}{\beta}t\right)dt = E_{\text{sub}}\left(\frac{\alpha}{\beta}s\right) \tag{3.55}$$

Example 3.3 ***System Modeled Through Two CSTRs in Parallel***
For a system of two reactors, like the one shown in Figure 3.5:

$$\bar{t}_1(s) = \frac{\alpha}{\beta}\bar{t}$$

$$\overline{t_2}(s) = \frac{1-\alpha}{1-\beta}\bar{t} \tag{3.56}$$

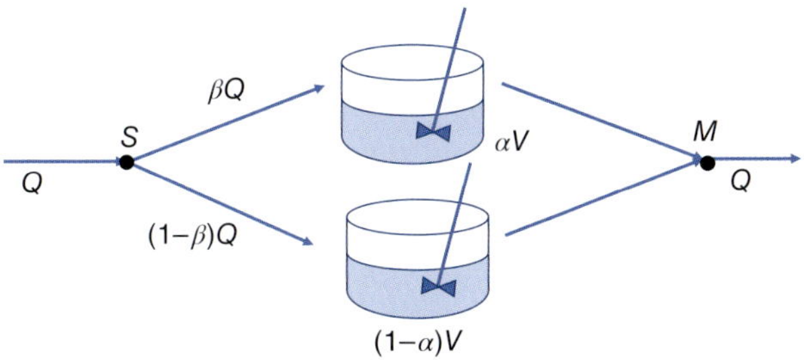

Figure 3.5 Two continuous stirred tanks in parallel.

where β and $(1-\beta)$ are the fractions of the flow and α and $(1-\alpha)$ are the fractions of the volume of each branch. In the same way:

$$E(s) = \beta E_1(s) + (1-\beta)E_2(s) \tag{3.57}$$

If the reactors of the system shown in the last figure were CSTRs:

$$E_1(s) = E_2(s) = \frac{1}{1+\bar{t}_j s} \qquad j = 1, 2 \tag{3.58}$$

and then:

$$E(s) = \beta \frac{1}{1+\frac{\alpha}{\beta}\bar{t}s} + (1-\beta)\frac{1}{1+\frac{1-\alpha}{1-\beta}\bar{t}s} \tag{3.59}$$

This means that:

$$E(t) = L^{-1}\{E(s)\} = \frac{\beta^2}{\alpha\bar{t}}e^{-\frac{\beta t}{\alpha\bar{t}}} + \frac{(1-\beta)^2}{(1-\alpha)\bar{t}}e^{-\frac{(1-\beta)t}{(1-\alpha)\bar{t}}} \tag{3.60}$$

3.5.3 Systems with Recycle

Now that we have learned the derivation of the transfer function for subsystems in parallel and in series, let us face systems with recirculation. The only new "rule" is the division rule: if a current is divided, each of the outgoing currents has the same transfer function. Consider a general recirculation system like the one shown in Figure 3.6.

Flow rate Q flows through a recycle system (the system within the dashed box is the system with recycle) of total volume V. Internally, at point M flow rate, Q

Figure 3.6 Combination of systems showing internal or external recycle current.

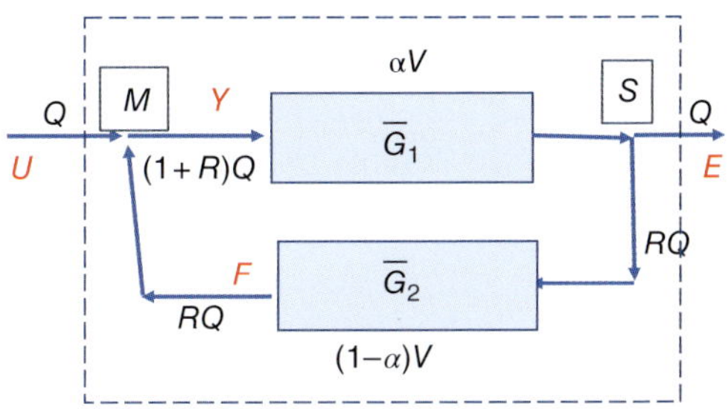

is joined by recycle flow rate, RQ, so that the flow rate of $(R+1)Q$ flows through the forward branch of the system that contains volume αV. At splitting point S, RQ is recycled through the recycle branch of volume $(1-\alpha)V$ when flow rate Q leaves the system.

Let F, U, Y, and E be the signals of the system in the branches shown in Figure 3.6. By definition, we can write

$$G_1 = \frac{E}{Y}$$

$$G_2 = \frac{F}{E}$$

G_1 and G_2 being the transfer functions of the systems 1 and 2, respectively. By addition of the signals, we have the following at the point M:

$$Q \cdot U + RQ \cdot F = (1+R)Q \cdot Y$$

Eliminating the value of Q:

$$U + R \cdot F = (1+R) \cdot Y$$

If we consider $U = 1$ (pulse input), and rearranging:

$$E = G_1 \cdot Y = \frac{1 + R \cdot F}{1+R} G_1 = \frac{1 + R \cdot G_2 \cdot E}{1+R} G_1$$

$$E = \frac{G_1}{1 + R - R \cdot G_1 \cdot G_2}$$

which can be also written as

$$E = \frac{1}{1+R} \frac{G_1}{1 - \frac{R}{R+1} G_1 G_2}$$

To get the impulse response in the time domain, inversion of the equation from the Laplace space is necessary. Often, it is interesting to expand the last equation by binomial theorem, to get

$$E = \frac{1}{1+R} G_1 \sum_{n=0}^{\infty} \left(\frac{R}{R+1} \right)^n G_1{}^n G_2{}^n \tag{3.61}$$

and obtain the impulse response in the time domain by inverting each term in the expression.

Example 3.4 ***Recirculation in a Packed Bed Reactor***

A reactor packed with solid catalyst is used to crack ethylene to its monomers at an elevated but constant temperature. A pulse input of C^{13}-marked tracer is fed, and the signal schematized in the figure is obtained. The flow rate was 1 l/min and the reactor volume (not occupied by the solid) is 5 l. Find the following: (a) the mean residence time of the gas, (b) a model valid for the reactor, and its $E(t)$ function.

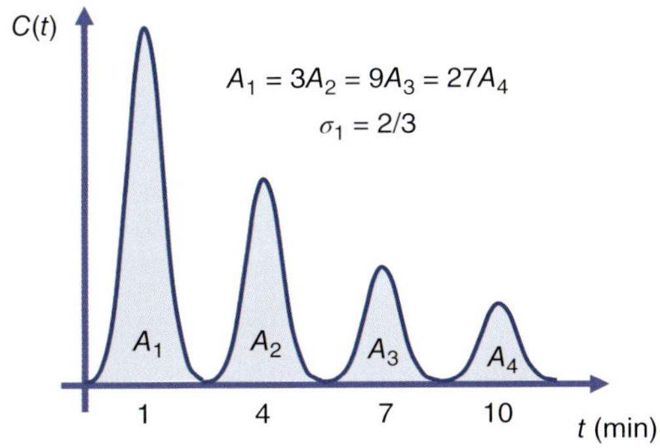

Solution

(a) The mean residence time can be calculated, according to Chapter 1, by

$$t_m = \frac{\int_0^\infty t \cdot C(t)dt}{\int_0^\infty C(t)dt} = \frac{1 \cdot A_1 + 4 \cdot \frac{A_1}{3} + 7 \cdot \frac{A_1}{9} + 10 \cdot \frac{A_1}{27}}{A_1 + \frac{A_1}{3} + \frac{A_1}{9} + \frac{A_1}{27}}$$

$$= \frac{94}{40} = 2.357 \text{ min}$$

(b) The RTD shown is typical of a system in which recirculation is occurring. Thus, we propose a model that is schematized in the figure. V_p represents a PFR. Let us call C_0, C_1, and C_2, respectively, to the signals of the tracer fed, entering the reactor model and in the recirculation current. Let us call R to the ratio of flow rate in the recirculation.

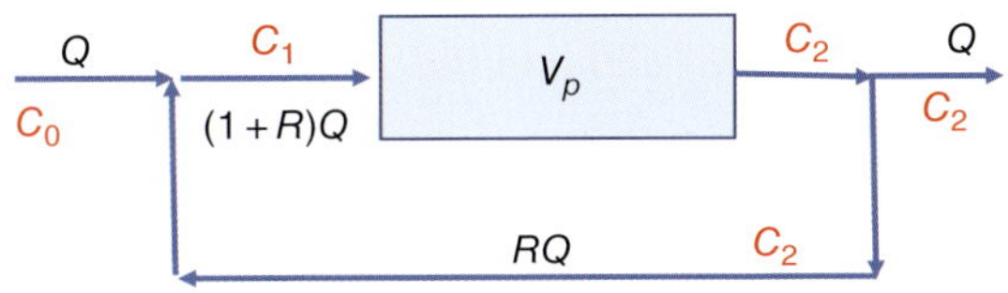

We can write, in the recirculation point:

$$Q \cdot C_0 + RQ \cdot C_2 = (1 + R)Q \cdot C_1$$

and then:

$$C_1 = \frac{C_0 + RC_2}{1 + R}$$

In the reactor itself, we have

$$G_1 = \frac{C_2}{C_1} = \frac{C_2}{\frac{C_0 + RC_2}{1 + R}} = \frac{(1 + R) \cdot C_2}{C_0 + RC_2}$$

"G_1" being the transfer function of the reactor vessel. We can clear, looking for signal exit:

$$(C_0 + RC_2) \cdot G_1 = (1 + R) \cdot C_2$$
$$C_0 G_1 + RC_2 G_1 = C_2 + RC_2$$
$$C_2 = \frac{C_0 G_1}{(1 + R) - RG_1}$$

Note that the transfer function of the whole system is given by

$$G_{\text{global}} = \frac{C_2}{C_0} = \frac{G_1}{(1 + R) - RG_1}$$

In this case, $G_1 = \exp(-\overline{t_p} s)$. The residence time in the vessel can be related to the average residence time in the system:

$$\overline{t_p} = \frac{V_p}{(1 + R) \cdot Q} = \frac{\bar{t}}{1 + R}$$

From this equation, we can calculate the recirculation flow:

$$\frac{5}{1 + R} = 2.357 \qquad \rightarrow R = 1.10$$

Taking into account these equations, finally:

$$G_{\text{global}} = \frac{\exp\left(-\frac{\bar{t}}{1+R} s\right)}{(1 + R) - R \exp\left(-\frac{\bar{t}}{1+R} s\right)}$$

As one can expect, it is very difficult to do the anti-transform of this function. For a transfer function such as $G_1 = \exp(-\overline{t_p} s) = \exp\left(-\frac{\bar{t}}{1+R} s\right)$ we can do the anti-transform, and it is

$$E_1(t) = \delta\left(t \quad \frac{\bar{t}}{1 + R}\right)$$

The anti-transform of G_{global} gives a $E(t)$ having a lot of peaks, as expected:

$$E(t) = \delta\left(t - \frac{\bar{t}}{1 + R}\right) + \delta\left(t - 2\frac{\bar{t}}{1 + R}\right) \delta + \left(t - 3\frac{\bar{t}}{1 + R}\right) + \ldots$$

Note that this reactor can also be modeled using a combination of the four PFRs of different volumes (V_1, V_2, V_3, and $1{-}V_1{-}V_2{-}V_3$) connected in parallel, each one with a flow rate Q_1, Q_2, Q_3, and $(Q{-}Q_1{-}Q_2{-}Q_3)$. The volume and flow rate in each PFR can be calculated bearing in mind that the reactor residence time is $t_i = V_i/Q_i$ and that the area of each peak is $A_i = Q_i/Q_{\text{total}}$. The result will give the same $E(t)$.

Example 3.5 ***Nonideal CSTR***

A second-order irreversible reaction ($A \rightarrow$ Products) is being produced in a tank reactor of volume $V = 25\ \text{m}^3$. The flow rate entering the reactor is $Q = 1\ \text{m}^3/\text{min}$, and the system was designed to operate as a CSTR with a high conversion, but now it is poorly operating at $X_A = 0.25$.

A pulse input of $M = 250$ g of tracer is used to characterize the reactor. At the exit, the concentration of tracer is measured, giving the following data:

t (min)	6	8	10	12	14	16	18	20
c (mg/l)	4.3	3.2	2.7	2.1	1.7	1.4	1.0	0.8

Besides this, initially we observe rapid fluctuations for five seconds, measuring a very high tracer concentration in this period of time.

(a) Propose a model for the reactor.
(b) If the reactor behaves as a perfect CSTR, what volume do we need for $X_A = 0.75$?
(c) What volume of a perfect CSTR do we need to get conversion that currently is produced by the well-mixed region?

Solution

(a) When a sharp increase of tracer is observed, the first possibility to test is the presence of a bypass. We cannot a priori discard the possibility of stagnancy either. We can propose the following model based on the available data:

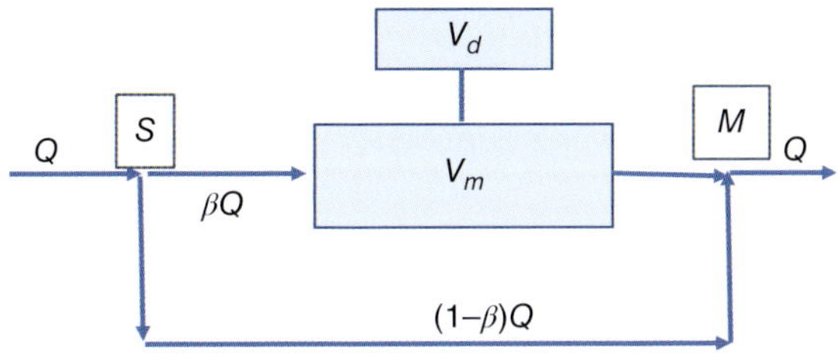

In the figure, V_m represents the volume of the perfectly mixed region, and V_d the "dead" volume, not accessible to flow. Let $V_m = \alpha \cdot V$ and $V = V_m + V_d$. The fraction of the flow in the bypass is $(1 - \beta)$. In this way, the transfer function for the whole system will be given by the sum of the bypass and the stirred tank:

$$E(s) = (1-\beta) + \frac{\beta}{1 + \frac{\alpha}{\beta}\bar{t}s}$$

And the impulse response is

$$E(t) = (1-\beta)\cdot\delta(t) + \frac{\beta^2}{\alpha\bar{t}}e^{-\frac{\beta t}{\alpha \bar{t}}}$$

where $\delta(t)$ is the Dirac delta function and $\bar{t} = \frac{V}{Q}$

If we have in mind that $C(t) = \frac{M\cdot E(t)}{Q}$, then:

$$C(t) = \frac{M}{Q}\left[(1-\beta)\delta(t) + \frac{\beta^2}{\alpha\bar{t}}e^{-\frac{\beta t}{\alpha\bar{t}}}\right]$$

We can plot in a semilog graph:

$$\ln(C(t)) = \ln\left(\frac{M\beta^2}{Q\alpha\bar{t}}\right) - \frac{\beta}{\alpha\bar{t}}t$$

The slope of the graph with the data given provides a value of −0.1177 and the extrapolation to $t = 0$ gives a value of 2.1549, this is, exp(2.1549) = 8.62 mg/l. Consequently, we can calculate the values of $\beta = 0.3$ and $\alpha = 0.1$. In this way, the active volume is approx. $0.3 \times 25 = 7.5\ \text{m}^3$.

In a CSTR for a second-order reaction, we have

$$\bar{t} = \frac{V}{Q} = \frac{C_{A0}X_A}{kC_{A0}^2(1 - X_A)^2}$$

where V and Q are the active volume and flow rate, respectively. It is important to note that X_A represents the conversion of the reactor, which does not equal the conversion of the isolated CSTR as there exists a bypass. In our example, we can write

$$\frac{\alpha V}{\beta Q} = \frac{C_{A0}X_{Av}}{kC_{A0}^2(1 - X_{Av})^2}$$

where X_{Av} represents the conversion just outside the reactor vessel. If a balance over the mixing point M is done:

$$(1 - \beta)n_{A0} + \beta n_{A0}(1 - X_{Av}) = n_{A0}(1 - X_A)$$

In our case, the system gives a conversion $X_A = 0.25$. Taking these considerations, one can see that:

$$X_{Av} = \frac{X_A}{\beta} = 0.83$$

And from all this, it is possible to calculate kC_{A0}, resulting in a value of $3.44\ \text{min}^{-1}$.

(b) $V_{new} = \dfrac{1000 \cdot 0.25}{3.44 \cdot (1 - 0.25)^2} = 128\ \text{l}$

(c) $V_{new} = \dfrac{1000 \cdot 0.83}{3.44 \cdot (1 - 0.83)^2} = 8333\ \text{l}$

Example 3.6 ***Deep Reactor***

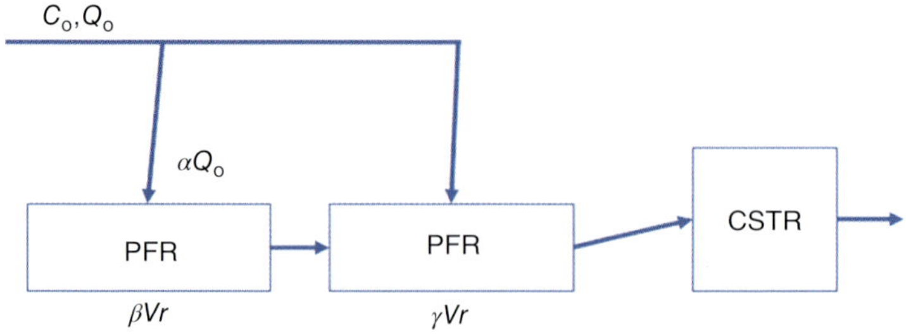

A very deep stirred reactor is agitated with two agitators on a single axis. A plausible model of the reactor is two PFRs partially in parallel and partially in series, followed by a CSTR, as indicated in the scheme. Let α be the fraction of the total flow that is fed into the first PFR and let β and γ be the fractions occupied by each PFR. Determine the values of the average residence times in each of the three reactors of the model, based on the total average time of residence. Calculate the transfer function of the system to a pulse input.

Solution
The volume of the CSTR is $(1-\beta-\gamma)\cdot V_r$ and the relation of the residence times with the total $\bar{t}$ would be

$$\overline{t_1} = \frac{\beta}{\alpha}\bar{t} \text{ (flow } \alpha Q_0 \text{ is flowing through this vessel)}$$

$$\overline{t_2} = \gamma\,\bar{t} \text{ (the total flow } Q_0 \text{ is flowing through this vessel)}$$

$$\overline{t_1} = (1-\beta-\gamma)\,\bar{t} \text{ (the total flow } Q_0 \text{ is flowing through this vessel)}$$

The transfer functions are

$$G_1 = \exp(-\overline{t_1}s)$$

$$G_2 = \exp(-\overline{t_2}s)$$

$$G_3 = \frac{1}{1+\overline{t_3}s}$$

By definition:

$$C_1 = G_1 \cdot C_0$$

Doing a balance, we can write

$$Q_0C_2 = G_2C_1\alpha Q_0 + C_0(1-\alpha)Q_0$$

and then:

$$[\alpha G_1 + (1-\alpha)]C_0G_2G_3 = C_3$$

From that:

$$\frac{C_3}{C_0} = [\alpha G_1G_2G_3 + (1-\alpha)G_2G_3]$$

And, finally:

$$G_{\text{total}} = \frac{C_3}{C_0} = \left[\frac{\alpha\exp(-\overline{t_1}s)\exp(-\overline{t_2}s)}{1+\overline{t_3}s} + (1-\alpha)\frac{\exp(-\overline{t_2}s)}{1+\overline{t_3}s}\right]$$

Example 3.7 ***Square Pulse of Tracer***
A square pulse of tracer of magnitude C_{in}/a and of duration "a" units of time is charged to a CSTR, starting at $t=0$. We are interested in simulating the response of that reactor vs. time for $a=0$, 1, and 2.

Solution

The input signal would have the following expression:

$$\text{Input} = \frac{C_{\text{in}}}{a}[u(t) - u(t - a)]$$

where $u(t)$ is the step input and $u(t - a)$ is the same input delayed "a" units. This equation gives the squared input signal. We can calculate the Laplace transform:

$$\begin{aligned}\text{Input}(s) &= \frac{C_{\text{in}}}{a}L(u(t)) - \frac{C_{\text{in}}}{a}L(u(t-a)) = \frac{C_{\text{in}}}{a}\frac{1}{s} - \frac{C_{\text{in}}}{a}\frac{\exp(-s \cdot a)}{s} \\ &= \frac{C_{\text{in}}}{a \cdot s}(1 - \exp(-s \cdot a))\end{aligned}$$

Bearing in mind that the CSTR has the transfer function shown in Eq. (3.32), at the output we will have

$$\text{Output}(s) = \text{Input}(s) \cdot E(s) = \frac{C_{\text{in}}}{a \cdot s}(1 - \exp(-s \cdot a)) \cdot \frac{1}{1 + \bar{t}s}$$

Rearranging:

$$\text{Output}(s) = \frac{C_{\text{in}}}{a \cdot s \cdot (1 + \bar{t}s)} - \frac{C_{\text{in}} \cdot \exp(-s \cdot a)}{a \cdot s \cdot (1 + \bar{t}s)}$$

And doing the inverse of Laplace transform:

$$\text{Output}(t) = \frac{C_{\text{in}}}{a}\left[1 - \exp\left(-\frac{t}{\bar{t}}\right)\right] - \frac{C_{\text{in}}}{a}\left[1 - \exp\left(-\frac{t-a}{\bar{t}}\right)\right]$$

With this equation, we can simulate the curve for the values of $a = 0$, 1, and 2:

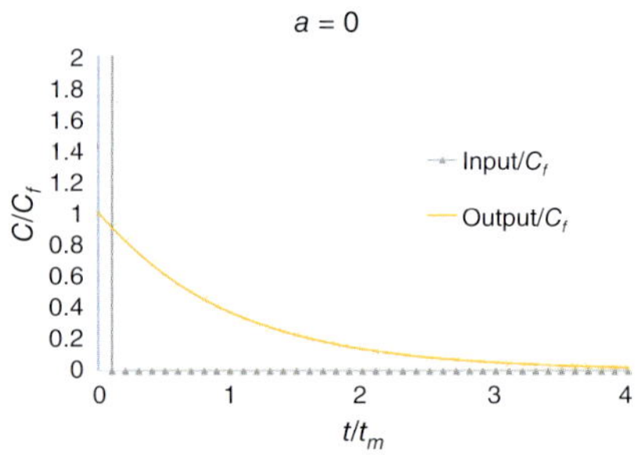

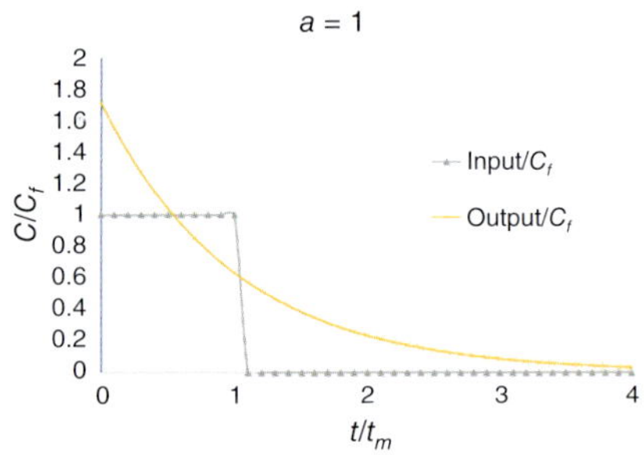

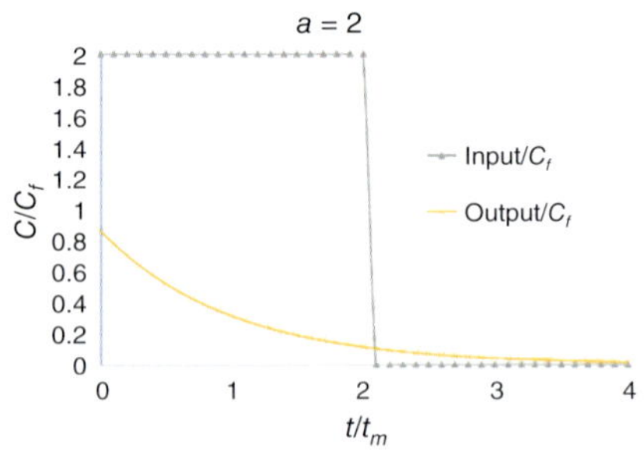

Example 3.8 ***Combining Mole Balance and Laplace Transform in a Reactor***
A test with a tracer enters a CSTR following the function:

$$C_{\text{in}} = 0.06 \cdot t \quad \text{when } 0 \le t \le 5$$
$$C_{\text{in}} = 0.06 \cdot (10 - t) \quad \text{when } 5 < t \le 10$$
$$C_{\text{in}} = 0 \quad \text{when } t > 10$$

Calculate the exit concentration as a function of time.

Solution
In the CSTR we have, for a non-reacting species:

$$\bar{t}\frac{dC_{\text{out}}}{dt} + C_{\text{out}} = C_{\text{in}} \quad \text{with } C_{\text{out}} = 0 \text{ at } t = 0$$

In the interval $0 \le t \le 5$:

$$\bar{t}\frac{dC_{\text{out}}}{dt} + C_{\text{out}} = 0.06t$$

Doing the Laplace transform:

$$\bar{t} \cdot s \cdot C(s) + C(s) = \frac{0.06}{s^2}$$

Solving for $C(s)$:

$$C(s) = \frac{\frac{0.06}{s^2}}{1 + \bar{t} \cdot s}$$

And doing the inverse of the Laplace transform:

$$C_{\text{out}}(t) = 0.06\bar{t}\left[\frac{t}{\bar{t}} - 1 + \exp\left(-\frac{t}{\bar{t}}\right)\right]$$

This equation gives at $t = 5 \rightarrow C_{\text{out}} = 0.11$
In the interval when $5 < t \le 10$:

$$\bar{t}\frac{dC_{\text{out}}}{dt} + C_{\text{out}} = 0.06 \cdot (10 - t) \quad \text{with } C_{\text{out}} = 0.11 \text{ at } t = 5$$

Let us call $C_{\text{out}}(t = 5) = C_5$. Transforming:

$$\bar{t} \cdot s \cdot (C(s) - C_5) + C(s) = 0.06\left[\frac{10}{s} + \frac{1}{s^2}\right]$$

Solving for $C(s)$ and doing the inverse of Laplace transform:

$$C_{\text{out}}(t) = 0.11 \cdot \exp\left(-\frac{t-5}{\bar{t}}\right) + 0.3 \cdot \left[1 - \exp\left(-\frac{t-5}{\bar{t}}\right)\right] - 0.06 \cdot \bar{t}\left[\frac{t-5}{\bar{t}} - 1 + \exp\left(-\frac{t-5}{\bar{t}}\right)\right]$$

This gives $C_{\text{out}}(t = 10) = C_{10} = 0.122$.

In the interval t > 10:

$$\bar{t} \cdot s \cdot C(s) + C(s) = 0$$

and

$$C_{out}(t) = 0.122 \cdot \exp\left(-\frac{t-10}{\bar{t}}\right)$$

Example 3.9 ***System of Two Plug Flow Reactors***

In a system of two reactors, designed as PFRs, certain deficiencies in its behavior are observed, so that it could be modeled with the following scheme:

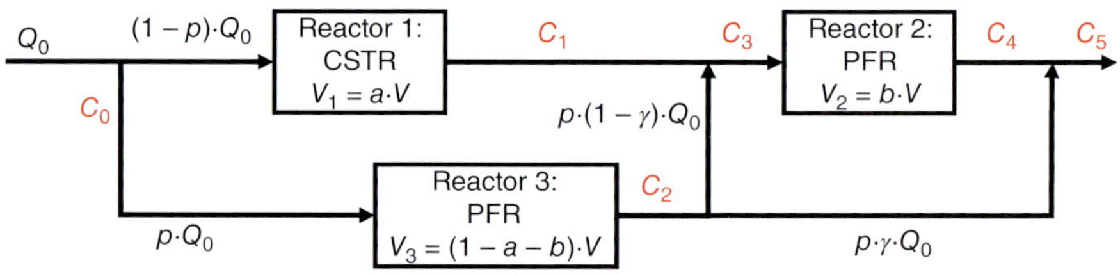

(a) Determine the values of the average residence times in each of the three reactors of the model, based on the average time of total residence.
(b) Taking into account the transfer function in the CSTR and PFR, find the transfer function of the proposed reactor network.

Solution

(a) Let us calculate the average residence times of each reactor as a function of the total residence time:

$$\overline{t_1} = \frac{V_1}{(1-p)Q_0} = \frac{aV}{(1-p)Q_0} = \frac{a}{(1-p)}\bar{t}$$

$$\overline{t_2} = \frac{V_2}{(1-p)Q_0 + p(1-\gamma)Q_0} = \frac{bV}{(1-p\gamma)Q_0} = \frac{b}{(1-p\gamma)}\bar{t}$$

$$\overline{t_3} = \frac{V_3}{pQ_0} = \frac{(1-a-b)V}{pQ_0} = \frac{1-a-b}{p}\bar{t}$$

(b) For calculating the global transfer function, let us use the definitions of the transfer functions in the three volumes:

$$C_1 = G_1 \cdot C_0$$
$$C_2 = G_3 \cdot C_0$$
$$C_4 = G_2 \cdot C_3$$

Let us write a balance at the mix point between reactors 1 and 2:

$$(1-p)C_1 + p(1-\gamma)C_2 = (1 - p + p - p\gamma)C_3$$

So:

$$\frac{(1-p)C_1 + p(1-\gamma)C_2}{(1-p\gamma)} = C_3$$

In the same way, at the final mixing point:

$$(1 - p\gamma)C_4 + p\gamma C_2 = C_5$$

We can say that the global transfer function is C_5/C_0; therefore:

$$G_{\text{global}} = p\gamma G_3 + (1 - p\gamma)G_2 \frac{(1 - p)G_1 + p(1 - \gamma)G_3}{(1 - p\gamma)}$$
$$= p\gamma G_3 + (1 - p)G_2 G_1 + p(1 - \gamma)G_2 G_3$$

As we know:

$$G_1 = \frac{1}{1 + \overline{t_1} s} \cdots G_2 = \exp(-\overline{t_2} s) \cdots G_3 = \frac{1}{1 + \overline{t_3} s}$$

So, finally:

$$G_{\text{global}} = p\gamma \frac{1}{1 + \overline{t_3} s} + (1 - p)\exp(-\overline{t_2} s)\frac{1}{1 + \overline{t_1} s} + p(1 - \gamma)\exp(-\overline{t_2} s)\frac{1}{1 + \overline{t_3} s}$$

3.6 Transfer Function for the Dispersion Model

It might be instructive to show how to obtain the solution to the ODE of the dispersion model (Eq. (1.52)) and the effect of the boundary conditions on that solution.

For the dispersion model, we have

$$D_e \frac{\partial^2 C_T}{\partial z^2} - u \cdot \frac{\partial C_T}{\partial z} = \frac{\partial C_T}{\partial t} \tag{1.52}$$

By taking Laplace transform respect to time:

$$D_e \cdot \frac{\partial^2 C(s)}{\partial z^2} - u \frac{\partial C(s)}{\partial z} = s \cdot C(s) \tag{3.62}$$

And then:

$$\frac{\partial^2 C(s)}{\partial z^2} - \frac{u}{D_e} \frac{\partial C(s)}{\partial z} - \frac{s}{D_e} \cdot C(s) = 0 \tag{3.63}$$

This is a homogeneous linear differential equation with constant coefficients, of the type:

$$A \cdot y'' + B \cdot y' + C \cdot y = 0 \tag{3.64}$$

whose characteristic equation is

$$A \cdot r^2 + B \cdot r + C = 0 \tag{3.65}$$

In our case:

$$r^2 + \left(-\frac{u}{D_e}\right) \cdot r + \left(-\frac{s}{D_e}\right) = 0 \tag{3.66}$$

In that equation, $(B^2 - 4 \cdot A \cdot C) > 0$, this is

$$\left(-\frac{u}{D_e}\right)^2 - 4 \cdot 1 \cdot \left(-\frac{s}{D_e}\right) = \left(\frac{u}{D_e}\right)^2 + 4 \cdot \left(\frac{s}{D_e}\right) > 0 \tag{3.67}$$

and then the values of r_1 and r_2 are distinct real numbers, with the following solution:

$$r_{1,2} = \frac{\frac{u}{D_e} \pm \sqrt{\left(\frac{u}{D_e}\right)^2 + \frac{4\cdot s}{D_e}}}{2} \tag{3.68}$$

The solution in the Laplace domain is then:

$$\begin{aligned} C(s) &= M_1 \cdot \exp\left[\left(\frac{\frac{u}{D_e} + \sqrt{\left(\frac{u}{D_e}\right)^2 + \frac{4\cdot s}{D_e}}}{2}\right) t\right] \\ &\quad + M_2 \cdot \exp\left[\left(\frac{\frac{u}{D_e} - \sqrt{\left(\frac{u}{D_e}\right)^2 + \frac{4\cdot s}{D_e}}}{2}\right) t\right] \\ &= M_1 \cdot \exp\left[\left(\frac{u}{2D_e} + \frac{u}{2D_e}\sqrt{1 + \frac{4 \cdot D_e \cdot s}{u^2}}\right) t\right] \\ &\quad + M_2 \cdot \exp\left[\left(\frac{u}{2D_e} + \frac{u}{2D_e}\sqrt{1 - \frac{4 \cdot D_e \cdot s}{u^2}}\right) t\right] \end{aligned} \tag{3.69}$$

The constants "M_1" and "M_2" should be calculated from the boundary conditions.

Example 3.10 ***Laplace Transform Solution to Example 2.8***

In Chapter 2 we have treated in Example 2.8 the case of an experimental flow reactor, initially designed as a CSTR, fed with small particles of solids. The response to an impulse signal (solid impregnated in a colored compound) has the function: $C(t) = a(10 - t)$ (with time in minutes). This reactor is followed by a fluidized bed reactor of mean residence time (of the solid) equal to three minutes; determine the concentration of tracer at the output of the system.

Let us solve the problem using Laplace transforms and transference functions.

Solution

At the exit of the second reactor, we have

$$C_{\text{out}} = E_1 \otimes E_2$$

In the Laplace space:

$$C_{\text{out}}(s) = E_1(s) \cdot E_2(s) = L\{0.02 \cdot (10 - t)\} \cdot L\left\{\frac{1}{\overline{t_2}} \cdot \exp\left(-\frac{t}{\overline{t_2}}\right)\right\}$$

Taking the transforms:

$$\begin{aligned} C_{\text{out}}(s) &= \left(\frac{0.2}{s} - \frac{0.02}{s^2}\right) \cdot \frac{1}{1 + \overline{t_2} \cdot s} = \frac{0.2}{s \cdot (1 + \overline{t_2} \cdot s)} - \frac{0.02}{s^2(1 + \overline{t_2} \cdot s)} \\ &= \frac{0.2}{(s + \overline{t_2} \cdot s^2)} - \frac{0.02}{(s^2 + \overline{t_2} \cdot s^3)} \end{aligned}$$

Doing the anti-Laplace transform:

$$C_{\text{out}}(t) = 0.2 \cdot \left(1 - \exp\left(-\frac{t}{\overline{t_2}}\right)\right) - 0.02 \cdot \left[\overline{t_2} \cdot \exp\left(-\frac{t}{\overline{t_2}}\right) - (\overline{t_2} - t)\right]$$

The following figure presents the data calculated for time between 0 and 12 minutes using the equation, which is completely equal to the result presented in Example 2.8.

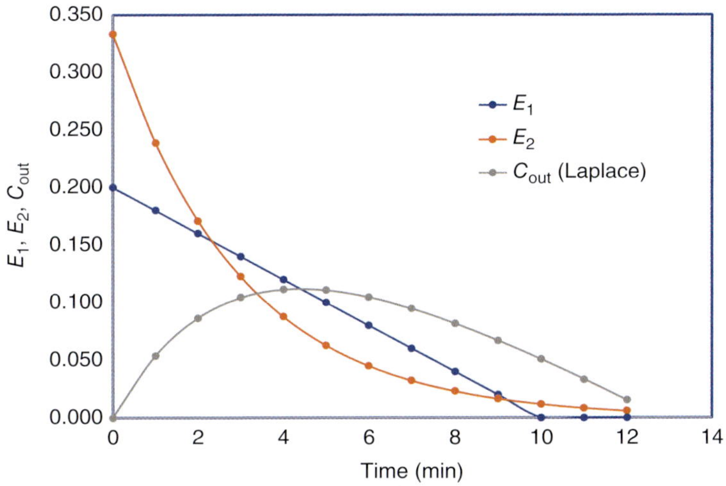

Example 3.11 ***Complex System of Three Reactors***
The behavior of a mixed flow reactor can be considered equivalent to three elements in parallel, with a fraction of flow for each one. The response of each branch of the system to a pulse input is given by (time in minutes):

$$C_1 = \exp(-5t)$$
$$C_2 = t \cdot \exp(-t)$$
$$C_3 = t^2 \cdot \exp(-3t)$$

(a) Calculate the mean residence time of the system, considering the same fraction of the flow to each reactor.
(b) Deduce the form of the $E(t)$ function, in the general case.

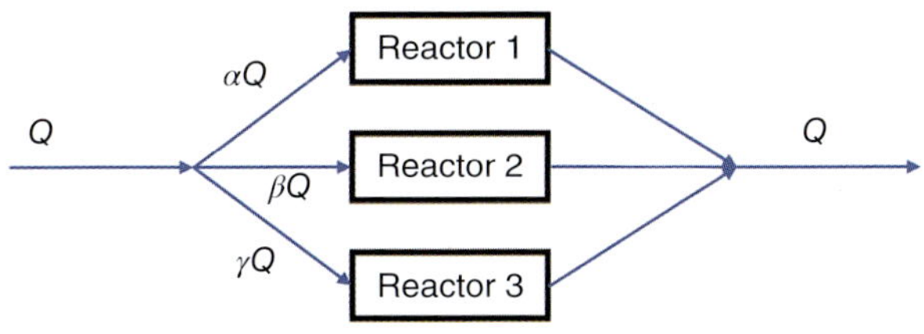

Solution
(a) Considering the system as a whole, we will write

$$C_{\text{out}} = \alpha C_1 + \beta C_2 + \gamma C_3 = \alpha \cdot \exp(-5t) + \beta \cdot t \exp(-t) + \gamma \cdot \text{t}^2 \exp(-3t)$$

We can then calculate the integrals to get the mean residence time:

$$\int_0^\infty C_{\text{out}} dt = \alpha \cdot \int_0^\infty \exp(-5t)dt + \beta \cdot \int_0^\infty t \cdot \exp(-t)dt + \gamma \int_0^\infty \text{t}^2 \cdot \exp(-3t)dt = 0.2 \cdot \alpha + 1 \cdot \beta + 0.075 \cdot \gamma$$

In the case proposed, where α=β=γ=0.33, the area is 0.4243.
On the other hand:

$$\int_0^\infty t \cdot C_{\text{out}} dt = \alpha \cdot \int_0^\infty \text{t} \cdot \exp(-5t)dt + \beta \cdot \int_0^\infty t^2 \cdot \exp(-t)dt + \gamma \int_0^\infty \text{t}^3 \cdot \exp(-3t)dt = 0.04 \cdot \alpha + 2 \cdot \beta + 0.075 \cdot \gamma$$

Using the proposed values of α, β and γ, we get 0.7 as the value of this integral. Then:

$$t_m = \frac{\int_0^\infty t \cdot C_{\text{out}} dt}{\int_0^\infty C_{\text{out}} dt} = \frac{0,7}{0.4243} = 1.64 \text{ min}$$

(b) For deducing the $E(t)$ function, we can use the definition of this curve when a pulse input is introduced:

$$E(t) = \frac{C(t)}{\int_0^\infty C_{\text{out}} dt} = \frac{\alpha \cdot \exp(-5t) + \beta \cdot t \exp(-t) + \gamma \cdot \text{t}^2 \exp(-3t)}{0.2 \cdot \alpha + 1 \cdot \beta + 0.075 \cdot \gamma}$$

Also, the system can be treated using transfer functions. In this case, the transfer functions of each reactor would be

$$G_1 = \frac{C_1(s)}{L\{\delta(t)\}} = C_1(s) = \frac{1}{s+5}$$

$$G_2 = \frac{C_2(s)}{L\{\delta(t)\}} = C_2(s) = \frac{1}{(s+1)^2}$$

$$G_3 = \frac{C_3(s)}{L\{\delta(t)\}} = C_3(s) = \frac{2}{(s+3)^3}$$

and then:

$$G = \alpha G_1 + \beta G_2 + \gamma G_3 = \frac{\alpha}{(s+5)} + \frac{\beta}{(s+1)^2} + \frac{2\gamma}{(s+3)^3}$$

That obviously will give the same $E(t)$ combining the three signals. Nevertheless, if the system were to be connected in series, the convolution of the three signals would be difficult without the aid of the transfer function, In that case, we would have

$$E = E_1 \otimes E_2 \otimes E_3$$

$$G = G_1 \cdot G_2 \cdot G_3 = \frac{1}{s+5} \cdot \frac{1}{(s+1)^2} \frac{2}{(s+3)^3} \cdot$$

and, doing the anti-transform, we get

$$E(t) = 2 \cdot (31\delta(t) + 10\delta'(t) + 16\delta''(t) + 6\delta'''(t) + \delta^{iv}(t) - 128e^{-5t})$$

Bibliography

Bourne, J. (1999). *Turbulent Mixing and Chemical Reactions*. New York, NY: Wiley.

Doraiswamy, L.K. (1984). *Recents Advances in the Engineering Analysis of Chemically Reacting Systems*. New York, NY: Wiley.

Kayode Coker, A. (2001). *Modelling of Chemical Kinetics and Reactor Design*. Elsevier.

Spiegel, M.R. (1965). *Theory and Problems of Laplace Transforms*, Schaum's Outline Series. New York, NY: McGraw-Hill.

Wallas, S.M. (1995). *Chemical Reaction Engineering Handbook*. Amsterdam: Gordon and Breach.

4

Partial Differential Equations in Reactor Design

4.1 Introduction

In mathematics an equation in partial derivatives (sometimes abbreviated as PDE) is a relation between a function F of several independent variables x, y, z, t, … and the partial derivatives of F with respect to those variables. The partial differential equations are used in the mathematical formulation of processes of physics and other sciences that are usually distributed in space and time. Typical problems are the propagation of sound or heat, electrostatics, electrodynamics, fluid dynamics, elasticity, quantum mechanics, and many others.

In chemical reaction engineering, this is especially important in the transient regime during the start-up and shut down of chemical reactors (Chapter 5). The method will permit to calculate, among others, the time to get a stationary regime in each case.

Also important is the application of the methods to the differential equations obtained in heterogeneous systems, where the reactant must be transferred from one phase to another, where it finally reacts.

We work also with models for nonideal flow that are based in PDE, where the initial values and boundary conditions must be proposed.

4.2 Classification of Partial Differential Equations

Consider the equation:

$$A\frac{\partial^2 f}{\partial x^2} + B\frac{\partial f}{\partial x \partial y} + C\frac{\partial^2 f}{\partial y^2} = 0 \tag{4.1}$$

Partial differential equations are classified according to the value of the discriminant D defined as

$$D = B^2 - 4AC$$

where

$$\begin{cases} D < 0 & \text{elliptical} \\ D = 0 & \text{parabolic} \\ D > 0 & \text{hiperbolic} \end{cases} \tag{4.2}$$

Chemical Reactor Design: Mathematical Modeling and Applications,
First Edition. Juan A. Conesa.

For this chapter, we are going to focus on the parabolic equations, although the results obtained can be extended without problems to equations of elliptical or hyperbolic type.

Examples of PDE related to chemical engineering are as follows:

Hyperbolic:

$$\frac{dC}{dt} - u\frac{dC}{dx} = 0 \quad \text{(Advection equation)}$$

$$\frac{d^2C}{dt^2} - a\frac{d^2C}{dx^2} = 0 \quad \text{(wave equation)}$$

Elliptical:

$$\frac{\partial^2 T}{\partial x^2} + \frac{\partial^2 T}{\partial y^2} = 0 \quad \text{(Laplace equation)}$$

Parabolic:

$$\frac{\partial C}{\partial t} - D_e\frac{\partial^2 C}{\partial x^2} = 0 \quad \text{(diffusion equation)}$$

4.3 Approximations by Finite Differences

The resolution of a differential equation in partial derivatives by finite differences tries to discretize the equations. In our case, we must obtain a difference approximation of the first and second derivatives using only values of the function we are studying. Let us see how we can obtain a first-order and a second-order approximation using a Taylor series development.

4.3.1 First-order Approximation

Developing the function $f(x)$ to first order around the point x_i:

$$f(x_{i+1}) = f(x_i) + \left(\frac{df(x_i)}{dx}\right)(x_{i+1} - x_i) + \dots \tag{4.3}$$

where $(x_{i+1} - x_i) = \Delta x$

Simply by isolating the derivative, we obtain

$$f'(x_i) = \left(\frac{f(x_{i+1}) - f(x_i)}{\Delta x}\right) \tag{4.4}$$

Equation (4.4) is a first-order forward approximation. Likewise, we could have obtained a backward approximation by simply using approximations in the points x_i and x_{i-1}.

To obtain the approximation of the second derivative, we use the previously obtained result:

$$f''(x_i) = \left(\frac{f'(x_{i+1}) - f'(x_i)}{\Delta x}\right) = \frac{\frac{f(x_{i+2}) - f(x_{i+1})}{\Delta x} - \frac{f(x_{i+1}) - f(x_i)}{\Delta x}}{\Delta x}$$

$$= \frac{f(x_{i+2}) - 2f(x_{i+1}) + f(x_i)}{\Delta x^2} \tag{4.5}$$

4.3.2 Approximation of Second Order

To obtain a second-order approximation of the first and second derivatives, let us now consider the Taylor series approximation of the function $f(x)$:

$$f(x_{i+1}) = f(x_i) + \left(\frac{\delta f(x_i)}{\delta x}\right)\Delta x + \frac{1}{2!}\left(\frac{\delta^2 f(x_i)}{\delta x^2}\right)\Delta x^2 + \cdots$$

$$f(x_{i+2}) = f(x_i) + \left(\frac{\delta f(x_i)}{\delta x}\right)2\Delta x + \frac{1}{2!}\left(\frac{\delta^2 f(x_i)}{\delta x^2}\right)(2\Delta x)^2 + \cdots \tag{4.6}$$

Multiplying by "a" and "b," respectively, the two previous equations and adding we get

$$a \cdot f(x_{i+1}) + b \cdot f(x_{i+2}) = (a+b)f(x_i) + \left(\frac{\delta f(x_i)}{\delta x}\right)\Delta x\,(a+2b) + \frac{1}{2!}\left(\frac{\delta^2 f(x_i)}{\delta x^2}\right)\Delta x^2(a+4b) + \cdots \tag{4.7}$$

These equations must be met for any value of "a" and "b." Therefore, if we want to obtain the value of the derivative, we can do that by

$$a + 2b = 1$$

Also, since we are interested in obtaining an expression for the derivative where only function values appear, we can do

$$a + 4b = 0$$

Solving the above-given system of equations results in $a = 2$, $b = 0.5$.

Thus:

$$2f(x_{i+1}) - \frac{1}{2}f(x_{i+2}) = \frac{3}{2}f(x_i) + \left(\frac{\delta f(x_i)}{\delta x}\right)\Delta x$$

and, isolating the derivative:

$$f'(x_i) = \left(\frac{4f(x_{i+1}) - 3f(x_i) - f(x_{i+2})}{2\Delta x}\right) \tag{4.8}$$

To obtain an approximation of the second derivative, we can follow the same procedure:

$$f(x_{i+1}) = f(x_i) + \left(\frac{\delta f(x_i)}{\delta x}\right)\Delta x + \frac{1}{2!}\left(\frac{\delta^2 f(x_i)}{\delta x^2}\right)\Delta x^2 + \frac{1}{3!}\left(\frac{\delta^3 f(x_i)}{\delta x^3}\right)\Delta x^3 + \cdots$$

$$f(x_{i+2}) = f(x_i) + \left(\frac{\delta f(x_i)}{\delta x}\right)(2\Delta x) + \frac{1}{2!}\left(\frac{\delta^2 f(x_i)}{\delta x^2}\right)(2\Delta x)^2 + \frac{1}{3!}\left(\frac{\delta^3 f(x_i)}{\delta x^3}\right)(2\Delta x)^3 + \cdots$$

$$f(x_{i+3}) = f(x_i) + \left(\frac{\delta f(x_i)}{\delta x}\right)(3\Delta x) + \frac{1}{2!}\left(\frac{\delta^2 f(x_i)}{\delta x^2}\right)(3\Delta x)^2 + \frac{1}{3!}\left(\frac{\delta^3 f(x_i)}{\delta x^3}\right)(3\Delta x)^3 + \cdots \quad (4.9)$$

Multiplying by a, b, c and adding:

$$a \cdot f(x_{i+1}) + b \cdot f(x_{i+2}) + c \cdot f(x_{i+3}) = (a+b+c)f(x_i) + f'(x_i)\,\Delta x\,(a+2b+3c) + \frac{1}{2!}f''(x_i)\Delta x^2(a+4b+9c) + \frac{1}{3!}f'''(x_i)\Delta x^3(a+8b+27c) \quad (4.10)$$

Taking:

$$a + 2b + 3c = 0$$
$$a + 4b + 9c = 1$$
$$a + 8b + 27c = 0$$

We get $a = -\frac{5}{2}; b = 2; c = -\frac{1}{2}$

With what you get:

$$f''(x_i) = \frac{2f(x_i) - 5f(x_{i+1}) + 4f(x_{i+2}) - f(x_{i+3})}{\Delta x^2} \quad (4.11)$$

The first-order approximation for the first and second derivative of a function is shown in Table 4.1. Table 4.2 shows the second-order approximation for the first and second derivatives of a function.

Table 4.1 First-order approximations of the first and second derivative.

Front differences:
$f'(x_i) = \frac{f(x_{i+1}) - f(x_i)}{\Delta x}$
$f''(x_i) = \frac{f(x_i) - 2f(x_{i+1}) + f(x_{i+2})}{\Delta x^2}$
Rear differences:
$f'(x_i) = \frac{f(x_i) - f(x_{i-1})}{\Delta x}$
$f''(x_i) = \frac{f(x_i) - 2f(x_{i-1}) + f(x_{i-2})}{\Delta x^2}$
Central differences: They do not exist.

Table 4.2 Second-order approximation of the first and second derivative.

Front differences:

$$f'(x_i) = \frac{4f(x_{i+1}) - 3f(x_i) - f(x_{i+2})}{2\Delta x}$$

$$f''(x_i) = \frac{2f(x_i) - 5f(x_{i+1}) + 4f(x_{i+2}) - f(x_{i+3})}{\Delta x^2}$$

Rear differences:

$$f'(x_i) = \frac{3f(x_i) - 4f(x_{i-1}) + f(x_{i-2})}{2\Delta x}$$

$$f''(x_i) = \frac{2f(x_i) - 5f(x_{i-1}) + 4f(x_{i-2}) - f(x_{i-3})}{\Delta x^2}$$

Central differences:

$$f'(x_i) = \frac{f(x_{i+1}) - f(x_{i-1})}{2\Delta x}$$

$$f''(x_i) = \frac{2f(x_{i+1}) - 2f(x_i) + f(x_{i-1})}{\Delta x^2}$$

4.4 Approaching the Problem Using Finite Differences

To facilitate the interpretation of the methodology, we propose the resolution scheme by means of an example, which we use throughout this chapter. The results obtained can be generalized directly to other types of problems.

We consider the diffusion of a species toward the center of a catalyst particle, which is completely explained in Chapter 9. The problem calculates the concentration distribution of the species along the particle. In the example, the substance "A" is contained in a gas and it diffuses and reacts inside of a solid catalyst. Before the contact between the phases ($t < 0$), the concentration of the species in the gas phase is known (C_{As}) and it is zero inside the solid. As time goes on, "A" is transferred to the inner part solid by a combination of diffusion and reaction of "A" in the catalyst: A (gas phase) → products (catalyzed reaction). Figure 4.1 shows a scheme of this situation.

If a mass balance for species "A" is done in a $dV = S \cdot dx$, "S" being the surface of the interface:

$$\text{Input} + \text{Generation} = \text{Output} + \text{Accumulation}$$

$$[n_A]_x \cdot S + r_A \cdot S \cdot dx = [n_A]_{x+dx} \cdot S + S \cdot dx \cdot \frac{dC_A}{dt} \tag{4.12}$$

n_A is the molar flow density (kmol/(s m^2)) and r_A the reaction rate, which will be considered to be of first order:

$$([n_A]_x - [n_A]_{x+dx}) \cdot S - k\, C_A \cdot S \cdot dx = S \cdot dx \cdot \frac{dC_A}{dt} \tag{4.13}$$

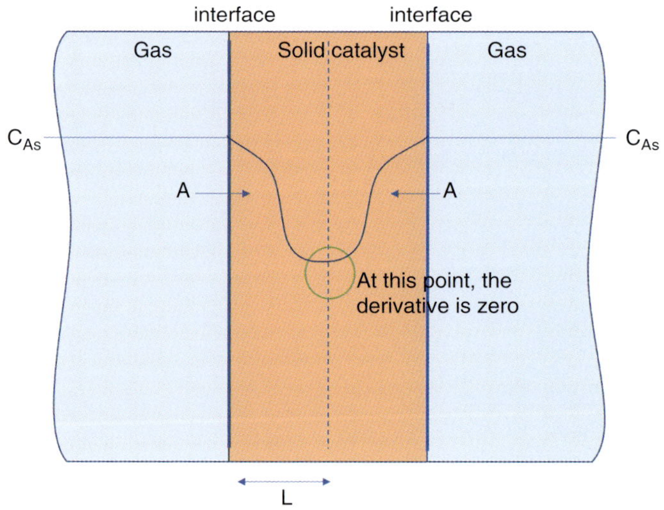

Figure 4.1 Diffusion of a gas inside a plane catalyst particle.

Using $dn_A = [n_A]_{x+dx} - [n_A]_x$ for the small increment of this variable and eliminating "S" we get

$$(-dn_A) - k\,C_A \cdot dx = dx \cdot \frac{dC_A}{dt} \tag{4.14}$$

This is

$$-\frac{dn_A}{dx} - kC_A = \frac{dC_A}{dt} \tag{4.15}$$

And, using Fick's law for the molar flow density:

$$n_A = -D_{AB} \cdot \frac{dC_A}{dx}$$

So,

$$\frac{dn_A}{dx} = -D_{AB} \cdot \frac{d^2C_A}{dx^2}$$

We can rearrange (Eq. (4.14)) using partial derivatives:

$$D_{AB} \cdot \frac{\partial^2 C_A}{\partial x^2} - kC_A = \frac{\partial C_A}{dt} \tag{4.16}$$

The numerical methods of solving this type of problem calculate the value of the dependent variable (in our example, the concentration) only in discrete points, called nodes. In this way, the region existing between the limit points is divided into nodal points. On the other hand, the time variable is discretized, resulting in a two-dimensional network, in which an axis represents the spatial variable and the other the temporary variable.

4.4.1 Explicit Method

The explicit method, for the parabolic equations, combines a forward difference in time with a normally centered approximation of order 2 of the spatial derivatives always using values of the dependent variable at the time instant before the one considered, which produces an explicit formula. It is possible to use other differences schemes; we only have to take into account the stability conditions that are discussed later.

Let us consider an interior point shown in Figure 4.2. Since all the values at the time "t" are assumed to be known, the spatial derivatives can be approximated using central differences:

$$\frac{\partial C_A}{\partial x} = \frac{C^t_{Ai+1} - C^t_{Ai-1}}{2 \cdot \Delta x} \tag{4.17}$$

$$\frac{\partial C^2_A}{\partial x^2} = \frac{C^t_{Ai+1} - 2 \cdot C^t_{Ai} + C^t_{Ai-1}}{\Delta x^2} \tag{4.18}$$

However, the first-order time derivative can be approximated by the forward difference:

$$\frac{\partial C_A}{\partial t} = \frac{C^{t+1}_{Ai} - C^t_{Ai}}{\Delta t} \tag{4.19}$$

The resolution procedure consists of substituting these approximations in the differential equation that governs the problem and resolving explicitly for C^{t+1}_{Ai}.

This equation will contain on the right side only terms of C^t_A. The boundary conditions will be used to calculate the limit points and the problem can be solved forward in time. As considerations, we must point out that the error in the resolution of this problem is proportional to $(\Delta t + \Delta x^2)$.

Applying the procedure to the previous example and using the indices "i" for the position and "t" for the time:

$$\frac{C^{t+1}_{Ai} - C^t_{Ai}}{\Delta t} = D_{AB}\frac{C^t_{Ai+1} - 2 \cdot C^t_{Ai} + C^t_{Ai-1}}{\Delta x^2} - k \cdot C^t_{Ai} \tag{4.20}$$

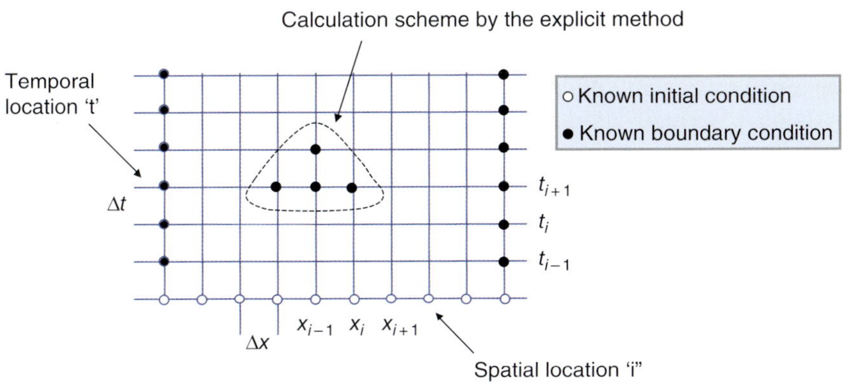

Figure 4.2 Resolution scheme for the explicit method.

Rearranging:

$$C_{Ai}^{t+1} = \frac{\Delta t \cdot D_{AB}}{\Delta x^2} C_{Ai+1}^{t} - 2 \cdot \frac{\Delta t \cdot D_{AB}}{\Delta x^2} C_{Ai}^{t} - \frac{\Delta t \cdot D_{AB}}{\Delta x^2} C_{Ai-1}^{t} + k \cdot \Delta t \cdot C_{Ai}^{t} + C_{Ai}^{t} \tag{4.21}$$

We finally get

$$C_{Ai}^{t+1} = \frac{\Delta t \cdot D_{AB}}{\Delta x^2} C_{Ai+1}^{t} + \left(1 - k \cdot \Delta t - 2 \cdot \frac{\Delta t \cdot D_{AB}}{\Delta x^2}\right) C_{Ai}^{t} + \frac{\Delta t \cdot D_{AB}}{\Delta x^2} C_{Ai-1}^{t} \tag{4.22}$$

With Eq. (4.22), we can obtain the values of the concentration in the nodes in time ($t + \Delta t$) depending on the value at time "t," which, together with the initial and boundary conditions, allows us to solve the problem.

4.4.2 Initial and Boundary Conditions

As can be seen, the problems of differential equations in partial derivatives require initial and limit conditions. The initial condition provides the value of the dependent variable over the entire spatial domain for a fixed time (although conditions at initial time are usually considered, the same methodology can be used for conditions at any other time).

For the case mentioned earlier, a valid initial condition is

> $t = 0 \rightarrow C_A = 0$, for all points but the one contacting the other phase, where $C_A = C_{As}$

In chemical engineering, boundary conditions are usually particularized in three types: constant value, insulation, and contours exposed to a transport boundary layer. In the case of the diffusion presented, a good assumption can be the following:

> At the interface ($x = 0$): $C_A = C_{A_s}$ (known and constant value), $\forall t$
>
> At the center of the particle ($x = L$), by simmetry: $\dfrac{\partial C_A}{\partial x} = 0$

In this sense, the first condition is of the constant value limit type. This type of condition is normally used as an approximation to the real situation, in which there are fluctuations on the constant value, but they are of small magnitude and, for practical purposes, the fluctuations can be neglected without introducing significant errors in the problem.

The boundary condition at the center of the catalyst, $x = L$, is similar to other conditions used in chemical engineering and reflects the fact that, in the present example, the derivative in the center is zero because it represents a minimum in the concentration of the species reacting.

The boundary condition $\frac{\partial C_A}{\partial x} = 0$ is like those used for "insulating surfaces," representing surfaces that do not allow the transport of a certain magnitude. For example, thermal insulators prevent the passage of heat, while solid surfaces act as insulators when passing matter. The common feature of a perfect insulator is that it does not allow passage in a direction perpendicular to it.

Other types of boundary conditions are also present in chemical technology, such as the transport boundary layer condition, which is usually modeled using an individual transport coefficient.

For the problem of the example, the initial conditions can be written as

$$t = 0; \quad C_A = 0; \qquad \text{for } \forall x \neq 0$$

$$t = 0; \quad C_A = C_{AS}; \qquad \text{for } \forall x = 0$$

And the limit conditions will be given by (constant limit value):

$$C_A = C_{As}; \quad \text{for } x = 0$$

$$\left[\frac{C^t_{Ai} - C^t_{Ai-1}}{\Delta x}\right]_{x=L} = 0 \quad \text{i.e. } C^t_{AN} = C^t_{AN-1} \tag{4.23}$$

Note that the derivative in the last point must be done in the form of a rear derivative, i.e. using the points "$i - 1$" and "i." If central or front differences are used, the information in the point "$N + 1$" is needed, and the system is not solvable.

We can also use a second-order approximation of the first derivative, and in this case for the last node (rear differences) we will have

$$\left[\frac{3C^{t+1}_{Ai} - 4C^t_{Ai-1} + C^t_{Ai-2}}{2 \cdot \Delta x}\right]_{i=N} = 0 \text{ i.e. } C^{t+1}_{AN} = \frac{1}{3}(4C^t_{AN-1} + C^t_{AN-2}) \tag{4.24}$$

The equation represents the relationship between the concentrations at the final node and the two nodal points closest to it. The extension of this result to other types of insulating surfaces is simple.

Another interesting example may be the transport of a quantity (heat, mass, or momentum) from the sine of a fluid to a solid surface, which can be described by a transfer coefficient. For example, the heat flow from a surface to the fluid can be represented by

$$q_S = h(T_S - T_0) \tag{4.25}$$

where T_0 is the temperature of the wall, T_S the temperature of the sine of the fluid, and "h" is the heat transfer coefficient. Applying the energy balance, the equation is given by

$$q_s = -k \cdot \left.\frac{dT}{dx}\right]_S = h \cdot (T - T_0) \tag{4.26}$$

Using a first-order approximation of the derivative:

$$-k \cdot \left.\frac{T_{i+1} - T_i}{\Delta x}\right]_{i=0} = h \cdot (T_S - T_0) \tag{4.27}$$

we can get

$$T_1 = \frac{h \cdot \Delta x}{k} \cdot (T_S - T_0) + T_0$$

$$T_0 = \frac{T_1 - \dfrac{h \cdot \Delta x}{k} \cdot T_S}{\left(1 - \dfrac{h \cdot \Delta x}{k}\right)}$$

If the derivative were approximated to the second order:

$$-k \cdot \left.\frac{4T_{i+1} - 3T_i - T_{i+2}}{2\Delta x}\right]_{i=0} = h \cdot (T_S - T_0)$$

we would obtain

$$T_0 = \frac{-4T_1 + T_2 - \dfrac{2 \cdot h \cdot \Delta x}{k} \cdot T_S}{\left(-3 - \dfrac{2h \cdot \Delta x}{k}\right)}$$

4.4.3 Stability

In order to correctly solve a system of differential equations by the finite difference method, three conditions must be met: that the difference scheme be consistent, that it be stable, and that it converge to the correct solution. If a scheme is used in differences of those commented on earlier, the system will be consistent. However, guaranteeing stability and convergence (that is, converging to the solution of the differential equation and not to another solution) is not trivial.

There are different alternatives to establish a scheme in finite differences. However, as noted, the increments of the independent variables (position and time in our example) cannot be chosen independently. Even taking very small values of Δx – forcing a great precision – it can happen that the system is not stable.

One way to ensure stability is through the following theorem. Given a scheme like the following:

$$C_i^{t+1} = A \cdot C_{i+1}^t + B \cdot C_i^t + D \cdot C_{i-1}^t \tag{4.28}$$

If A, B, D are all positive and also $(A + B + D) \leq 1$, the scheme is stable; the errors tend to decrease as we proceed with the iterations and converge to the solution of the differential equation that we are trying to solve.

So, for our example on the gas diffusing in a solid catalyst:

$$A = \frac{\Delta t \cdot D_{AB}}{\Delta x^2} \quad B = \left(1 - k \cdot \Delta t - 2 \cdot \frac{\Delta t \cdot D_{AB}}{\Delta x^2}\right) \quad D = \frac{\Delta t \cdot D_{AB}}{\Delta x^2}$$

Studying the first condition, the coefficients A and D are always positive. The coefficient B must, therefore, meet:

$$\left(1 - k \cdot \Delta t - 2 \cdot \frac{\Delta t \cdot D_{AB}}{\Delta x^2}\right) \geq 0$$

which leads us to

$$\Delta t \leq \frac{1}{k + \dfrac{2 \cdot D_{AB}}{\Delta x^2}} \tag{4.29}$$

In other words, you cannot choose the time and position increments completely independently of each other. Usually, the increment of position is fixed, and an increment of time that allows the system to be stable is chosen.

To check the second condition:

$$\left[\frac{\Delta t \cdot D_{AB}}{\Delta x^2} + \left(1 - k \cdot \Delta t - 2 \cdot \frac{\Delta t \cdot D_{AB}}{\Delta x^2}\right) + \frac{\Delta t \cdot D_{AB}}{\Delta x^2}\right] \leq 1 \tag{4.30}$$

In this example, it is always fulfilled.

4.4.3.1 Resolution of the Selected Problem and Programming

For the resolution of the selected examples, we use Matlab® programming. Matlab program for this example is the following one.

Program Listing 4.1

```
% DIFFERENTIAL EQUATION IN PARTIAL DERIVATIVES
% METHOD OF FINITE DIFFERENCES "EXPLICIT"
% NON-STATIONARY CONCENTRATION PROFILE OF A SUBSTANCE 'A'
% DIFFUSING TO A CATALYST
DAB = 0.01;% cm2 / s
L = 0.3; % cm
k = 1; % 1 / mol · s
CAs = 1; % mol / l
N = 50; % Number of values of x to integrate in x
incx = L / (N-1);
inct = 0.9 / (k + 2 * DAB / incx ^ 2);
C (1) = CAs;
C (2: N) = 0;
x = 0: incx: L;
A = DAB * inct / incx ^ 2;
plot (x, C)
t=0;

while t <50
t = t + inct;
Cn (1) = CAs; % BOUNDARY CONDITION SURFACE
Cn (N) = 1/3 * (4 * C (N-1) -C (N-2)); % BOUNDARY CONDITION CENTER
for i = 2: N-1
Cn (i) = A * C (i-1) + (1-k * inct-2 * A) * C (i) + A * C (i + 1);
end

C = Cn;
plot (x, C)
drawnow
end
```

In the first lines of the program, the variables of the system are defined, and the number of intervals "N" is fixed. The corresponding x-increment (length) is calculated by $L/(N-1)$, and the time increment is selected by multiplying a number between 0–1 (in this case 0.9, although this can be changed) by the condition given in Eq. (4.29). If the time increment does not fulfill this condition, the system will be unstable. This can be checked using 1.1 (or higher) instead of 0.9.

The program now defines the initial value of the vector "C" (concentration) and the values to be used of the vector "x," at time equal to zero. A loop is defined

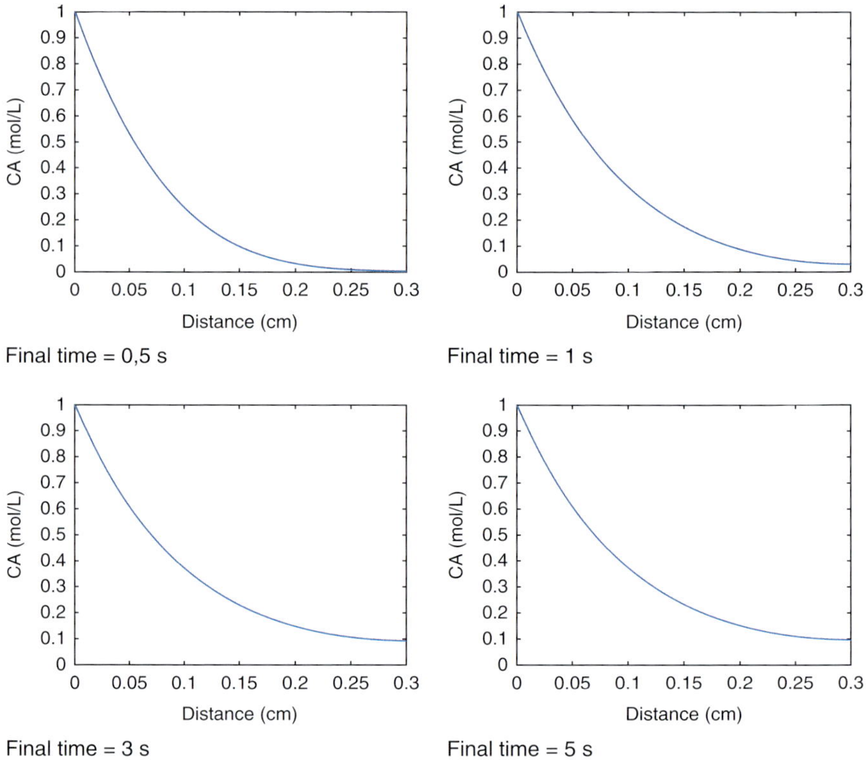

Figure 4.3 Graphical representation of the concentration of gas in the solid particle at different times.

until time reaches a certain value (that can be, obviously, changed). In the loop, a new vector of concentration "Cn" is calculated, corresponding to the value of the concentration in a time incremented by a value of "inct." Conditions for points at the input (Cn(1)) and the end (Cn(N)) are defined, and the points going from 2 to $N-1$ are calculated by an equation corresponding to Eq. (4.22). Finally, the new value of the vector concentration to be used as the last one calculated is changed, by doing $C = \text{Cn}$, and the graph is plotted.

This program permits to visualize the evolution of the concentration profile with time, until time reaches the fixed value.

Figure 4.3 shows some of the results obtained with this program at different values of the final time.

4.5 Other Applications of PDE – Numerical Methods

4.5.1 The RTD of a Complex System

Let us apply the finite difference method to the simulation of the flow in a complex system. In this example, we have an electrochemical reactor having a three-dimensional electrode made of carbonaceous matrix inserted between two

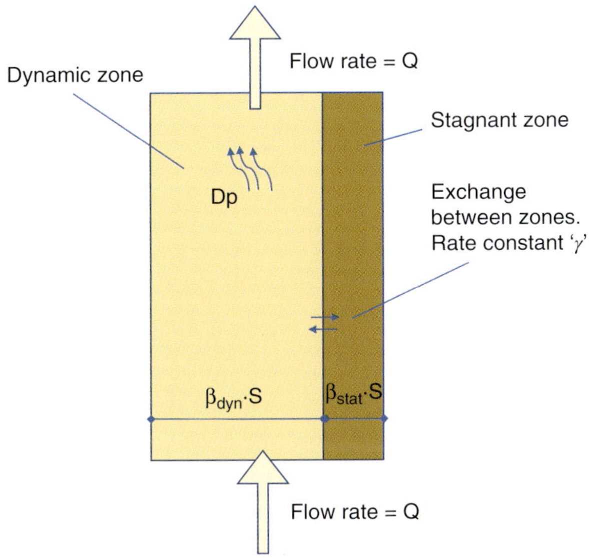

Figure 4.4 Scheme of the model proposed for an electrochemical reactor with a three-dimensional electrode.

metals. The system shows an unusual behavior, as the residence time distribution (RTD) would depend on the width of the carbonaceous felt used.

To develop a model for the observed behavior, different nonideal flow models have been tested. Different authors consider the hydrodynamic mixing in a porous medium with the help of a model containing a series of perfect mixers with stagnant areas. The porous reactor showed a distribution with a single peak in the RTD, but it can be represented as the sum of two distributions: a normal distribution and an exponentially decaying one. The proposed model for the flow inside this electrochemical reactor is represented in Figure 4.4.

In our case, the model will consider only one pathway, for simplicity. In that pathway we will assume that the electrolyte flows following a dispersed plug-flow model, and that a stagnant area exists (such as shown in Figure 4.4). In this way, the electrolyte in the stagnant areas will be refreshed by the electrolyte flowing in the region considered in an axially dispersed plug flow.

The electrolyte holdup in this zone, $\beta_{\text{tot}} = \text{m}^3$ liquid/m^3 column, is divided into a dynamic holdup, β_{dyn}, and a static or stagnant holdup, β_{stat} (see Figure 4.4). The rate of exchange between the static (s) and dynamic (d) zones is assumed to be proportional to the concentration difference. An exchange coefficient, γ (s^{-1}), is defined for the transport:

$$\frac{\text{moles exchanged}}{\text{m}^3\ \text{liquid s}} = \gamma(c_d - c_s)$$

where c_d and c_s are the concentrations in the dynamic and static phase, respectively. "γ" can be considered as the product of a mass transfer coefficient and the specific interfacial area between the flowing and the stagnant zones, $k_L a$ (in units of time^{-1}).

The model equations will be deduced applying the mass balance equation to the dynamic and the static zones over a thin slice perpendicular to the direction of the flow. We will assume the density and other variables to be constant over the length coordinates:

$$\beta_{\text{dyn}} \cdot S \cdot \frac{\partial C_d}{\partial t} = \beta_{\text{dyn}} \cdot S \cdot D_d \cdot \frac{\partial^2 C_d}{\partial z^2} - \beta_{\text{dyn}} \cdot S \cdot u_d \cdot \frac{\partial C_d}{\partial z} - \beta_{\text{tot}} \cdot S \cdot \gamma \cdot (C_d - C_s) \tag{4.31}$$

$$\beta_{\text{stat}} \cdot S \cdot \frac{\partial C_s}{\partial t} = -\beta_{\text{tot}} \cdot S \cdot \gamma \cdot (C_s - C_d) \tag{4.32}$$

where S = cross-sectional area of the reactor (m^2); D_d = dispersion coefficient in dynamic phase (m^2/s); u_d = velocity in the dynamic phase (m/s) of volume $V = Q/S\beta_{\text{dyn}}$; and Q = electrolyte flow through volume V (m^3/s).

For the calculation of the RTD due to this pathway, we will solve the system of Eqs. (4.31) and (4.32) with the appropriate boundary conditions.

As in other examples, we will use the finite differences method for solving the system of equations. If we develop Eq. (4.32) in terms of this method, we find that:

$$\beta_{\text{stat}} \cdot S \cdot \frac{C_{s,i}^{t+1} - C_{s,i}^{t}}{\Delta t} = -\beta_{\text{tot}} \cdot S \cdot \gamma \cdot (C_{s,i}^{t} - C_{d,i}^{t}) \tag{4.33}$$

where the subscript "i" refers to a position in the length dimension, and the superscript "t" to the positions in time. Rearranging:

$$\begin{aligned} C_{s,i}^{t+1} &= C_{s,i}^{t} - \frac{\beta_{\text{tot}}}{\beta_{\text{stat}}} \cdot \Delta t \cdot \gamma \cdot (C_{s,i}^{t} - C_{d,i}^{t}) \\ &= \left(1 - \frac{\beta_{\text{tot}}}{\beta_{\text{stat}}} \cdot \Delta t \cdot \gamma\right) C_{s,i}^{t} + \frac{\beta_{\text{tot}}}{\beta_{\text{stat}}} \cdot \Delta t \cdot \gamma \cdot C_{d,i}^{t} \end{aligned} \tag{4.34}$$

Using the following groups of constants:

$$A = 1 - \frac{\beta_{\text{tot}}}{\beta_{\text{stat}}} \cdot \Delta t \cdot \gamma$$

$$B = \frac{\beta_{\text{tot}}}{\beta_{\text{stat}}} \cdot \Delta t \cdot \gamma$$

We can rewrite the equation by

$$C_{s,i}^{t+1} = A \cdot C_{s,i}^{t} + B \cdot C_{d,i}^{t} \tag{4.35}$$

For Eq. (4.31), following the same procedure, it is found that:

$$\begin{aligned} \beta_{\text{dyn}} \cdot \frac{C_{d,i}^{t+1} - C_{d,i}^{t}}{\Delta t} &= \beta_{\text{dyn}} \cdot D_d \cdot \frac{C_{d,i-1}^{t} - 2C_{d,i}^{t} + C_{d,i+1}^{t}}{\Delta z^2} \\ &\quad - \beta_{\text{dyn}} \cdot u_d \cdot \frac{C_{d,i+1}^{t} - C_{d,i}^{t}}{\Delta z} - \beta_{\text{tot}} \cdot \gamma \cdot (C_{d,i}^{t} - C_{s,i}^{t}) \\ C_{d,i}^{t+1} &= \Delta t \cdot D_d \cdot \frac{C_{d,i-1}^{t} - 2C_{d,i}^{t} + C_{d,i+1}^{t}}{\Delta z^2} \\ &\quad - \frac{\Delta t \cdot u_d}{\Delta z} \cdot (C_{d,i+1}^{t} - C_{d,i}^{t}) \\ &\quad - \frac{\beta_{\text{tot}} \cdot \Delta t \cdot \gamma}{\beta_{\text{dyn}}} \cdot (C_{d,i}^{t} - C_{s,i}^{t}) + C_{d,i}^{t} \end{aligned}$$

$$\begin{aligned} C_{d,i}^{t+1} &= C_{d,i+1}^{t} \cdot \left(\frac{\Delta t \cdot D_d}{\Delta z^2} - \frac{\Delta t \cdot u_d}{\Delta z} \right) \\ &+ C_{d,i}^{t} \cdot \left(1 - \frac{2 \cdot \Delta t \cdot D_d}{\Delta z^2} + \frac{\Delta t \cdot u_d}{\Delta z} - \frac{\beta_{\text{tot}} \cdot \Delta t \cdot \gamma}{\beta_{\text{dyn}}} \right) \\ &+ C_{d,i-1}^{t} \left(\frac{\Delta t \cdot D_d}{\Delta z^2} \right) + \frac{\beta_{\text{tot}} \cdot \Delta t \cdot \gamma}{\beta_{\text{dyn}}} \cdot C_{s,i}^{t} \end{aligned} \tag{4.36}$$

and renaming:

$$\begin{aligned} A' &= \left(\frac{\Delta t \cdot D_d}{\Delta z^2} - \frac{\Delta t \cdot u_d}{\Delta z} \right) \\ B' &= \left(1 - \frac{2 \cdot \Delta t \cdot D_d}{\Delta z^2} + \frac{\Delta t \cdot u_d}{\Delta z} - \frac{\beta_{\text{tot}} \cdot \Delta t \cdot \gamma}{\beta_{\text{dyn}}} \right) \\ D' &= \left(\frac{\Delta t \cdot D_d}{\Delta z^2} \right) \end{aligned}$$

we get

$$C_{d,i}^{t+1} = A' \cdot C_{d,i+1}^{t} + B' \cdot C_{d,i}^{t} + D' \cdot C_{d,i-1}^{t} + \frac{\beta_{\text{dyn}}}{\beta_{\text{stat}}} B \cdot C_{s,i}^{t} \tag{4.37}$$

To study the stability of the equations, we must check for the stability conditions, which, as mentioned before, are

$$A > 0 \quad \text{(from Eq. (4.34))}$$

$$A' > 0; \quad B' > 0; \quad D' > 0; \quad (A' + B' + D') \leq 1 \quad \text{(from Eq. (4.36))}$$

If the theorem is applied to Eq. (4.35), bearing in mind the terms that accompany each value of C_d and C_s, we find that, if $A > 0$, then:

$$\Delta t < \frac{\beta_{\text{stat}}}{\beta_{\text{tot}} \cdot \gamma} \equiv \text{Condition 1}$$

For the analysis of Eq. (4.36), we have that $(A' + B' + D')$ is always minor or equal to one, and also always $D' > 0$. Considering only $A' > 0$ and $B' > 0$, then:

$$\Delta z < \frac{D_d}{u_d} \equiv \text{Condition 2}$$

and also:

$$\Delta t < \frac{1}{\left(\frac{2 \cdot D_d}{\Delta z^2} - \frac{u_d}{\Delta z} + \frac{\beta_{\text{tot}} \cdot \gamma}{\beta_{\text{dyn}}} \right)} = \frac{\Delta z}{\left(\frac{2 \cdot D_d}{\Delta z} - 1 + \frac{\beta_{\text{tot}} \cdot \gamma \cdot \Delta z}{\beta_{\text{dyn}}} \right)}$$

$$\equiv \text{Condition 3}$$

So, for the system of differential equations to be stable, we will need to compare the time increments given by conditions 1 and 3, and choose the lower one.

4.5.1.1 Boundary Conditions for Partial Differential Equations

For a "closed" reactor (i.e. when a portion of fluid that enters the reactor cannot leave it by the same way), the consideration of flux balances at the entrance and at the exit provide what are usually termed the "Danckwerts' boundary conditions" (see Chapter 1):

$$C_{A0} = C_A(0^+) - \left(\frac{D}{u}\right)\frac{\partial C_A(0^+)}{\partial z} \tag{4.38}$$

where C_{A0} is the input concentration of species A and the point 0^+ represents the first differential point inside the reactor. If we apply Eq. (4.38) to our case, we have

$$C^t_{A0} = C^t_{d,1} - \left(\frac{D_d}{u_d}\right)\frac{C^t_{d,1} - C^t_{A0}}{\Delta z} \tag{4.39}$$

that rearranging results in:

$$C^t_{d,1} = C^t_{A0} \tag{4.40}$$

That will constitute the boundary condition at the input for the problem. On the other side, at the exit, we can write (following Eq. (1.68))

$$C^t_{d,N} = C^t_{d,N-1} \tag{4.41}$$

For the calculation of the E curve (or the RTD curve), we will simulate an input signal as a step experiment:

$$\text{Concentration} = 0 \quad \text{if } t < 0$$

$$\text{Concentration} = 1 \quad \text{if } t \geq 0$$

These relations constitute the other boundary condition of the problem, at the input of the reactor. This would give what is called the F curve, i.e. the integral residence time curve. For calculating the E curve corresponding to the reactor, the derivative of F will be taken.

In the following lines, a Matlab program is given to simulate a particular case of the system shown in Figure 4.4, with a group of constants with a given value (listed in the program).

Program Listing 4.2

```
%PDE DIFFERENTIAL EQUATIONS
%ELECTROCHEMICAL REACTOR RESIDENCE TIME DISTRIBUTION

%DEFINITIONS OF CONSTANTS:
Bdyn=0.38; % m^3/m^3
Btot=0.98; % m^3/m^3
Bstat=Btot-Bdyn; % m^3/m^3
gamma=0.1; % 1/s
Dd=0.01;% m^2/s
ud=1; % m/s
L=1; % m Total length

N= 150; % Number of length intervals

Az=L/(N-1); % Length-increment

% Stability conditions:
At1 = 0.5 * Bstat/Btot/gamma; % Following Condition 1
At2 = 0.5 * Az / (2*Dd/Az - 1 + Btot*gamma*Az/Bdyn);
     % Following condition 3
```

```
At =min(At1, At2) % Choosing the minor value of At1 and At2

A = 1+ Btot/Bstat*At*gamma;
B = Btot/Bstat*At*gamma;
Aprime  = At*Dd/Az^2-At*ud/Az;
Bprime  = 1-2*At*Dd/Az^2+At*ud/Az-Btot*At*gamma/Bdyn;
Dprime  = At*Dd/Az^2;

t=0;
tend=2; % Ending time to be used, secs
% Initial values:
Cdo= 1; %mol/m3 (step experiment simulation)
Cd(1:N+1)=zeros(size(1:+1));
Cd(1) =Cdo;
Cs(1:N+1)=zeros(size(1:N+1)) ;
tv=[]; % All time values to be used
Cout=[]; % All concentration values
Z=0:Az:L; % Values of z-variable

while t<tend
t = t+At
Cdn(1)=Cdo; %Boundary condition at the input
Csn(N)=Cs(N-1); %Boundary condition at the exit
Cdn(N)=Cd(N-1); %Boundary condition at the exit
for i=2:N-1
Cdn(i) =Aprime*Cd(i+1) + Bprime*Cd(i) + Dprime*Cd(i-1)
     + B*Bdyn/Bstat*Cs(i);
Csn(i) = A*Cs(i) + B*Cd(i);
end
Cd=Cdn;
Cs=Csn;
tv=[tv t]; % Initially is void but it is filled with the
     corresponding value
Cout=[Cout Cd(N)];% Initially is void but it is filled with the
     corresponding value
end

plot(Z,Cd(1:N),Z,Cs(1:N))
legend('Cd','Cs')
xlabel('Z (m)')
ylabel('Cd (mol/m^3); Cs (mol/m^3)')
drawnow
% Calculation of the RTD
E=diff(Cout)/At ; %E(t) of the system (derivative exit concentra-
tion of the step-input run)
figure
plot(tv, [0 E])
xlabel('t(s)')
ylabel('E(t)')
E=[0 E];
% Calculation of the average time and variance:
tE= tv.*E ;
t_m=trapz(tv,tE)
var = trapz(tv,(((tv - tm).^2 ) .* E ))
```

The structure of the program is similar to those explained before. First, the different variables of the system are fixed, the value of space intervals is fixed, and the corresponding space increment is calculated. Then, the minimum time increment is calculated, by using conditions 2 and 3. The initial values of the concentrations and times are then defined. Note that a step tracer input is considered, as a value of $C = 1$ is defined. Void vectors "tv" and "C_{out}" are defined, which will be used to store the values of the time and concentration at the exit of the reactor that will be used in the calculation. At this point, the time loop begins, where we use two vectors Cdn and Csn to represent the concentration in dynamic and stagnant zones at an incremented time. Finally, the new value of the vector concentration to be used as the last ones calculated are changed, by doing Cd = Cdn and Cs = Csn. The concentration at the exit of the reactor is given by the last numbers of these vectors and are stored in the C_{out} variable.

In order to calculate the RTD vector of the reactor, at the final part of the program a numerical derivation of the concentration found is done. This is because the simulation is done using a step input for the tracer, so the $E(t)$ would correspond to the derivative of the obtained signal.

At the last part of the program, the mean residence time t_m and the variance of the RTD obtained is calculated. The graphs shown in Figure 4.5 are obtained with different values of β_{dyn}, i.e. with different fractions of the reactor vessel with a static behavior. As we can see, as the value of β_{stat} increases, the RTD presents a long tail, due to the slow mass transfer to the dynamic phase.

4.5.2 Concentration Profile in a Reactor in Which There Is Flow and Dispersion

PDE numerical methods explained before can be applied to many engineering situations. In this example, a stream of water is circulating along a tube. At a given moment, we begin to introduce a certain substance at one end, so that, keeping the flow constant, the concentration varies with the dimensionless time, according to the following expression:

$$C_0 = \exp\left(-\frac{(t-0.4)^2}{0.005}\right)$$

where t refers to units of dimensionless time and C is the concentration that enters the pipe at a given time. The variation of the concentration as a function of time along the axial direction within the tube can be calculated by the

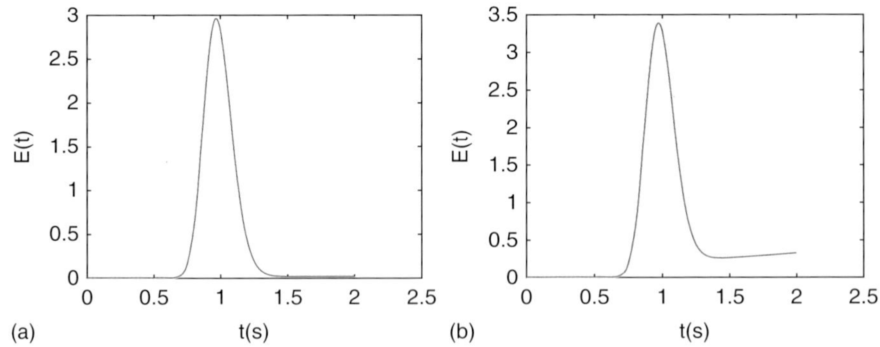

Figure 4.5 RTD calculated in the reactor, for (a) $\beta_{dyn} = 0.38$ and (b) $\beta_{dyn} = 0.68$.

expression

$$\frac{\partial C}{\partial t} = M\frac{\partial^2 C}{\partial x^2} - \frac{\partial C}{\partial x}$$

where C is the concentration, t is dimensionless time, M is the diffusion module, and x is a dimensionless length.

With contour condition at the point of departure, $\frac{\partial C}{\partial x} = 0$ at $x = 1$

Using the explicit finite difference method, we can calculate the profile of concentrations that the tube would have after a dimensionless time unit for three different values of M, for example, 0.01, 0.001, and 0.0001. We would like, in this example, to represent the profiles obtained in a single figure with different colors.

Using the first-order approximation of the derivatives, we can write

$$\frac{C_i^{t+1} - C_i^t}{\Delta t} = M\frac{C_{i+1}^t - 2\cdot C_i^t + C_{i-1}^t}{\Delta x^2} - \frac{C_{i+1}^t - C_i^t}{\Delta x}$$

Rearranging to solve for the concentration at the time "$t + 1$," we get

$$C_i^{t+1} = \left(\frac{M\Delta t}{\Delta x^2} - \frac{\Delta t}{\Delta x}\right)C_{i+1}^t + \left(1 - \frac{2M\Delta t}{\Delta x^2} - \frac{\Delta t}{\Delta x}\right)C_i^t + \frac{M\Delta t}{\Delta x^2}C_{i+1}^t$$

From this equation we can calculate the stability conditions:

$$\left(\frac{M\Delta t}{\Delta x^2} - \frac{\Delta t}{\Delta x}\right) > 0$$

$$\left(1 - \frac{2M\Delta t}{\Delta x^2} - \frac{\Delta t}{\Delta x}\right) > 0$$

where the first one is equivalent to:

$$\Delta x < M$$

and the second one:

$$\Delta t < \left(\frac{1}{\frac{2M}{\Delta x^2} + \frac{1}{\Delta x}}\right)$$

In the example, the initial condition of the system is clear, as all the tube is empty of the substance (all $C(x,0) = 0$ except in the first increment that will be C_0).

The contour conditions are mentioned in the statement, and are

$C_{i=1} = C_0$ In the first interval

$\frac{C_{N-1}^t - C_N^t}{\Delta x} = 0$ In the last interval, i.e.$C_{N-1}^t = C_N^t$

The corresponding Matlab program is the following one, which is based on the same structure as that of the previous examples.

Program Listing 4.3

```
% DIFFERENTIAL EQUATION IN PARTIAL DERIVATIVES
% METHOD OF FINITE DIFFERENCES "EXPLICIT"
% "FLOW + DISPERSION IN A PIPE"
M = 0.01;
L = 1;
N = 200; % Number of values of x to make the integration in x
```

```
incx = L / (N-1);
inct = 0.9 / (2 * M / incx ^ 2-1 / incx); %PONER AQUI LO QUE SEA
     EXPLICANDO
C (1) = exp (0);
C (2: N) = 0;
x = 0: incx: L;
r = M * inct / incx ^ 2;
t = 0;
while t <1
t = t + inct;
Cnnew (1) = exp (- (t-0.4) ^ 2 / 0.005);
Cnnew (N) = C (N-1); % Implies that dC / dx = 0 in i = N
for i = 2: N-1
Cnnew (i) = r * C (i + 1) + (1 + inct / incx-2 * r) * C (i)
     + (- inct / incx + r) * C (i-1);
end
C = Cnew;
plot (x, C)
axis ([0 0.1 0 1])
    drawnow
end
```

4.5.3 Reaction on a Catalytic Flat Wall

In a catalytic reactor, a fluid of concentration "C" and linear velocity "u" is flowing and reacting between two wide plates, with rate equation $-r_b = k_b \cdot C_b^\alpha$ ("b" refers to the bulk fluid). The walls are coated with a catalyst on which the reaction rate is first order $-r_w = k_w C_w$ ("w" refers to the wall). Diffusion in the axial direction is small in comparison with the mass transfer by bulk flow, but it has been observed as a certain diffusion in the direction from wall to wall. Make a material balance over an elemental prism of unit width between y and $y + dy$ laterally and $z + dz$ axially, in order to determine the concentration of the reacting species over the complete system. Apply the finite differences method and determine the stability conditions, as well as the boundary and initial conditions of the system.

Apply the solution to the following case: $\alpha = 1$; $k_b = 1 \cdot 10^{-4}$ mol/(l s); $C_{A0} = 1$ mol/l; $k_w = 1 \cdot 10^{-6}$ l/(mol s); $2w = 0.02$ m; $u = 0.1$ m/s; $D = 10^{-3}$ m^2/s. Length of the (micro)reactor 0.01 m.

A scheme of the system considered is given in Figure 4.6. Assume that the separation between the plates is "$2w$" and then $y = 0$ at the center and $y = w$ in both catalytic plates.

Considering the elemental prism, the volume of that prism is $dV = 1 \cdot dy \cdot dz$. The current is flowing in the "z" direction, and then the area of the prism in the axial direction is $S = 1 \cdot dy$. Nevertheless, the lateral area of the prism, through which has been observed diffusion, is $S' = 1 \cdot dz = dz$.

With this premise, we can write

$$\text{Input} = (\text{Convection} + \text{diffusion in the}' y' \text{direction})$$
$$= u \cdot C_b \cdot S - D \cdot S' \cdot \frac{\partial C_b}{\partial y}$$

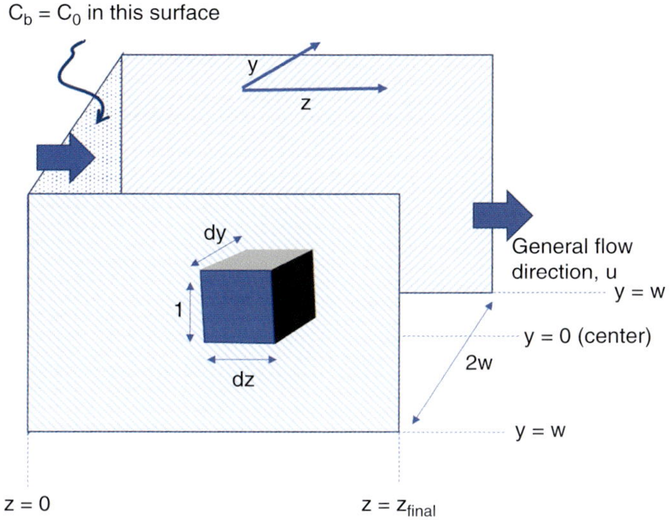

Figure 4.6 Catalytic flat wall.

$$= u \cdot C_b \cdot dy - D\frac{\partial C_b}{\partial y}dz$$

$$\text{Output} = \text{Input} + d\left[-D\frac{\partial C_b}{\partial y}dz + u \cdot C_b \cdot dy\right]$$

$$\text{Generation} = r_b dV = k_b C_b^{\propto}\, dy\, dz$$

$$\text{Accumulation} = -\frac{\partial C_b}{\partial t}\, dy\, dz$$

Putting them together and dividing by "*dy dz*," we get

$$-D\frac{\partial^2 C_b}{\partial y^2} + u\frac{\partial C_b}{\partial z} + k_b C_b^{\propto} + \frac{\partial C_b}{\partial t} = 0$$

Taking the steady state:

$$-D\frac{\partial^2 C_b}{\partial y^2} + u\frac{\partial C_b}{\partial z} + k_b C_b^{\propto} = 0$$

the boundary conditions are as follows:

- At the inlet: $C(y,0) = C_{A0}$ (all the section is at the same concentration)
- At the center, by symmetry: $y = 0$, $\left[\frac{\partial C_b}{\partial y}\right]_{y=0} = 0$
- At the wall, $y = 0$ and $y = w$, the rate of diffusion equals rate of reaction: $-D\left(\frac{\partial C}{\partial y}\right)_{y=0} = k_w C_w$

We can apply the finite differences method in this way:

$$\frac{\partial C_b}{\partial z} = \frac{C_{b,m}^{n+1} - C_{b,m}^{n}}{\Delta z}$$

$$\frac{\partial^2 C_b}{\partial y^2} = \frac{C^n_{b,m+1} - 2C^n_{b,m} + C^n_{b,m-1}}{\Delta y^2}$$

where subscript "m" refers to the position on the y-axis, and superscript "n" to the position on the z-axis.

Considering the mole balance:

$$-D\frac{\partial^2 C_b}{\partial y^2} + u\frac{\partial C_b}{\partial z} + k_b C_b^{\propto} = 0$$

it is easy to get

$$-D\frac{C^n_{b,m+1} - 2C^n_{b,m} + C^n_{b,m-1}}{\Delta y^2} + u\frac{C^{n+1}_{b,m} - C^n_{b,m}}{\Delta z} + k_b(C^n_{b,m})^{\propto} = 0$$

Rearranging:

$$C^{n+1}_{b,m} = \frac{\Delta z}{u}\left[\frac{D}{\Delta y^2}(C^n_{b,m+1} - 2C^n_{b,m} + C^n_{b,m-1}) - k_b(C^n_{b,m})^{\propto}\right] + C^n_{b,m}$$

And finally, for first-order ($\alpha = 1$) reaction:

$$C^{n+1}_{b,m} = \frac{\Delta z \cdot D}{u \cdot \Delta y^2}C^n_{b,m+1} + \left(-2\frac{\Delta z \cdot D}{u \cdot \Delta y^2} - \frac{\Delta z}{\Delta y^2}k_b + 1\right)C^n_{b,m} + \frac{\Delta z \cdot D}{u \cdot \Delta y^2}C^n_{b,m-1}$$

Renaming:

$$\gamma = \frac{\Delta z \cdot D}{u \cdot \Delta y^2}$$

we get

$$C^{n+1}_{b,m} = \gamma C^n_{b,m+1} + \left(-2\gamma - \frac{\Delta z}{\Delta y^2}k_b + 1\right)C^n_{b,m} + \gamma C^n_{b,m-1}$$

For the system to be stable, $\gamma \geq 0$ (always fulfilled), and $\left(-2\gamma - \frac{\Delta z}{\Delta y^2}k_b + 1\right) \geq 0$, so:

$$\Delta y \geq \sqrt{\Delta z\left(\frac{2D}{u} + k_b\right)}$$

At the center ($y = 0$), we will have

$$\frac{\partial C_b}{\partial z} = \frac{C^n_{b,\text{center}} - C^n_{b,\text{center}-1}}{\Delta z} = 0$$

where subscripts refer to the center and the position just before the center("center-1"). And then:

$$C^n_{b,\text{center}} = C^n_{b,\text{center}-1}$$

This is true for any position on the z-axis.

On the other hand, at the wall ($y = w$):

$$\frac{\partial C_w}{\partial y} = \frac{C^n_{b,\text{wall}} - C^n_{b,\text{wall}-1}}{\Delta y} = \frac{k_w C^n_{b,\text{wall}}}{D}$$

as the position $z = 0$ coincides with the wall. Then:

$$C^n_{b,\text{wall}} = \frac{C^n_{b,\text{wall}-1}}{1 + \frac{k_w \Delta y}{D}}$$

We can use the later equations to program in Matlab the case required. The following program can be used.

Program Listing 4.4

```
clear all
close all

kb=0.0001; % mol/l·s
CA0=1; % mol/l
kw=0.000001; % l/molS
w=0.01; % m half of the total lenght
u=0.1; % m/s
D=1e-3; %m2/s

N=50; % Number of increment in the y-axis

incy=w/((N-1)); % Increment of the variable y - half of the total
incz=0.95*incy^2/(2*D/u+kb); % Must be fulfilled
     incz<incy^2/(2*D/u+kb)

y=0:incy:w; % y-values to be used in the calculation
CAv=[]; % Initialization of a matrix that will contain all CA calcu-
lated
% Initial condition (entering the reactor):
z=0;
CA=ones(1,N)*CA0;

zfinal=0.01; % Length of the reactor
gamma=incz*D/(u*incy^2);
A=gamma;
B=-2*gamma-incz*kb/incy^2+1;

while z<zfinal
z=z+incz;
CAn(N)=CA(N)/(1+kw*incy/D); % Boundary condition at the walls
CAn(1)=CA(2); % Boundary condition in the center
for i=2:N-1
CAn(i)=A*CA(i+1)+B*CA(i)+A*CA(i-1);
end
CA=CAn;
CAv=[CAv CA']; % To save the CA vector in a matrix

end

z=0:incz:zfinal;
mesh(z,y,CAv)
xlabel('z')
ylabel('y')
zlabel('C_A')
```

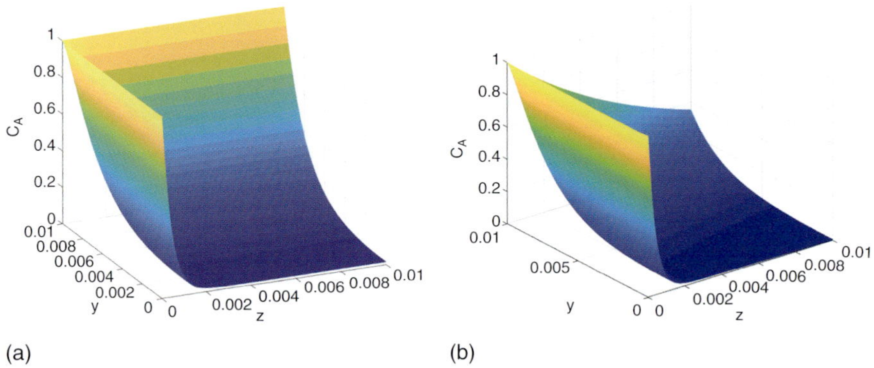

Figure 4.7 Concentration profile obtained with different values of the constants. (a) $k_w = 10^{-6}$ l/(mol s), (b) $k_w = 10^{-4}$ l/(mol s).

The program plots a 3D figure that can be rotated representing half of the reactor, where we can check the evolution of the concentration with positions. Figure 4.7a shows the results.

In this figure, we can see that the input concentration of reactant "A" (in the position $z = 0$) is quickly consumed in the center of the reactor ($y = 0$). Nevertheless, in the walls the reaction is being produced, but at a lower rate giving a characteristic profile. If the kinetic constant of the catalytic reaction is increased, for example, at $k_w = 1 \cdot 10^{-3}$ l/(mol s), the reaction will also take place at the wall, giving a concentration profile similar to that in Figure 4.7b.

Bibliography

Carrillo Ledesma, A., González Rosas, K.I., and Mendoza Bernal, O. (2019). Introducción al Método de Diferencias Finitas & su Implementación Computacional. http://132.248.182.159/acl/Textos/ (accessed 2 May 2019).

LeVeque, R.J. (2007). *Finite Difference Methods for Ordinary and Partial Differential Equations: Steady-State and Time-Dependent Problems*. Society for Industrial and Applied Mathematics.

Hjortso, M.A. and Wolenski, P. (1954). *Linear Mathematical Models in Chemical Engineering*, 2e. World Scientific.

Shankar, P.M. (2016). Pedagogy of second order homogeneous differential equations: a holistic approach using a Matlab workbook. *Computer Applications in Engineering Education* 24 (1): 114–121.

Wallas, S.M. (1995). *Chemical Reaction Engineering Handbook*. Amsterdam: Gordon and Breach.

5

Unsteady State Regime Simulation in Reactor Design

5.1 Introduction

In this chapter, we discuss the behavior of ideal reactors during the start-up and shutdown of the flow, i.e. the unsteady state behavior of such systems.

We will increase the complexity of the differential equations to be solved, and in this way we will begin with a continuous stirred tank reactor (CSTR) working in a dynamic regime. Later we discuss the plug flow reactor (PFR) without and with dispersion, and finish the chapter discussing the existence of multiple states in CSTR in which an exothermal reaction is produced.

5.2 CSTR Working in Unsteady State

Let us first discuss the system shown in Figure 5.1. During an unsteady state, we will have

Input = n_{A0} (mol/time)
Output = n_A(mol/time)
Accumulation = $V{\cdot}dC_A/dt$ (volume·(mol/volume)/time)
Generation = $r_A{\cdot}V$ ((mol/volume·time)·volume)

So that, in the mass balance:

$$\text{Input} + \text{Generation} = \text{Output} + \text{Accumulation}$$

$$n_{A0} + r_A \cdot V = n_A + \frac{dC_A}{dt} \cdot V \tag{5.1}$$

If we assume a first-order reaction:

$$Q \cdot C_{A0} - k \cdot C_A \cdot V = Q \cdot C_A + \frac{dC_A}{dt} \cdot V$$

$$\frac{dC_A}{dt} = \frac{C_{A0} - C_A}{V/Q} - k \cdot C_A = \frac{C_{A0} - C_A}{\bar{t}} - k \cdot C_A \tag{5.2}$$

This equation is an ordinary differential equation, not in partial derivatives, and can be solved numerically with no problem using the finite differences method:

$$\frac{C_A^{t+1} - C_A^t}{\Delta t} = \frac{C_{A0} - C_A^t}{\bar{t}} - k \cdot C_A^t \tag{5.3}$$

Chemical Reactor Design: Mathematical Modeling and Applications,
First Edition. Juan A. Conesa.

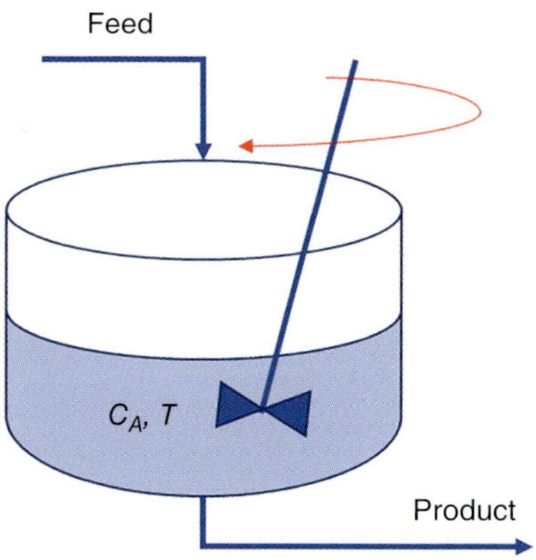

Figure 5.1 Continuous stirred tank reactor (CSTR).

and solving for the concentration:

$$C_A^{t+1} = \Delta t \left[\frac{C_{A0} - C_A^t}{\bar{t}} - k \cdot C_A^t \right] + C_A^t \tag{5.4}$$

For simulating this behavior, a simple Matlab® program can be used. For example, in case we want to simulate a CSTR reactor with $\bar{t} = 10\,\text{s}$ and $C_{A0} = 1\,\text{mol/m}^3$, where $k = 1\,\text{s}^{-1}$, the corresponding Matlab program can be that shown in Program Listing 5.1.

Program Listing 5.1

```
clear all
tresid=10; % sec, average residence time
k=1; % sec-1, kinetic first-order constant
CA0=1; % Input concentration, mol/m3
t=0;
i=1;
CA(1)=CA0; % Boundary condition at the input
tfinal=8; % sec, final time
Nt=100; % Number of intervals of time
inct=tfinal/Nt; % Sec, increment of time
tv=0:inct:tfinal; % Values of time to be used in the calculation
%CSTR concentrations are the same in all points of the reactor
while t<tfinal
    i=i+1
    t=t+inct;
    CAn(i)=inct*((CA0-CA(i-1))/tresid-k*CA(i-1))+CA(i-1);
End
CA=CAn;
plot(tv,CA(1:Nt+1))
xlabel('Time (sec)')
ylabel('C_A (mol/m^3)')
```

The main result is given in Figure 5.2. As you can see, the concentration at the exit varies with time, while the steady state is established. In the CSTRs, by definition, the concentration is the same in all points of the reactor and equals the exit concentration.

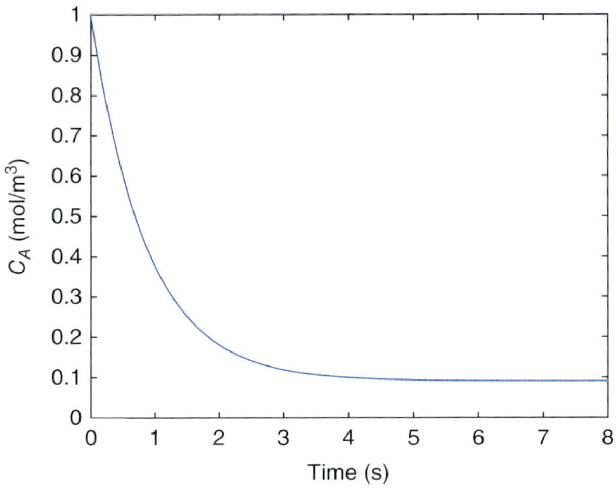

Figure 5.2 CSTR working in unsteady state regime. The concentration at the exit of the CSTR is constant in steady-state regime, but varies when considering the start-up or shutdown.

5.3 PFR Working in Dynamic Regime (No Dispersion)

As in the PFR, different reaction rates are present, and the mass balance should be done over a differential volume, dV, as shown in Figure 5.3.

As different variables are used, differential forms are partial differential, and introducing into the balance:

$$\text{Input} + \text{Generation} = \text{Output} + \text{Accumulation}$$

$$n_A + r_A \cdot \partial V = (n_A + \partial n_A) + \frac{\partial C_A}{\partial t} \cdot \partial V \tag{5.5}$$

Bearing in mind, for a constant-density system:

$$Q \cdot \partial C_A = \partial n_A \tag{5.6}$$

which means that:

$$\frac{\partial C_A}{\partial t} = r_A - Q\frac{\partial C_A}{\partial V} \tag{5.7}$$

If we assume a first-order kinetics:

$$\frac{\partial C_A}{\partial t} = -kC_A - Q\frac{\partial C_A}{\partial V} \tag{5.8}$$

Figure 5.3 The mole balance in a PFR.

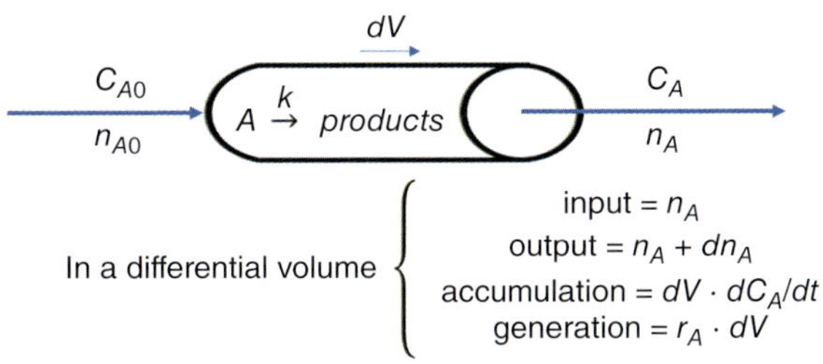

Using the finite differences approximation, first order:

$$\frac{C_{Ai}^{t+1} - C_{Ai}^{t}}{\Delta t} = -kC_{Ai}^{t} - Q\frac{C_{Ai+1}^{t} - C_{Ai}^{t}}{\Delta V} \tag{5.9}$$

Rearranging:

$$C_{Ai}^{t+1} = \left(1 - k \cdot \Delta t + Q\frac{\Delta t}{\Delta V}\right) C_{Ai}^{t} - Q\frac{\Delta t}{\Delta V} C_{Ai+1}^{t} \tag{5.10}$$

Applying the stability criteria (already mentioned in Chapter 4), the system is stable (errors do not increase), if:

$$Q\frac{\Delta t}{\Delta V} > 0$$

$$\left(1 - k \cdot \Delta t - Q\frac{\Delta t}{\Delta V}\right) > 0$$

$$\left(1 - k \cdot \Delta t - Q\frac{\Delta t}{\Delta V}\right) + Q\frac{\Delta t}{\Delta V} \leq 1$$

First and third conditions are always fulfilled, and from the second one:

$$\Delta t < \frac{1}{\left(k + \frac{Q}{\Delta V}\right)} \tag{5.11}$$

In the initial state:

$$\forall V \text{ if } t = 0 \rightarrow C_A = 0 \tag{5.12}$$

The boundary conditions of this system are

- At the input ($V = 0$): $C_A = C_{A0}$ (known concentration)
- At the exit ($V = V_{\text{total}}$) $\rightarrow C_A(V^-) = C_A(V^+)$ (closed vessel), so that at this section:

$$\left[\frac{\partial C}{\partial V}\right]_{V_{\text{total}}} = 0 \tag{5.13}$$

The procedure can be applied to different designs. We will apply the method for the simulation of a 20 m^3 reactor where a feed of 1 mol/m^3 of reactive is fed, with a flow rate of 100 m^3/s and $k = 17$ $mol/(m^3\ s)$. The Matlab program in this case is that shown here.

Program Listing 5.2

```
clear all

CA0=1; % Input concentration mol/m3
Qv0=100; % Volumetric flow rate m3/s
Vtotal=20; % m3 Total volume
k=17; % Reaction Kinetic constant , mol/m3·s
N=50; % Number of increment in x (position)

incV=Vtotal/(N-1); % Increment V
inct=0.95*incV/(Qv0+k*incV); % Must be lesser than defined in Eq 5.11.
V=0:incV:Vtotal; % Values of V that Will be used in the calculation
```

```
% Initial values:
tend=[];
CAend=[];
t=0;
CA=zeros(1,N);
CA(1)=CA0;
tfinal=1; % secs, Final time
A=1-k*inct-Qv0*inct/incV;

while t<tfinal
    t=t+inct
    CAn(1)=CA0; % Boundary condition for the first increment
    CAn(N)=CA(N-1); % Boundary condition for the last increment.
      Eq. 5.13
    for i=2:N-1
        CAn(i)=A*CA(i)+(Qv0*inct/incV)*CA(i-1); % PDE given in Eq. 5.10
    end
    CA=CAn;
    CAend=[CAend CA(end)]; % Saving the concentration at the exit
    tend=[tend t]; % Saving the values of time used
    plot(V,CA)
    xlabel('Volume (m^3)')
    ylabel('C_A (mol/m^3)')
    drawnow
end

figure
plot(tend,CAend)
xlabel('Time (s)')
ylabel('C_A at the exit (mol/m^3)')
```

In the program, the last value of the concentration at each time, i.e. the concentration at the exit, is saved in a vector (C_{end}) in order to see the evolution of this concentration with time.

Some results are presented in Figure 5.4, using different values of final time in order to see the evolution of the concentration profile.

5.4 PFR Working in Dynamic Regime (with Dispersion)

As pointed out in previous section, the mass balance in the PFR system in the dynamic regime can be expressed as

$$n_A + r_A \cdot \partial V = (n_A + \partial n_A) + \frac{\partial C_A}{\partial t} \cdot \partial V \tag{5.14}$$

But in this case:

$$Q \cdot \partial C_A \neq \partial n_A \tag{5.15}$$

As:

$$n_A = Q \cdot C_A - D_e \cdot S \cdot \frac{\partial C_A}{\partial z} \tag{5.16}$$

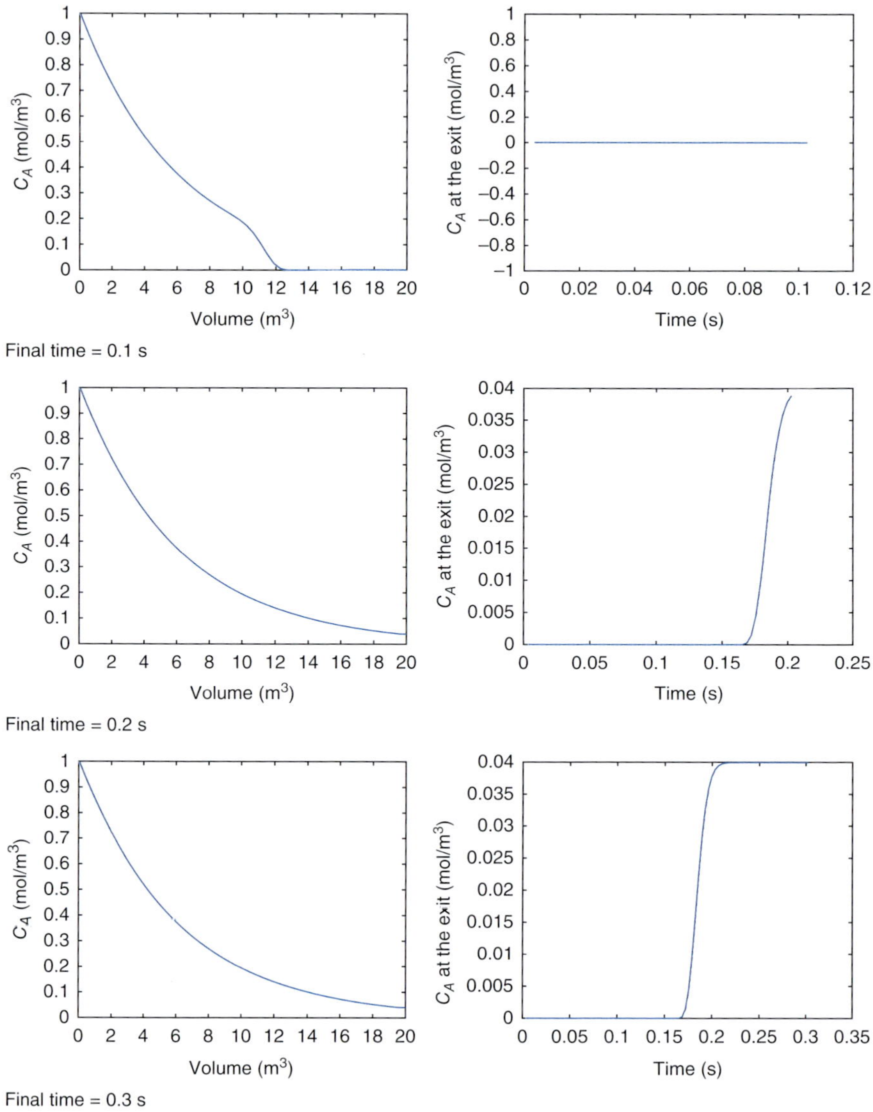

Figure 5.4 PFR working in unsteady state regime (without dispersion) at different times.

In the previous equation, two terms are present, the first one corresponding to the convection (general movement of the flow) and the second one corresponding to the diffusion by the Fick's law. Deriving and doing $Q = (u \cdot S)$, then:

$$\frac{\partial n_A}{\partial z} = u \cdot S \cdot \frac{\partial C_A}{\partial z} - D_e \cdot S \cdot \frac{\partial^2 C_A}{\partial z^2} \tag{5.17}$$

and introducing in the mole balance:

$$D_e \cdot \frac{\partial^2 C_A}{\partial z^2} - u \cdot \frac{\partial C_A}{\partial z} + r_A = \frac{\partial C_A}{\partial t} \tag{5.18}$$

Using the finite differences approximation, first order, and assuming a first-order reaction:

$$D_e \cdot \frac{C^t_{Ai+1} - 2C^t_{Ai} + C^t_{Ai-1}}{\Delta z^2} - u \cdot \frac{C^t_{Ai+1} - C^t_{Ai}}{\Delta z} - k \cdot C^t_{Ai} = \frac{C^{t+1}_{Ai} - C^t_{Ai}}{\Delta t} \tag{5.19}$$

and rearranging:

$$C^{t+1}_{Ai} = D_e \frac{\Delta t}{\Delta z^2} \cdot C^t_{Ai-1} + \left(1 - 2\frac{\Delta t}{\Delta z^2} D_e + u\frac{\Delta t}{\Delta z} - k\Delta t\right) C^t_{Ai} + \left(\frac{\Delta t}{\Delta z^2} D_e - u\frac{\Delta t}{\Delta z}\right) C^t_{Ai+1} \tag{5.20}$$

The system is stable (errors do not increase) if:

$$D_e \frac{\Delta t}{\Delta z^2} > 0$$

$$\left(1 - 2\frac{\Delta t}{\Delta z^2} D_e + u\frac{\Delta t}{\Delta z} - k\Delta t\right) > 0$$

$$\left(\frac{\Delta t}{\Delta z^2} D_e - u\frac{\Delta t}{\Delta z}\right) > 0$$

$$\left[D_e \frac{\Delta t}{\Delta z^2} + \left(1 - 2\frac{\Delta t}{\Delta z^2} D_e + u\frac{\Delta t}{\Delta z} - k\Delta t\right) + \left(\frac{\Delta t}{\Delta z^2} D_e - u\frac{\Delta t}{\Delta z}\right)\right] \leq 1$$

As in other equations previously discussed, first and last conditions are always fulfilled, but from the third one:

$$\Delta z < \frac{D_e}{u} \tag{5.21}$$

and from the second one:

$$\Delta t < \frac{1}{\left(\frac{2}{\Delta z^2} D_e - \frac{u}{\Delta z} + k\right)} \tag{5.22}$$

Obviously, these two conditions must fulfill, so the lower value of Δt should be selected.

The procedure can be applied to different designs. We will apply the method for the simulation of a 1 m length reactor where a feed of 0.01 mol/m^3 of reactive is fed, with a gas velocity of 0.01 m/s and values of $k = 1.7 \cdot 10^{-5}$ mol/(m^3 s) and De = 10^{-4} m^2/s. Program Listing 5.3 shows a possible program for Matlab.

Program Listing 5.3

```
clear all

CA0=0.01; % Input concentration mol/m3
u=0.01; % Gas Velocity m/s
Ltotal=1; % m
k=0.000017; % Reaction kinetic constant, mol/m3·s
De=1e-4; % Difusivity in m2/s
N=150; % Number of increments in Z (position)
incZ=Ltotal/(N-1); % Z increment
if incZ>De/u
```

```
    incZ
    De/u
    disp('Something is wrong')
    return % This condition must be fulfilled (Eq. 5.21)
end
inct=0.95/(-u/incZ+2*De/incZ^2+k); % Must be lower according
     to Eq. 5.22.
alfa=De*inct/incZ^2
beta=1-2*De*inct/incZ^2+u*inct/incZ-k*inct
gamma=De*inct/incZ^2-u*inct/incZ
L=0:incZ:Ltotal; % Values of V to be used in the calculation
% Initial condition:
t=0;
CA=zeros(1,N);
CA(1)=CA0;
tfinal=100; % Secs, final time
alfa=De*inct/incZ^2;
beta=1-2*De*inct/incZ^2+u*inct/incZ-k*inct;
gamma=De*inct/incZ^2-u*inct/incZ;

while t<tfinal
    t=t+inct
    CAn(1)=CA0; % Boundary condition for the first increment
    CAn(N)=CA(N-1); % Boundary condition for the last increment
    for i=2:N-1
        CAn(i)=alfa*CA(i-1)+beta*CA(i)+gamma*CA(i+1);
    end
    CA=CAn;
    plot(L,CA)
    drawnow
end
```

The program is similar to those used in Chapter 2 and in the previous section. Figure 5.5 shows some interesting results.

5.5 Multiple Steady States in CSTR with Exothermal Reaction

Let us assume we are working in a CSTR with a cooling jacket as that shown in Figure 5.6. Let us assume we are working with an exothermal first-order reaction: $A \rightarrow B$. Let us assume also a negligible heat loss, and constant density. We will use the symbols included in Figure 5.6.

In the cooling jacket, we will assume that the heat removed is

$$q_c = U \cdot A \cdot (T - T_c) \tag{5.23}$$

The mass balance of reacting species in the reactor, in the unsteady state, is

$$QC_{A0} - (kC_A) \cdot V = QC_A + V\frac{dC_A}{dt} \tag{5.24}$$

The last term represents the variation of the amount of reactant with time, i.e. the accumulation term of the mole balance. This is usually assumed to be zero,

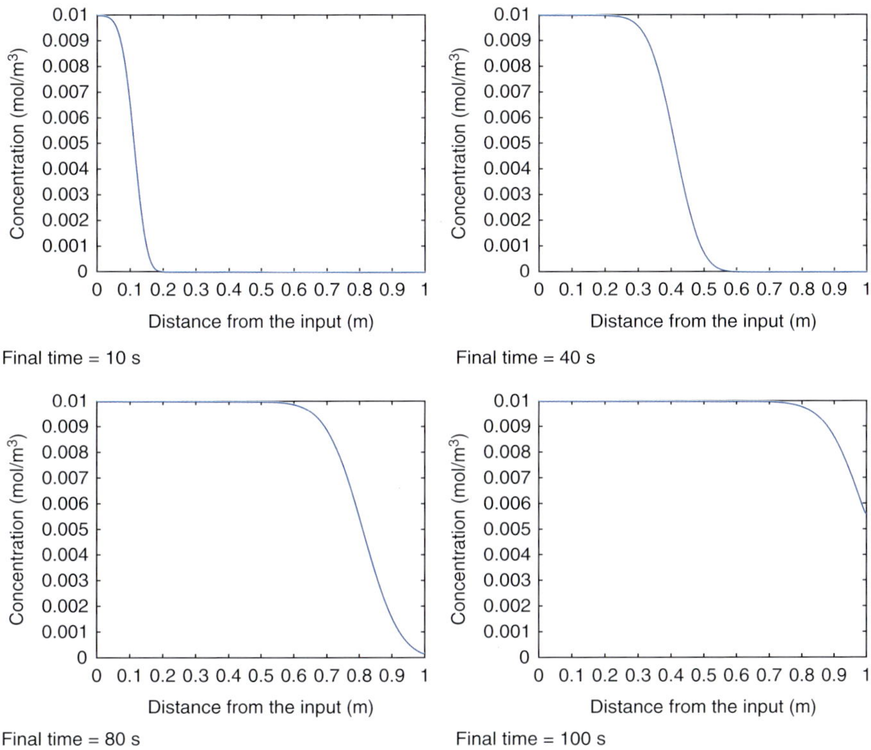

Figure 5.5 PFR working in unsteady state regime (with dispersion) at different times.

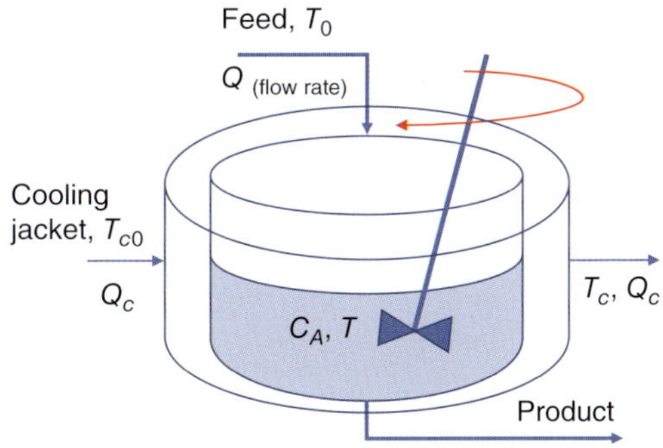

Figure 5.6 CSTR with cooling jacked. T is the temperature of the reactor; T_c is the temperature of the cooling fluid. Q and Q_c are the flow rates of reactants and cooling fluid, respectively.

as steady conditions are usually applied. V represents the volume of the reactor, which is constant.

In the same way, an energy balance at the reactor can be done:

$$\rho c_p Q(T_0 - T) - (\Delta H_r)(kC_A) \cdot V = UA(T - T_c) + \rho c_p V \frac{dT}{dt} \tag{5.25}$$

In this case, the heat of reaction is ΔH_r and the heat capacity of the reacting fluid is C_p. Energy balance at the cooling jacket can be written as

$$\rho_c c_{pC} Q_c (T_{c0} - T_C) + UA(T - T_c) = \rho_c c_{pC} V_c \frac{dT_c}{dt} \tag{5.26}$$

where V_c is the volume of the fluid in the jacket that is also constant.

In many textbooks and manuals, the variation of C_A, T, and T_c with time are not considered and only the steady state is studied. In that case, the balance equations "loss" the term depending with time, in such a way that:

$$QC_{A0} - (kC_A) \cdot V = QC_A \tag{5.27}$$

$$\rho c_p Q(T_0 - T) - (\Delta H_r)(kC_A) \cdot V = U \cdot A \cdot (T - T_c) \tag{5.28}$$

$$\rho_c c_{pc} Q_c (T_{c0} - T_c) + U \cdot A \cdot (T - T_c) = 0 \tag{5.29}$$

In this way, Eqs. (5.27)–(5.29) represent the usual CSTR at steady state. We can solve these equations for C_A and T_c. This gives:

$$C_A = \frac{Q_0 C_{A0}}{Q_0 + V \cdot k} \tag{5.30}$$

$$T_c = \frac{T_{c0} + \dfrac{UA}{\rho_c C_{pc} Q_c} \cdot T}{1 + \dfrac{UA}{\rho_c C_{pc} Q_c}} \tag{5.31}$$

We can define the heat generated (q_G) and the negative of the heat removed ($-q_R$) as

$$q_G = -\Delta H_r \cdot (kC_A) \cdot \mathrm{V} \tag{5.32}$$

$$-q_R = -[\rho c_p Q_0 (T - T_0) - UA(T - T_c)] \tag{5.33}$$

We can easily simulate the behavior of the reactor with a simple Matlab program. For illustrating this situation, we will use the following constants:

$Q_0 = 0.01\ \mathrm{m^3/s}$	$k_0 = 7\mathrm{e}11\ \mathrm{s^{-1}}$	$T_0 = 300\ \mathrm{K}$
$Q_{c0} = 0.01\ \mathrm{m^3/s}$	$E/R = 10\,500\ \mathrm{K^{-1}}$	$\Delta H_r = -10\,000\ \mathrm{J/mol}$
$C_{A0} = 1\ \mathrm{mol/m^3}$	$U = 40\ \mathrm{W/m^2{\cdot}K}$	$c_p\ \rho_o = 720\ \mathrm{J/K}$
$V = 1\ \mathrm{m^3}$	$A = 2\ \mathrm{m^2}$	$c_{pc}\ \rho_{oc} = 1000\ \mathrm{J/K}$
$V_c = 0.01\ \mathrm{m^3}$	$T_{c0} = 300\ \mathrm{K}$	

Program Listing 5.4

```
%%% CSTR multiple states

Qv0=0.010; % m3/s
Qr=0.01;% m3/s
CA0=1; %mol/m3
V=1; % m3
VR=0.01; % m3
k0=7e11; % seg-1
E_R=10500; % K-1
R=8.314; %J/mol·K
U=40; % W/m2·K
A=2; % m2
TC0=300; % K
T0=300; % K
```

```
DHr=-10000; % J/mol
rCp=720; % J/K
rCpR=1000; % J/K

% First part: steady state assumption. Looking for the solutions
     of the balance equations

QR=[];
QG=[];
Tv=280:430; % Temperature range, in K

% Heat generated and removed (steady state):
for T=Tv
    k=k0*exp(-E_R/T);
    CA=Qv0*CA0/(Qv0+V*k);
    beta=U*A/(rCpR*Qr);
    Tc=(TC0+beta*T)/(1+beta);
    QG=[QG -DHr*V*k*CA];
    QR=[QR -(rCp*Qv0*(T-T0)-U*A*(T-Tc))];
end

plot (Tv,QG,'x',Tv,QR)
xlabel ('Temperature (K)')
ylabel ('J/s')
legend ('Q_g_e_n_e_r_a_t_e_d','Q_r_e_m_o_v_e_d')
```

In the program, the variables are fixed to the mentioned values, and the temperature of the reactor is varied from 280 to 430 K. The corresponding C_A and T_c are calculated using the expressions (5.30) and (5.31). As a result of the program, the following figure is obtained.

As we can see, three different solutions to Eqs. (5.27)–(5.29) are found, i.e. three different steady states are possible. The exact values of the temperatures and composition are presented in Table 5.1.

In principle, the one giving a lower concentration of C_A is the best, as it implies a higher conversion.

But it is important to be aware of the stability of these three solutions, because the solutions may not be stable, and then the reactor will work in an unintended state. If we look only at the results in Figure 5.7, and the reactor is started at, for example, 330 K (between intermediate and upper steady states), the heat generated by the reaction is greater than the heat removed. This will cause the temperature in the reactor to rise, and the rise will continue until the upper

Table 5.1 Steady-state conditions of the example, without considering their stability.

Steady state	T **(K)**	T_c **(K)**	C_A **(mol/m^3)**
Upper	352.9	347.02	0.1066
Intermediate	328.8	325.6	0.5137
Lower	303.8	303.37	0.9360

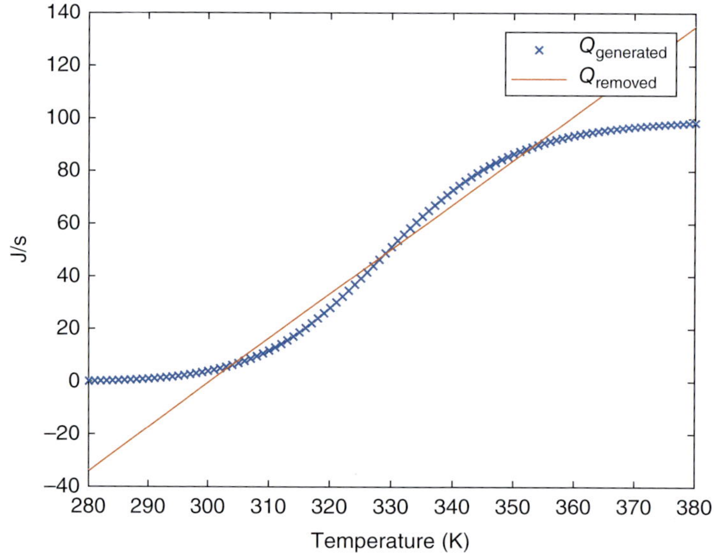

Figure 5.7 Three possible steady states in the exothermal cooled CSTR.

steady state is reached. It is easy to show, using similar arguments, that the upper and lower states are stable, but it is not the intermediate. Let us show that this is not completely true.

In order to simulate the dynamical behavior of the reactor, we will solve numerically Eqs. (5.24)–(5.26). To do that, let us use the first-order approximations of the derivatives:

$$QC_{A0} - (kC_A^t) \cdot V = QC_A^t + V\frac{C_A^{t+1} - C_A^t}{\Delta t}$$

(Eq.(5.34), from Eq.(5.24))

$$\rho c_p Q(T_0 - T^t) - (\Delta H_r)(kC_A^t) \cdot V = UA(T^t - T_c^t) + \rho c_p V\frac{T^{t+1} - T^t}{\Delta t}$$

(Eq.(5.35), from Eq.(5.25))

$$\rho_c c_{pC} Q_c(T_{c0} - T_c^t) + UA(T^t - T_c^t) = \rho_c c_{pC} V_c \frac{T_c^{t+1} - T_c^t}{dt}$$

(Eq.(5.36), from Eq.(5.26))

We must solve to find the calculated C_A^{t+1}, T^{t+1}, and T_c^{t+1}:

$$C_A^{t+1} = \frac{Q(C_{A0} - C_A^t) - (kC_A^t) \cdot V}{V}\Delta t + C_A^t \tag{5.34}$$

$$T^{t+1} = \frac{\rho c_p Q(T_0 - T^t) - (\Delta H_r)(kC_A^t) \cdot V - UA(T^t - T_c^t)}{\rho C_p V} + T^t \tag{5.35}$$

$$T_c^{t+1} = \frac{\rho_c c_{pC} Q_c(T_{c0} - T_c^t) + UA(T^t - T_c^t)}{\rho_c c_{pC} V_c} + T_c^t \tag{5.36}$$

We can now simulate the behavior starting at the different steady states. In the simulation, we will start at a steady state. Note that the values of concentration at the input and temperatures are the same, as the system has not changed. The initial state of the system is the steady state, and the evolution with time is followed. The Matlab program can be the following one.

Program Listing 5.5

```
%%% CSTR multiple states
% Second part: non-steady state
% Values of the variables at the input:
 CA0=1;
 T0=300;
TC0=300;
T=[];
CA=[];
TC=[];
% Initial values of the variables at the reactor == steady state
     (in this case, lower steady state)
  CAss=0.9360;
  Tss=303.8;
  TCss=303.3778;
CA(1)=CAss;
T(1)=Tss;
TC(1)=TCss;
t=0;
i=1;
tfinal=200 ; % s
inct=0.1; % s
tv=0:inct:tfinal; % Values of time to be used in the calculation

while t<=tfinal
    i=i+1;
    t=t+inct;
    k=k0*exp(-E_R/T(i-1));
    CA(i)=inct*((Qv0*(CA0-CA(i-1))-k*CA(i-1)*V)/V)+CA(i-1);
    T(i)=inct*(rCp*Qv0*(T0-T(i-1))-DHr*V*k*CA(i-1)-U*A*(T(i-1)
        -TC(i-1)))/(rCp*V)+T(i-1);
    TC(i)=inct*(rCpR*Qr*(TC0-TC(i-1))+U*A*(T(i-1)
        -TC(i-1)))/(rCpR*VR)+TC(i-1);
end

plot(tv,T(1:i-1),'b-')
xlabel('Time (s)')
ylabel('Temperature in the reactor (K)')
```

The calculation gives the evolution of temperature seen in Figure 5.8a–c. "a" corresponds to the upper steady state, "b" to the intermediate, and "c" to the lower one. As can be seen, actually the upper and intermediate states are not stable at all, and the temperature tends to reach the only stable state, which is the lower one. Figure 5.8c represents this stable state, and, as can be checked, a very small temperature change is observed. In this sense, the study of the stationary states is useless, as it can give unreal steady states. Actually, instead of talking about three

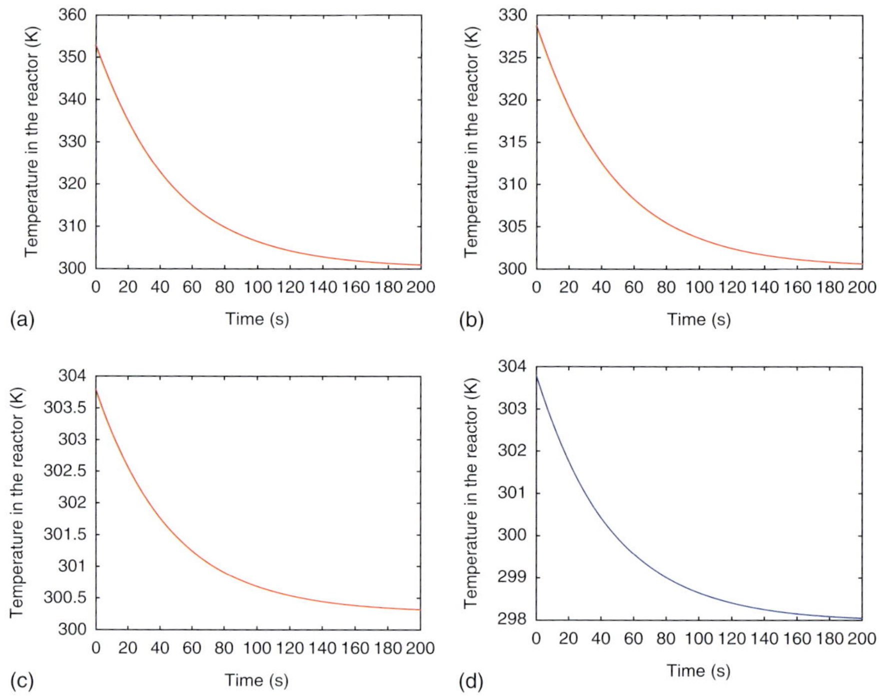

Figure 5.8 Temperature evolution with time in the system starting at different conditions. (a) Upper steady state. (b) Intermediate steady state. (c) Lower steady state. (d) Introducing a perturbation in the lower steady state.

"steady states," we would better off talking about three "different solutions of the balances in stationary conditions."

We can check for the robustness of the stationary state with another simulation. Starting from the stationary state, we can introduce a small perturbation in one of the input conditions, for example, $T_{\text{input}} = (T_0 - 5)$ in such a way that the feed is 5 °C cooler. In this case, the $(-Q_r)$ curve will move to higher values (Figure 5.9) and the new steady states will be somewhat different (states 1′, 2′, and 3′ in the figure). The usual behavior is then that, if we start from steady states 1, 2, and 3, the system will evolve to states 1′, 2′, and 3′, respectively. Nevertheless, with the data proposed in this example, as we have seen, states 2 and 3 are unstable and all cases finish in the state 1′.

Matlab programming of this case is quick and simple. The same program as that in the previous case can be used, but the perturbation changes the temperature of the feed:

```
T0=300-5; % Perturbation of (-5) K in the
    temperature at the input
```

The result of the simulation, starting at the stable lower steady state, is shown in Figure 5.8d. As we can see, the system evolves to the new steady state that has been changed as the term "$Q_{removed}$" has changed.

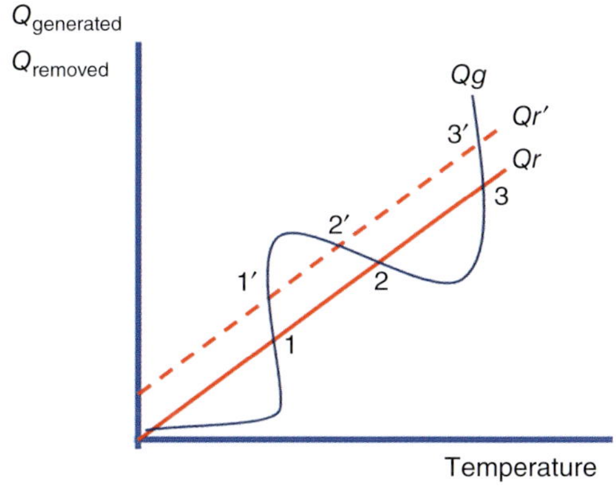

Figure 5.9 Small changes in the input temperature will move the $Q_{removed}$ and define new steady states.

Bibliography

LeVeque, R.J. (2007). *Finite Difference Methods for Ordinary and Partial Differential Equations: Steady-State and Time-Dependent Problems*. Society for Industrial and Applied Mathematics.

Hjortso, M.A. and Wolenski, P. (1954). *Linear Mathematical Models In Chemical Engineering*, 2e. World Scientific.

6

Scaling and Stability of Chemical Reactors

6.1 Introduction

A key problem in the development of a new process is scaling. A definition of scaling could be "correct operation and start up a commercial size unit whose design and operating procedures are partly based on experimentation and demonstration in the operation of a smaller system."

The reactor is the nucleus of many chemical plants. It is also the part of the plant where there are most scaling problems, particularly when the reaction involves several phases, because all phenomena are not equally affected by the dimensions of the plant.

Until around World War II, scaling was an empirical process. The common way of doing this was to enlarge the process in small steps, as shown schematically in Figure 6.1 for a chemical reactor. Figure 6.1 also shows the approximate production rate and the time intervals associated with the design/construction and operation. Depending on the nature of the process, these time intervals may differ greatly from those shown.

Initially, large safety margins were incorporated, often producing plants with a much higher capacity than designed. Clearly, scaling in small steps is very expensive and insecure, as long as a predictive model is not used. Today, however, the scaling process uses the data of the laboratory and/or pilot plant, supplemented by studies in models and mathematical studies, to determine the size and dimensions of the industrial units.

Due to the different predictive power of different mathematical models within the production plant, there are different equipment for which the maximum scale factor, from which the operation is unreliable, is different. Table 6.1 shows this maximum factor for several equipment.

Let us assume that we have a reactor in which 30% conversion is obtained. We want to design a reactor with 90% conversion. Someone could think about making the reactor three times larger, but we know that many reactions show a rate that decreases with increasing conversion. Also, thermodynamic limitations could exist, preventing higher conversion of, for example, 30%. In fact, a conversion of 30% may be achievable in a three times smaller reactor. In many cases, the scaling is complicated by the existence of lateral reactions, thermal effects, phase separation, etc.

Chemical Reactor Design: Mathematical Modeling and Applications,
First Edition. Juan A. Conesa.

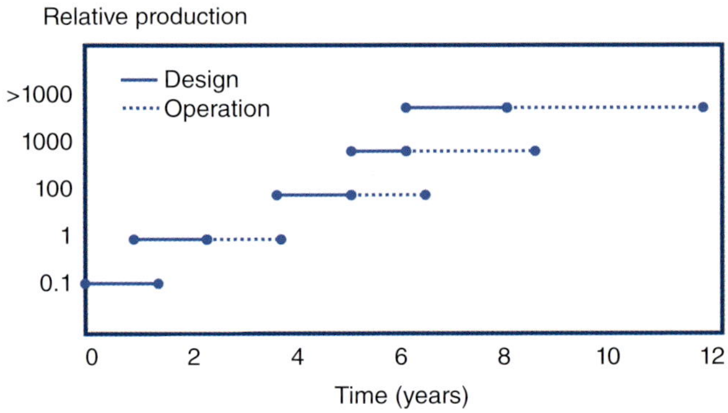

Figure 6.1 Conventional scaling process. Source: Moulijn et al. (2013). Adapted with permission from John Wiley & Sons.

Table 6.1 Maximum typical values of the scale factor of several equipment.

Equipment	Maximum factor of scale
Reactors	
Multitubular	>10 000
Homogeneous plug flow and stirred tank	>10 000
Columns of bubbles	<1000
Fluidized bed	50–100
Separation processes	
Distillation and rectification	1000–50 000
Absorption	1000–50 000
Extraction	500–1000
Drying	20–50
Crystallization	20–30

A scaling factor with a low level of risk can only be achieved in situations in which there are known correlations or when an engineering approach is possible. Thus, the unit operations of distillation, absorption, and rectification can be scaled without an intermediate step. The behavior of the gases and the liquid–gas equilibrium can be explained well in physical terms, for which the calculations are direct. For solids, the values of the maximum scale factor are much lower, due to complications such as deposits and abrasion.

One of the common objectives during scaling is to convert a batch process (tested in the laboratory) to continuous operation. In this sense, a familiar concept in terms of scaling is to assume the residence time in the continuous reactor identical to the reaction time of the batch reactor.

Moreover, if we have a continuous process with a flow rate Q_1 that we want to increase to a $Q_2 \gg Q_1$ producing a material with the same characteristics, the scale factor is $S = Q_2/Q_1$ and volume of the system will be given by $S = V_2/V_1$ so

Table 6.2 Standards for scaling based on residence time.

Data from reactor	Extrapolation to large reactor	Scaling criterion	Precautions/ commentary
Batch isotherm, Ci $= f(t)$	Batch isotherm of greater volume	$t_1 = t_2$	– Control heat transfer in the large reactor – Ensure same conditions of agitation
Batch isothermal with constant density, Ci $= f(t)$	Isothermal tubular reactor of constant density	$t_1 = \bar{t}_2$	– Heat transfer – Plug flow deviations
Batch isothermal isobaric with variable density, Ci $= f(t)$	Isothermal tubular reactor with variable density, $\Delta P \approx 0$	$t_1 = \bar{t}_2$	– Heat transfer – Plug flow deviations
Isothermal PFR, Ci $= f(\bar{t})$	Isothermal PFR	$\bar{t}_1 = \bar{t}_2$	– Heat transfer – Deviations in RTD
CSTR isothermal, Ci $= f(\bar{t})$	CSTR isothermal larger size	$\bar{t}_1 = \bar{t}_2$	– Heat transfer – RTD deviations

that the average residence time is maintained:

$$\bar{t}_1 = V_1/Q_1 = V_2/Q_2 = \bar{t}_2$$

$\bar{t}$ constancy during scaling is natural for chemical reactors, since any alteration will change the yields and can also change the selectivity.

In processes where heat transmission is important, it would also be desirable for the external area of the system to follow the same scale factor, but this is rarely possible. In fact, it is impossible if you want to maintain geometric similarity.

Table 6.2 shows the rules that we could deduce for the scaling of homogeneous reactors, based on the different concepts studied in previous topics.

As can be seen in Table 6.2, in the scaling of isothermal reactors, it is important to know the heat transfer phenomena. More important will be in adiabatic reactors, which are abundant in the industry. In this type of reactors, the phenomena associated with energy changes are so important that direct scaling from the reaction time is very difficult in practice.

Thus, this chapter discusses the study of the phenomena related to the heat transfer during the process of scaling a chemical reactor.

6.2 Scaling the Batch Tank Stirred Reactor

Batch reactors are frequently used for production of specialty chemicals such as cosmetics, medicines, etc. The hardest problems occur when we are dealing with exothermic reactions.

6.2.1 Temperature Control. Heat Transmission

Often, batch reactors operate as shown in Figure 6.2: at first, the temperature of the heat transfer media (T_c) is high in order to heat the reaction mixture to the

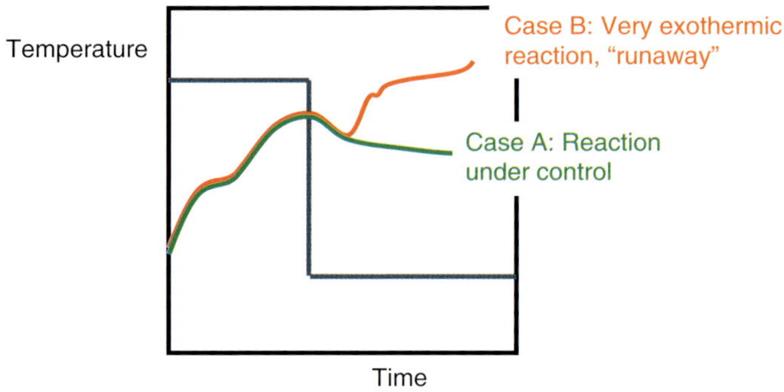

Figure 6.2 Temperature control in batch reactors manipulates the temperature of the external fluid.

reaction temperature (T). When the exothermic reaction begins, the temperature of the medium that transfers heat is decreased, often to the lowest possible level. After the reaction is under control (case A), the temperature can be increased to maintain the reaction mixture at the optimum temperature to achieve a reaction rate sufficiently high. If a reaction is faster or very exothermic, it may be impossible to keep the temperature under control even with maximum cooling (case B). Obviously, such an event is not desirable, and a thermal runaway is produced.

Thermal control is so important in many reactors that, usually, they have the appearance of heat exchangers. The reaction rate and the equilibrium are greatly affected by temperature, and therefore also are affected by side reactions, formation of other products, yields, selectivities, etc.

Many of the agitated reactors found in the chemical industry are of the jacketed type. As a consequence, the area of heat transfer per unit volume decreases with increasing the reactor volume. Generally, the heat transfer could be improved with a higher agitation; however, usually, this does not produce a dramatic increase in the heat transference.

The heat transfer can also be improved using a coolant with lower temperature, but the use of a medium at very low temperature is expensive, and the components of the reaction mixture could precipitate on the walls.

The solution to this problem is to separate the reaction volume and the heat exchange area, i.e. using external heat exchangers or by introducing heating/cooling coils in the reactor. Thus, the batch reactor may be scaled from laboratory to industrial, simply based on the reaction time. In addition, if necessary, the rate of heat production is lowered by reducing the concentrations of the reactants or catalysts (the operation in semi-batch could be attractive, as discussed in Section 6.2.2).

6.2.2 Example of Scaling a Batch and Semi-batch Reactor

In general, the histories of temperature, concentration, and mixing conditions in a large reactor differ significantly from those obtained in a laboratory reactor. Therefore, it is not surprising that selectivities also change during scaling.

We illustrate the influence of heat transfer performance and selectivity during scaling in a semi-batch batch reactor by means of a simulation using a reaction in series. This reaction is composed of two irreversible elementary steps, both exothermic and with first-order kinetics:

$$A \xrightarrow{k_1} P \xrightarrow{k_2} S$$

A is the reactant, P the desired product, and S an unwanted side product. k_1 and k_2 are the kinetic constants of the reaction (s^{-1}). For the batch reactor, the material and energy balances have the following form:

$$\frac{dC_A}{dt} = r_1 = -k_1 \cdot C_A \tag{6.1}$$

$$\frac{dC_P}{dt} = r_2 - r_1 = -k_2 \cdot C_P - (-k_1 \cdot C_A) \tag{6.2}$$

$$\rho \cdot C_P \cdot \frac{dT}{dt} = (-r_1) \cdot \Delta H_1 + (-r_2) \cdot \Delta H_2 + U \cdot (T_c - T) \cdot \frac{A}{V} \tag{6.3}$$

Remember that in the given expression of the heat energy balance, it is assumed that enthalpies are not dependent on the reaction temperature, and that the values of the density and capacity of reactants and products are the same. Thus, the first term is derived from the enthalpy change of reagents due to change in temperature, terms with ΔH_1 and ΔH_2 are associated with changing enthalpy occurring as a result of the reaction, and the last term accounts for the heat transfer through the walls.

The overall coefficient of heat transfer U is evaluated using the appropriate expressions. The ODE system (1–3) can be solved using numerical methods.

The values that we use to simulate the reactor are the following:

$k_{10} = 5\,s^{-1}$, $k_{20} = 2.5 \cdot 10^8\,s^{-1}$, pre-exponential factors of the constants k_1 and k_2
$E_1 = 20$ kJ/mol, $E_2 = 90$ kJ/mol, activation energies
$\Delta H_1 = -400$ kJ/mol, $\Delta H_2 = -300$ kJ/mol, reaction enthalpies
$c_p = 3$ kJ/(kg K), specific heat
$\rho = 1000\,kg/m^3$, density
$C_{A0} = 1500\,mol/m^3$, $C_{P0} = 0\,mol/m^3$, initial concentration of A and P
$T_0 = 295$ K, input temperature of the reaction mixture
$T_c = 345$ K if $0 < t < 360$ s
$= 295$ K if $t > 360$ s, temparature of cooling medium
$U = 5\,kJ/(s\,m^2\,K)$, global coefficient of heat transfer.

We assume the reactor to be cylindrical and use two values of A/V runs corresponding to a reactor 42 cm in diameter and one with 400 cm. The corresponding quotients are

$$(A/V)_{small} = 9.50\,m^{-1} \quad (A/V)_{large} = 1.00\,m^{-1}$$

A valid Matlab® program is presented here. Figure 6.3 shows the results of the calculations for a small and large reactor.

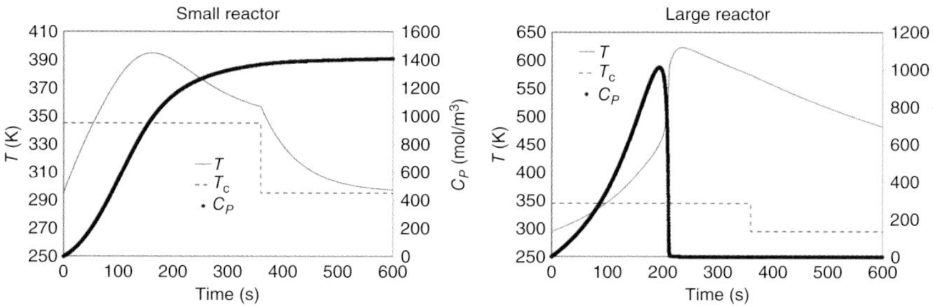

Figure 6.3 Concentration and temperature profiles in the small and the large reactor.

Program Listing 6.1

```
k10 = 5;% s -1
k20 = 250000000;% s -1, pre-exponential factors of the
constants k1 and k2
E1 = 20;% kJ mol -1,
E2 = 90;% kJ mol -1, activation energies
DH1 = -400;% kJ mol -1,
DH2 = -300;% kJ mol -1, reaction enthalpies
cp = 3;% kJ kg-1 K-1, specific heat
d = 1000;% kg m-3, density
CA0 = 1500;% mol m-3,
CP0 = 0;% mol m-3, initial concentration of A and P
T0 = 295;% K, input temperature of the reaction mixture
Tc = 345;% K if 0 <t <360 s
% 295 K if t > 360 s, temperature of cooling medium
U = 5;% kJ s -1 m -2 K -1, global coefficient of heat
transfer

A_V=9.5 ;% m-1 small reactor area-to-volume ratio,
should be changed to see the effect

% Initial conditions:
CA(1)=CA0;
CP(1)=CP0;
T(1)=T0;
t=0;
N=1000; % Number of time increments
tfinal=600; % sec, final time to be used
inct=tfinal/(N-1);
% Integration by finite increments:
for i=2:N
    t(i)=t(i-1)+inct;
    if t(i)<360
        Tc=345;
```

```
    else
        Tc=295;
    end
    k1=k10*exp(-E1/(0.008314*T(i-1)));
    k2=k20*exp(-E2/(0.008314*T(i-1)));
    R1=-k1*CA(i-1);
    R2=-k2*CP(i-1);
    CA(i)=CA(i-1)+R1*inct;
    CP(i)=CP(i-1)+(R2-R1)*inct;
    T(i)=T(i-1)+((R1*DH1)+(R2*DH2)+(U*A_V*(Tc-T(i-
1))))*inct/(d*cp);
end

plot(t,CA,t,CP,t,T)
```

In the previous figure, we can check that the selectivity of the process (and the resultant quantity of the desired compound for a total time of 10 min [600 s]) in a large-scale reactor is significantly lower than that of a pilot plant reactor. The reason is that the relationship between the area of exchange and the reactor volume, in the large-scale reactor is much smaller than that in the small reactor, which produces incorrect control of the temperature. Thus, the value of the reaction rate at the second reaction increases, thus lowering the selectivity.

It is possible to opt for a semi-batch operation when the temperature is difficult to control. Initially, only a small part of the reaction is charged, and when the desired temperature is reached, more reactant is added over time. The heat production rate can be controlled effectively by careful dosing so that a loss of control temperature is prevented.

To simulate the semi-batch mode, we will assume the same conditions as in the batch reactor. However, the initial concentration of A will be minor, and the remaining amount A is added at a rate of Q_0 (m^3/s) starting at time t_d. It is assumed for simplicity that the physical properties of the mixture are unchanged during dosing. Equations (6.1)–(6.3) describe the course of the process before the dosing of the additional amount of A begins. The equation of the molar balance during the period $t > t_d$ was modified as

$$\frac{dC_A}{dt} = r_1 + \frac{Q_o}{V} \cdot (C_{A0} - C_A) \tag{6.4}$$

where Q_0 is the addition flow rate of A and V is the volume at each instant. We will assume the following values:

$$Q_0 = 1\ \text{m}^3/\text{h} \quad t_d = 100\ \text{s} \quad C_{A0} = 150\ \text{mol/m}^3$$

Figure 6.4 shows the results of the calculations for a large semi-batch reactor with $A/V = 1.00\ \text{m}^{-1}$ with an initial volume of 500 l with a concentration of 1000 (mol A)/m^3. Table 6.3 compares the results obtained in the batch and semi-batch reactors. A valid program for computing the conditions is given.

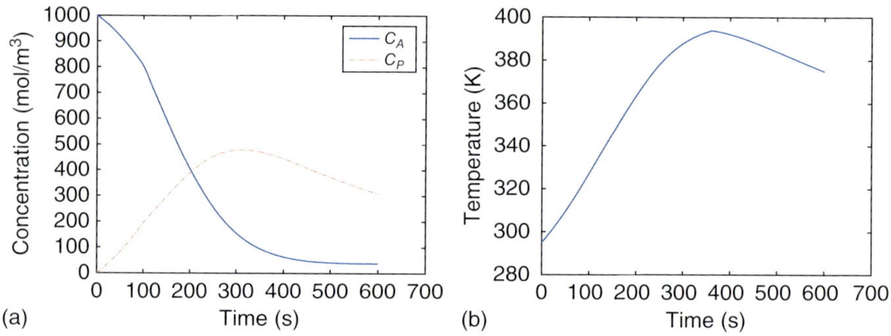

Figure 6.4 Concentration of reactive A (a), product P, and reactor temperature (b).

Table 6.3 Results in the batch and semibatch reactor after 600 seconds.

	A/V (m^{-1})	Final C_p (mol/m^3)	T_{max} (K)
Small	9.50	1409.1	394.7
Large	1.00	0.0	622.9
Large semi-batch	1.00	308.8	393.7

Program Listing 6.2

```
k10 = 5;% s -1
k20 = 250000000;% s -1, pre-exponential factors of the
constants k1 and k2
E1 = 20;% kJ mol -1,
E2 = 90;% kJ mol -1, activation energies
DH1 = -400;% kJ mol -1,
DH2 = -300;% kJ mol -1, reaction enthalpies
cp = 3;% kJ kg-1 K-1, specific heat
d = 1000;% kg m-3, density
CA0 = 150;% mol m-3, Current entering the system
CP0 = 0;% mol m-3, initial concentration of A and P
Q0=10/3600; %m3 s-1
T0 = 295;% K, input temperature of the reaction mixture
Tc = 345;% K if 0 <t <3600 s
% 295 K if 3660 <t <5400 s, temperature of cooling medium
U = 5;% kJ s -1 m -2 K -1, global coefficient of heat
transfer

% A_V=9.5 ;% m-1 small reactor area-to-volume ratio
A_V=1 ;% m-1 large reactor area-to-volume ratio

% Initial conditions:
CA(1)=1000; % Assuming initial condition in the reactor
CP(1)=CP0;
```

```
T(1)=T0;
V(1)=0.5; % Initial volume, m3
t=0;
N=10000; % Number of time increments
tfinal=600; % sec, final time to be used
tchange=360; % sec, time where Tc is changed
td=100; % sec, time when adding the input current
inct=tfinal/(N-1);
% Integration by finite increments:
for i=2:N
    t(i)=t(i-1)+inct;

    if t(i)<tchange
        Tc=345;
    else
        Tc=295;
    end

    k1=k10*exp(-E1/(0.008314*T(i-1)));
    k2=k20*exp(-E2/(0.008314*T(i-1)));
    R1=-k1*CA(i-1);
    R2=-k2*CP(i-1);

    if t(i)<td
        CA(i)=CA(i-1)+R1*inct;
        CP(i)=CP(i-1)+(R2-R1)*inct;
        T(i)=T(i-1)+((R1*DH1)+(R2*DH2)+(U*A_V*(Tc-T(i-
1))))*inct/(d*cp);
        V(i)=V(1);
    end
    if t(i)>=td
        CA(i)=CA(i-1)+(R1+Q0*(CA0-CA(i-1)))*inct;
        CP(i)=CP(i-1)+(R2-R1-Q0*CP(i-1))*inct;
        T(i)=T(i-1)+((R1*DH1)+(R2*DH2)+(U*A_V*(Tc-T(i-
1))))*inct/(d*cp);
        V(i)=V(i-1)+Q0*inct; % Volume changing with time
    end
end


figure(1)
plot(t,CA,t,CP,'-.')
legend('C_A', 'C_P')
ylabel('Concentration, mol/m^3')
xlabel('Time, s')
figure(2)
plot(t,T)
ylabel('Temperature, K')
xlabel('Time, s')
```

The data presented show that even in the case of an incorrect cooling process, the yield and selectivity in a large semi-batch reactor are better than those obtained in a smaller batch reactor. With this example it is also clear that the use of a model helps in the design and operation process of a reactor of this type.

6.3 Rapid Exothermic Reaction in a Tubular Reactor

Sometimes a continuous operation is desired, even for small productions. For example, batch reactors are not suitable for very fast reactions or purposes requiring short response times of the order of seconds or minutes.

We will consider the reaction presented earlier, $A \rightarrow P \rightarrow S$, but now with values of k_1 and k_2 much higher than before. This means that the required reaction time will be quite low. A tubular reactor operating as a plug flow reactor (PFR) will be appropriate in this case. However, the overall heat transfer coefficient, U, must be larger than that in the previous example due to the better transfer that occurs as a result of the movement of the liquid stream in the tubular reactor. We will use a value of 1 kJ/(s m^2 K).

Remember that the equations of balances of energy and mass for a PFR are very similar to those corresponding to a batch reactor, but with the dependent variables expressed as a function of volume or distance. In the case treated before, we have

$$u\frac{dC_A}{dL} = r_1 \tag{6.5}$$

$$u\frac{dC_P}{dL} = r_2 - r_1 \tag{6.6}$$

The energy balance can be deduced by considering the transfer in a differential volume:

(Input–Output) + (Generation by reaction)
  – (Transfer of heat to exterior) = 0

$$Q \cdot \rho \cdot C_P \cdot dT + (-r_A) \cdot (-\Delta H_r) \cdot S \cdot dL - U \cdot (T - T_c) \cdot dA = 0$$

In this expression, S refers to the section of the reactor and "A" is the exterior area of the tubular reactor, where the heat transfer takes place. Using the relation $Q = S{\cdot}u$ and dividing by S:

$$u \cdot \rho \cdot C_P \cdot dT + (-r_A) \cdot (-\Delta H_r) \cdot dL - U \cdot (T - T_c) \cdot \frac{dA}{S} = 0$$

Dividing by the dL:

$$u \cdot \rho \cdot C_P \cdot \frac{dT}{dL} + (-r_A) \cdot (-\Delta H_r) - U \cdot (T - T_c) \cdot \frac{dA}{S \cdot dL} = 0$$

If the tubular reactor has no irregularities, we can consider $\frac{dA}{S \cdot dL} = \frac{A}{V}$, and then, for the system with two reactions:

$$u \cdot \rho \cdot C_P \cdot \frac{dT}{dL} = (-r_1) \cdot \Delta H_1 + (-r_2) \cdot \Delta H_2 + U \cdot (T_c - T) \cdot \frac{A}{V} \tag{6.7}$$

To ensure that a piston flow is obtained in the tubular reactor, two criteria have to be met:

$$Re = \frac{\rho u D}{\mu} > 10\,000 \quad \text{and} \quad L/D > 100 \tag{6.8}$$

The typical liquid velocities in the pipes oscillate between 2 and 5 m/s. In this example, we choose a velocity (u) of 2.5 m/s, so the first criterion is met if $D > 0.004$ m. By choosing a maximum residence time of 100 s, a reactor length of 250 m results. Thus, the second criterion is satisfied for $D < 2.5$ m.

A Matlab program valid for this simulation is presented.

Program Listing 6.3

```
k10 = 50;% s -1
k20 = 25000000;% s -1, pre-exponential factors of the
constants k1 and k2
E1 = 20;% kJ mol -1,
E2 = 60;% kJ mol -1, activation energies
DH1 = -300;% kJ mol -1,
DH2 = -250;% kJ mol -1, reaction enthalpies
cp = 4;% kJ kg-1 K-1, specific heat
d = 1000;% kg m-3, density
CA0 = 1000;% mol m-3,
CP0 = 0;% mol m-3, initial concentration of A and P
T0 = 295;% K, input temperature of the reaction mixture
Tc = 345;% K if 0 <t <3600 s
% 295 K if 3660 <t <5400 s, temperature of cooling medium
U = 1;% kJ s -1 m -2 K -1, global coefficient of heat
transfer
u=2.5;% m/s Linear velocity of the flow
Diameter=0.1; % m It can be changed to visualize the effect
A_V=4*3.1416/Diameter ;% m-1 area-to-volume ratio

% Conditions at the input:
CA(1)=CA0;
CP(1)=CP0;
T(1)=T0;
L(1)=0;
N=10000; % Number of length increments
Lfinal=250; % m, length of the reactor
incL=Lfinal/(N-1);

% Integration by finite increments:
for i=2:N
    L(i)=L(i-1)+incL;
    if L(i)<25
        Tc=345;
    else
        Tc=295;
    end
    k1=k10*exp(-E1/(0.008314*T(i-1)));
```

```
      k2=k20*exp(-E2/(0.008314*T(i-1)));
      R1=-k1*CA(i-1);
      R2=-k2*CP(i-1);
      CA(i)=CA(i-1)+R1*incL;
      CP(i)=CP(i-1)+(R2-R1)*incL;
      T(i)=T(i-1)+((R1*DH1)+(R2*DH2)+(U*A_V*(Tc-T(i-
1))))*incL/(u*d*cp);
end

plot(L,CA,L,CP,L,T)
```

Firstly, the PFR can operate in an isothermal or adiabatic mode, or with a temperature profile along the reactor. Figure 6.5 shows the concentration and temperature profiles in reactors of various diameters. When there is no heat transfer, i.e. in an adiabatic reactor, the diameter is not important. The optimum reactor length (for a higher selectivity) decreases as the diameter increases. The yield of the product P also decreases when the diameter of the reactor increased, because the cooling is less efficient. Table 6.4 resumes the results.

Figure 6.5 and Table 6.4 show that the larger the diameter of the reactor, the more it resembles the adiabatic operation. Therefore, instead of increasing the diameter of the reagent or in order to increase the production capacity, it is better to use a large number of parallel tubes of smaller diameter, so that the temperature control is improved.

As mentioned before, the isothermal operation is also possible, in principle, in a tubular reactor. Figure 6.6 shows the concentration of P and the temperature of the hot medium necessary to maintain the reactor at the desired temperature for large and small diameter reactors.

Figure 6.6 shows that, in practice, the cooling temperature to achieve the isothermal operation of the reactor would have to be unacceptably low

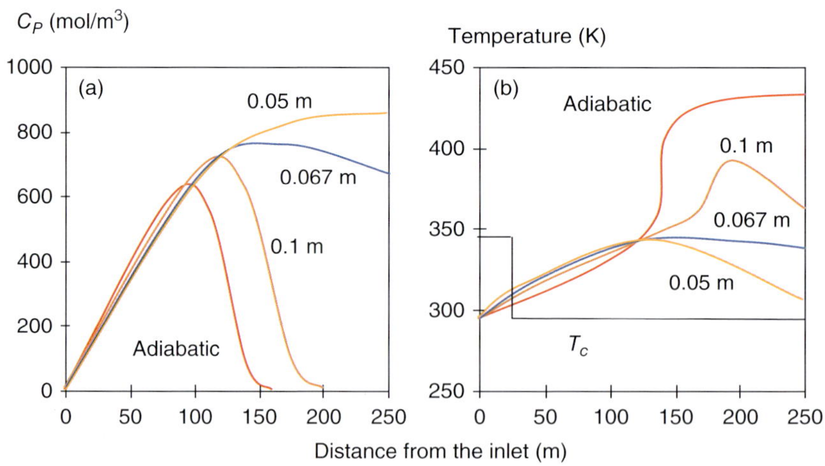

Figure 6.5 Concentration of P (a) and reactor temperature (b) in adiabatic and jacketed piston flow reactors of different diameters.

Table 6.4 Optimum length, conversion and performance for tubular reactors.

Diameter (m)	A/V (m^{-1})	Optimal L (m)	Conversion (% mol)	Yield of P (% mol)
0.05	80	250	96.2	85.1
0.067	60	155	88.7	76.2
0.1	40	127	82.3	71.1
0.2	20	115	77.3	68.9
(Adiabatic)	(0)	110	74.5	66.9

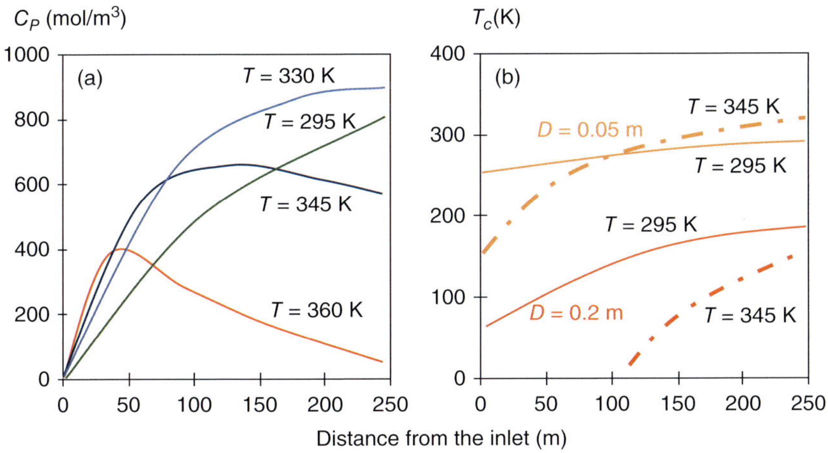

Figure 6.6 (a) Concentration of P as a function of the temperature of the reactor (T) in isothermal tubular reactors (diameter not relevant). (b) Required temperature of the refrigerant to maintain isothermal reactor at temperature for a large reactor and a small one.

(even impossible in large diameter reactors). The reason for needing this extreme cooling in the first part of the reactor is that the reaction rate is very high initially, with a large production of heat in the exothermic reaction.

6.3.1 Study of the Stability of the Process

Chemical reactors, especially those in which one exothermic reaction is going on, are often difficult to control. This is the case when large temperature changes are produced due to variations in one or more operational parameters. These zones of operation are known as "parametrically sensitive areas" and "areas of runaway." In the last ones, a small change the initial input conditions produces results in the system not returning to the initial state, whereas in the parametrically sensitive areas, the system is able to return to the initial conditions.

Operating the reactor in an unstable region can produce a defective product, a not well-controlled temperature, rapid deterioration of the catalyst, unwanted side reactions, etc.

This section tries to calculate the maximum temperature ($T_{\max}$) that will be achieved in a tubular reactor system where an exothermic reaction occurs. Consider that a single reaction is occurring in a constant density system. The energy balance in this case would be, according to Eq. (6.7):

$$u \cdot \rho \cdot C_P \cdot \frac{dT}{dL} = (-r_A) \cdot \Delta H_r + U \cdot (T_c - T) \cdot \frac{A}{V} \tag{6.9}$$

The mole balance in the reactor tells us that:

$$(-r_A) = -u \cdot \frac{dC_A}{dL} = u \cdot C_{A0} \cdot \frac{dX_A}{dL}$$

and then:

$$dL = u \cdot C_{A0} \cdot \frac{dX_A}{(-r_A)} \tag{6.10}$$

where X_A represents the molar conversion. Clearing dT from the first equation and making use of the second, one can easily be reached:

$$dT = \frac{C_{A0} \cdot (-\Delta H_r)}{\rho \cdot C_P} \cdot dX_A + \frac{U \cdot (T_c - T) \cdot \frac{A}{V}}{u \cdot \rho \cdot C_P} \cdot dL \tag{6.11}$$

The term that accompanies dX_A would be the temperature increase that would occur if the reactor were adiabatic, so it is usually called ΔT_{ad}:

$$\Delta T_{\text{ad}} = \frac{C_{A0} \cdot (-\Delta H_r)}{\rho \cdot C_P} \tag{6.12}$$

Therefore, we can write

$$\frac{dT}{dX_A} = \Delta T_{\text{ad}} + \frac{U \cdot C_{A0} \cdot (T_c - T) \cdot \frac{A}{V}}{(-r_A) \cdot \rho \cdot C_P} \tag{6.13}$$

From this equation, if $T = T_{\max}$, we would have $dT = 0$; therefore:

$$T_{\max} = T_h - \frac{\Delta T_{\text{ad}} \cdot (-r_A) \cdot \rho \cdot C_P}{U \cdot C_{A0} \cdot \frac{A}{V}} \tag{6.14}$$

In this equation, $(-r_A)$ should be calculated at the temperature and the molar conversion obtained at the $T_{\max}$. These quantities are determined by the whole history of the reaction mixture, so that Eq. (6.14) does not allow the direct calculation of $T_{\max}$. However, it is possible to give an approximate solution to Eq. (6.13) using some approximations.

To do this, let us express Eq. (6.13) using dimensionless numbers:

$$N_c = \frac{U \cdot \frac{A}{V}}{k_c \cdot \rho \cdot C_P}$$
$$N_{\text{ad}} = \frac{E \cdot \Delta T_{\text{ad}}}{R \cdot T_c^2} \tag{6.15}$$

and defining a dimensionless temperature difference:

$$\Delta \upsilon = \frac{E}{RT_c^2}(T - T_c) \tag{6.16}$$

At the reactor inlet, the conversion is zero and the difference in dimensionless temperature will be

$$\Delta\upsilon_0 = \frac{E}{RT_c^2}(T_0 - T_c)$$

Only when the inlet temperature (T_0) equals the temperature of the cooling medium (T_c) will we have that $\Delta\upsilon_0 = 0$.

The dimensionless number N_{ad} takes into account the adiabatic temperature increase, while N_c is a measure of the cooling capacity of the medium. In these numbers, k_c represents the kinetic constant at the temperature of the cooling medium. The number N_c is very sensitive to changes in the cooling temperature, as it contains a term exponentially dependent on T_c, while N_{ad} is not much temperature dependent. The relationship between k_c and the kinetic constant at any temperature is

$$\begin{aligned} k &= k_0 \cdot \exp\left(-\frac{E}{RT}\right) = k_0 \cdot \exp\left(-\frac{E}{RT_c}\right) \cdot \exp\left(\frac{E}{RTT_c}(T - T_c)\right) \\ &\approx k_c \cdot \exp\left(\frac{E}{RT_c^2}(T - T_c)\right) = k_c \cdot \exp(\Delta\upsilon) \end{aligned} \tag{6.17}$$

We can use this approach with first-order kinetics:

$$(-r_A) = k_c \cdot \exp(\Delta\upsilon) \cdot C_{A0} \cdot (1 - X_A) \tag{6.18}$$

Making use of the definitions (Eq. (6.15)) and of Eqs. (6.13) and (6.18):

$$\frac{E}{RT_c^2} \cdot \frac{dT}{dX_A} = N_{ad} + N_c \frac{E \cdot (T_c - T) \cdot \exp\left(-\frac{E}{RT_c^2} \cdot (T - T_c)\right)}{(1 - X_A) \cdot RT_c^2} \tag{6.19}$$

Finally, Eq. (6.19) is expressed as follows:

$$\frac{d(\Delta\upsilon)}{dX_A} = N_{ad} - N_c \frac{\Delta\upsilon \cdot \exp(-\Delta\upsilon)}{(1 - X_A)} \tag{6.20}$$

Qualitative information about reactor stability can be obtained by examining the Eq. (6.20), which we recall is a simplified form of the energy balance in a cooled tubular reactor. For positive values of $\Delta\upsilon$, the slope $\frac{d(\Delta\upsilon)}{dX_A}$ can never be greater than N_{ad}. As the conversion approaches unity, $\frac{d(\Delta\upsilon)}{dX_A}$ becomes a growing negative number, so $\Delta\upsilon$ will tend to zero. So, except in the case where $\Delta\upsilon$ is always negative, there must be a point for each solution of (Eq. (6.20)) where there is a $\Delta\upsilon_{max}$. The value of the maximum temperature difference that will occur is what is called the "hot spot" of the reactor.

We do note that this is true only if taken as independent variable the conversion of the reactive A, that is, eliminating the differential length (dL) of reactor, as has been done previously. If the treatment is done using dL, it would not necessarily be a maximum. This is what is called a "safe approach," criticized by some authors for not being very real.

Figure 6.7 shows, as an example, temperature profiles obtained for the case $N_c/N_{ad} = 2$, for various values of N_{ad} and assuming that the initial temperature matches the temperature of the refrigerant (that is $\Delta\upsilon = 0$) and a single curve obtained in the case $\Delta\upsilon_0 = 2$.

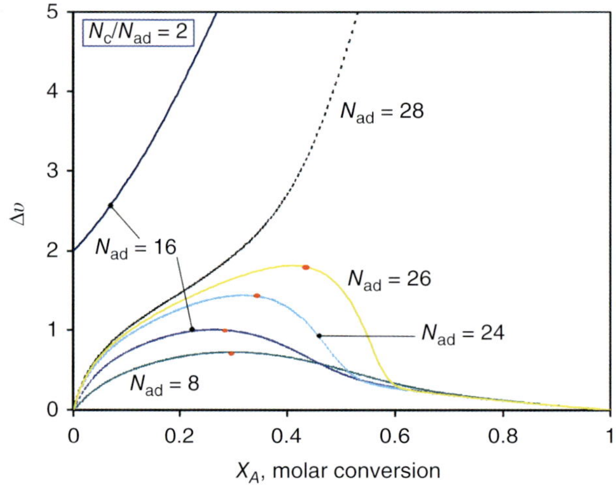

Figure 6.7 Temperature increase for exothermic reactions in a tubular reactor. $N_c/N_{ad} = 2 < e$.

We see in the graph that above a certain value of N_{ad}, (in Figure 6.7, between 26 and 28) the curve $\Delta\upsilon - X_A$ completely loses the lower part observed at high conversion values and the temperature increase is close to that obtained in the adiabatic regime (in the graph it is not observed, but all the curves return to $\Delta\upsilon = 0$ for $X_A = 1$).

Let us calculate the locus of the maxima of the paths given by Eq. (6.20). To do this, we equal the expression (Eq. (6.20)) to zero and clear:

$$X_A = 1 - \frac{N_c}{N_{ad}} \cdot \Delta\upsilon_{max} \cdot \exp(-\Delta\upsilon_{max}) \tag{6.21}$$

For each value of conversion, Eq. (6.21) has either two solutions (one with $\Delta\upsilon_{max} > 1$ and another with $\Delta\upsilon_{max} < 1$) or none (in the interval $X_A = 0$–1). Figure 6.7 shows (in red) the locus of the points represented by (6.21) in the case $N_c/N_{ad} = 2$.

Figure 6.8 shows (in red) examples of loci given by Eq. (6.21), for $N_c/N_{ad} = 5$, and 2, with arrows (blue) representing the value of the slope $d(\Delta\upsilon)/dX_A$ given by Eq. (6.20). The value of this slope is always positive to the right of the locus given by Eq. (6.21), and negative to the left. In Figure 6.8, the position of the "mm" points is also indicated.

When $N_c/N_{ad} > e$, the place of the maxima intersects the axis $X_A = 0$ (no conversion) in two points, for example, the values 2.52 and 0.26 for $N_c/N_{ad} = 5$, as shown in Figure 6.8. In this case, if a reaction begins with $\Delta\upsilon_0 < 0.26$, it will follow a path similar to that shown in Figure 6.9 (blue line): the value of $\Delta\upsilon$ will not increase very much (in fact, $\Delta\upsilon$ will always be less than 0.26) because it must decrease as soon as the solid line is passed, which marks the maximums in the derivative, and the reaction can be easily controlled. The same happens if $0.26 < \Delta\upsilon_0 < 2.52$ (black line), since the temperature must decrease monotonically. However, if $\Delta\upsilon_0 > 2.52$ (see green line), the temperature initially

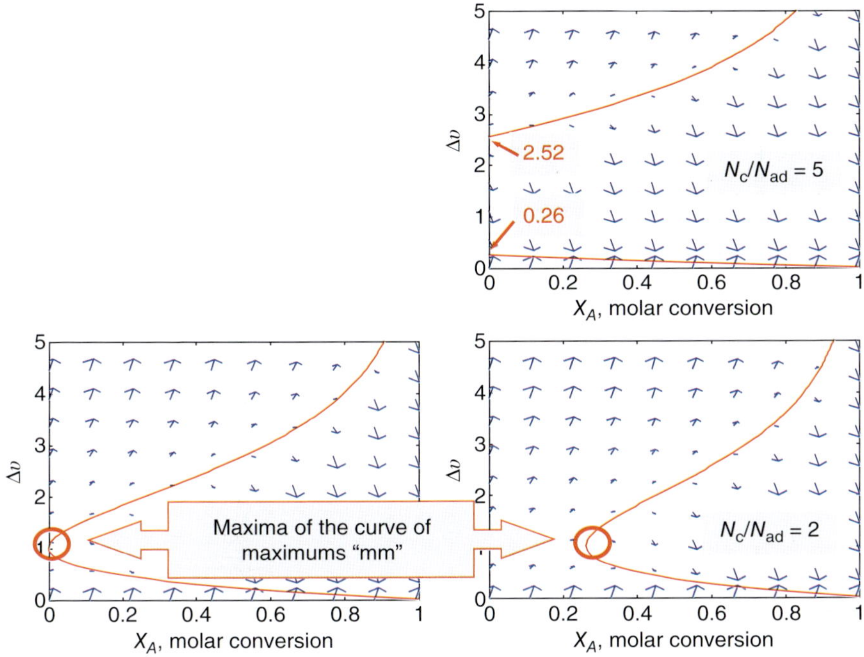

Figure 6.8 Geometric place of the maximum temperature differences for several values of N_c/N_{ad}.

increases and will continue to grow until the maximum area is crossed. Clearly, the point $\Delta \upsilon_0 = 2.52$ is a stability limit (for the case $N_c/N_{ad} = 5$).

As shown in Figure 6.8, by decreasing the value of N_c/N_{ad} the curve presenting $\Delta \upsilon_{max}$ is closer to the right, so that the possibility that the reactor temperature runaway increases. If $N_c/N_{ad} = e$, the place of maxima is tangent to the vertical axis.

Let us analyze carefully the case $N_c/N_{ad} = 2$ (Figure 6.7). We see that the temperature shoots up when the path crosses the line after its maximum (after the point we call "mm"). Thus, the path that passes through "mm" can be considered as critical, in the sense that at the moment that some point of the reactor is located above this path, there are many possibilities for the temperature in the reactor to be triggered. For example, if in the case of Figure 6.7 the value of the temperature at the inlet makes $\Delta \upsilon_0 = 2$, probably the reactor to be thermally unstable and temperature increase indefinitely.

Let us see what equation defines the loci of the maximums of the curve of maximum values (points "mm"). For this, we derive Eq. (6.21) and equate to zero:

$$\frac{dX_A}{d\Delta \upsilon_{max}} = -\frac{N_c}{N_{ad}}[\exp(-\Delta \upsilon_{mm}) - \Delta \upsilon_{mm} \cdot \exp(-\Delta \upsilon_{mm})] \tag{6.22}$$

The solution to this equation is $\Delta \upsilon_{mm} = 1$ (it can be perfectly seen in Figures 6.7–6.9), no matter the value of N_c/N_{ad}, or what is the same:

$$\Delta \upsilon_{mm} = 1 = \frac{E}{RT_c^2}(T_{mm} - T_c) \Rightarrow T_{mm} = T_c + \frac{RT_c^2}{E} \tag{6.23}$$

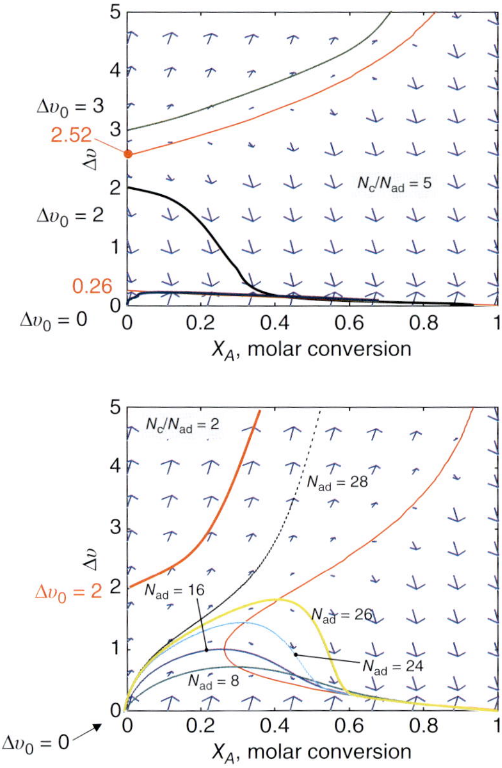

Figure 6.9 System behavior for $N_c/N_{ad} = 5$ and $N_{ad} = 8$.

The corresponding value of the conversion is obtained by substituting in Eq. (6.21):

$$X_{A,\mathrm{mm}} = 1 - \frac{N_c}{e \cdot N_{\mathrm{ad}}} \tag{6.24}$$

Other studies use the temperature T_{mm} as the reference temperature instead of the temperature of the cooling medium, T_c. Thus, the condition $\Delta \upsilon_{\mathrm{mm}} \leq 1$ is converted into

$$\frac{E}{RT_{\mathrm{mm}}^2}(T_{\mathrm{mm}} - T_c) \leq 1 \tag{6.25}$$

That is equivalent to

$$\frac{R}{E}T_{\mathrm{mm}}^2 - T_{\mathrm{mm}} + T_c \leq 0$$

The application of Eq. (6.25) gives greater values of T_{mm}. Solving the second-order degree equation (6.25) you get to

$$\frac{RT_{\mathrm{mm}}}{E} \leq 0.5\left[1 - \sqrt{\left(1 - \frac{4RT_c}{E}\right)}\right] \tag{6.26}$$

This is what is called the first stability criterion, and it provides the conditions inside the reactor that produce a risk situation. But what matters is knowing what conditions at the reactor inlet will produce those situations. Van Welsenare and Froment empirically obtained a second criterion for thermal stability, carefully studying the trajectory of the curves T–X_A, valid only in the case $T_c = \mathrm{T}_0$ (i.e. $\Delta\upsilon_0 = 0$):

$$\frac{N_{\text{ad}}}{\Delta\upsilon_{\text{mm}}} = \frac{\Delta T_{\text{ad}}}{T_{\text{mm}} - T_c} \leq 1 + \sqrt{\left(\frac{N_c}{e}\right)} + \frac{N_c}{e} \tag{6.27}$$

The condition (Eq. (6.27)) is helpful when trying to determine the maximum allowable concentrations of reagents as $C_{A0} = \rho C_P \Delta T_{\text{ad}}/(-\Delta H_r)$. The applicability of the method of van Welsenare and Froment is limited, since the second stability criterion is derived with the condition $T_0 = T_c$. In practice, for very exothermic reaction at elevated temperature, the feed is preheated by the refrigerant, so $T_0 \cong T_c$.

Equation (6.27) is also useful to estimate the value of minimum N_c necessary (or what is the same, the cooling conditions needed by the system) to keep the system stable given the conditions of entry.

Example 6.1 ***Batch Reactor***

In a batch tank stirred reactor in which the reaction $A \rightarrow P \rightarrow R$ occurs and in which the desired product is "P," the following data is valid:

$k_{10} = 0.5\,\text{s}^{-1}$, $k_{20} = 25\,000\,\text{s}^{-1}$, pre-exponential factors of the constants k_1 and k_2
$E_1 = 20\,\text{kJ/mol}$, $E_2 = 60\,\text{kJ/mol}$, activation energies
$\Delta H_1 = -300\,\text{kJ/mol}$, $\Delta H_2 = -250\,\text{kJ/mol}$, reaction enthalpies
$C_{A0} = 1000\,\text{mol/m}^3$, $C_{P0} = 0$, initial concentration of A and P
$U = 0.5\,\text{kJ/(s m}^2\,\text{K)}$, global coefficient of heat transfer.

Calculate the cooling temperature profiles necessary to maintain the reactor at 300 K throughout the reaction, for a cylindrical reactor of 42 cm in diameter and another of 400 cm.

Solution

For the batch reactor, the material and energy balances have the following form:

$$\frac{dC_A}{dt} = r_1 = -k_1 C_A$$

$$\frac{dC_P}{dt} = r_2 - r_1 = -k_2 C_P - (-k_1 C_A)$$

$$\rho C_P \frac{dT}{dt} = (-r_1)(\Delta H_1) + (-r_2)(\Delta H_2) + U(T_c - T)\frac{A}{V}$$

Solving previous balances knowing that $dT/dt = 0$, T_c is cleared:

$$0 = k_1 C_{A0}(1 - X_A)(\Delta H_1) - k_2 C_{A0} X_A (\Delta H_2) + U(T_c - T)\frac{A}{V}$$

Knowing that the temperature must be T=300 K, we can calculate k_1 and k_2:

$$k_1 = 0.5 \cdot \exp\left(-\frac{20}{0.008\,314 \cdot 300}\right) = 1.56 \cdot 10^{-4}\,\text{s}^{-1}$$

$$k_2 = 25\,000 \cdot \exp\left(-\frac{60}{0.008\,314 \cdot 300}\right) = 8.9255 \cdot 10^{-7}\ \text{s}^{-1}$$

and, finally:

$$T_c = \frac{\begin{array}{c} -1.56 \cdot 10^{-4} \cdot \text{s}^{-1} 1000\ \text{mol/m}^3 (1 - X_A)(-300\,000\ \text{J/mol}) \\ +8.9255 \cdot 10^{-7}\ \text{s}^{-1}\ 1000\ \text{mol/m}^3 \cdot X_A(-250\,000\ \text{J/mol}) \end{array}}{500\ \text{J/(s m}^2\ \text{K)} \cdot \frac{A}{V}\ \text{m}^{-1}} + 300\ \text{K}$$

and the following profiles are obtained, using a spreadsheet:

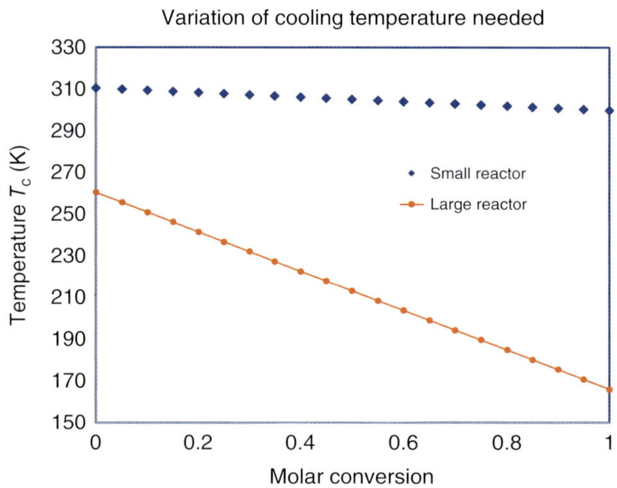

In the previous figure, it can be seen that in the large-scale reactor the refrigeration to be supplied is greater than that in the smaller reactor and that in the large reactor the A/V ratio is lower.

Example 6.2 ***Jacketed Piston Flow Reactor***
In a jacketed piston flow reactor, the first-order reaction $A \rightarrow R$ is carried out, in which the following constants are present:

$k_0 = 1 \cdot 10^9\ \text{s}^{-1}$
$U = 150\ \text{W/(m}^2\ \text{K)}$
Diameter = 250 mm
$E/R = 15\,300\ \text{K}$
$\rho \cdot C_p = 1300\ \text{J/(m}^3\ \text{K)}$
$\Delta H_r = -1800\ \text{kJ/mol}$

(a) If it is intended that at the entrance the food is introduced at the temperature of the refrigerant, which is 300 °C, what temperature in the reactor makes the system thermally unstable? What concentration of maximum at the inlet makes the reactor stable?
(b) Maintaining the concentration calculated before, and estimating that the maximum stability temperature is 800 K, what value of the heat transfer coefficient makes it possible? What if it was 900 K?

Solution

(a) Applying the second Welsenare criterion, the T_{mm} that makes the system is thermally unstable can be obtained by

$$\frac{RT_{\mathrm{mm}}}{E} \leq 0.5\left[1 - \sqrt{\left(1 - \frac{4RT_c}{E}\right)}\right]$$

Using the known data:

$$\frac{T_{\mathrm{mm}}}{15\,300} \leq 0.5\left[1 - \sqrt{\left(1 - \frac{4 \cdot 300}{15\,300}\right)}\right]$$

We obtain $T_{\mathrm{mm}} = 596.2$ K. The maximum concentration can be obtained from

$$C_{A0} = \frac{\rho C_P \cdot \Delta T_{\mathrm{ad}}}{(-\Delta H_r)}$$

where ΔT_{ad} can be obtained from the relationship:

$$\frac{\Delta T_{\mathrm{ad}}}{T_{\mathrm{mm}} - T_c} \leq 1 + \sqrt{\left(\frac{N_c}{e}\right)} + \frac{N_c}{e}$$

And the value of N_c is

$$N_c = \frac{U \cdot \frac{A}{V}}{k_c \cdot \rho \cdot C_P}$$

In a tubular reactor, $A/V = 2/R$, which equals 16 m^{-1} in the present example. The value of N_c is then 728.9, and a value of $\Delta T_{\mathrm{ad}} = 6633$ K is obtained and therefore a $C_{A0} = 4.8\ \mathrm{mol/m^3}$.

(b) First, the new T_c is calculated from the following equation:

$$\frac{E}{RT_c^2}(T_{\mathrm{mm}} - T_c) \leq 1 \quad \text{i.e.} T_c = 758.2\ \mathrm{K}$$

The ΔT_{ad} is obtained from a $C_{A0} = 4.8\ \mathrm{mol/m^3}$

$$C_{A0} = \frac{\rho C_P \cdot \Delta T_{\mathrm{ad}}}{(-\Delta H_r)} \quad \text{i.e.} \Delta T_{\mathrm{ad}} = 6633.3\ \mathrm{K}$$

Next, N_c is calculated from

$$\frac{\Delta T_{\mathrm{ad}}}{T_{\mathrm{mm}} - T_c} \leq 1 + \sqrt{\left(\frac{N_c}{e}\right)} + \frac{N_c}{e}$$

Finally, U can be obtained from

$$N_c = \frac{U \cdot \frac{A}{V}}{k_c \cdot \rho \cdot C_P}$$

Therefore, for $T_{\mathrm{mm}} = 800$ K, a value of $N_c = 4.36$ is obtained, what implies a $U = 79.4\ \mathrm{W/(m^2\ K)}$ and for $T_{\mathrm{mm}} = 900$ K we obtain $N_c = 4.17$ and $U = 99.5\ \mathrm{W/(m^2\ K)}$.

Example 6.3 ***Using Matlab to Calculate Stability Conditions***
In a tubular reactor system, the $A \rightarrow P$ reaction is occurring with the following constants:

$k = 3.94 \cdot 10^{12} \cdot \exp(-11\,400/T)\ \text{s}^{-1}$
$\Delta T_{ad} = 146\,°\text{C}$
$4U/(dt \cdot \rho \cdot C_P) = 0.2\ \text{s}^{-1}$

The inlet temperature of the reagents is, in any case, 340 K.

(a) Calculate the temperature-conversion profiles for various refrigerant temperatures between 335 and 350 K. Determine the maximum temperature that will be obtained in the reactor at each of these temperatures.
(b) Determine the variation of N_c/N_{ad} with the cooling temperature.
(c) Determine the $\Delta\nu$-conversion curves for different values of the quotient N_c/N_{ad}.
(d) Calculate the curve $\Delta\nu_{max}$-conversion for this system.
(e) If the temperature of the refrigerant is given by the conditions of the equipment and is 335 K, determine what will happen when changing to a system whose diameter is four times higher.
(f) Using the "quiver" function of Matlab, draw the relative size and the direction of the derivatives of the representation $\Delta\nu$-conversion, in the case that $N_c/N_{ad} = e$.

Solution
(a)–(d) In this system, Eq. (6.20) can be applied:

$$\frac{d(\Delta\upsilon)}{dX_A} = N_{ad} - N_c \frac{\Delta\upsilon \cdot \exp(-\Delta\upsilon)}{(1 - X_A)} \tag{6.20}$$

Using the finite differences method, we can write

$$\Delta\upsilon^{t+1} = \Delta\upsilon^{t} + \left(N_{ad} - N_c \frac{\Delta\upsilon \cdot \exp(-\Delta\upsilon^t)}{(1 - X_A^t)} \right) \cdot \Delta X_A$$

For calculating the requested curves, we can use the following Matlab program.

Program Listing 6.4

```
clear all
close all
% Data given in the example:
To=340; %K
dTad=146; %K
%4U/dtdCp=0.2; %s-1

dv=[]; % Vector for saving values of dv
T=[];% Vector for saving values of T
DVmax=[]; % Vector for saving the maximum values of dv
NcdivNad=[]; Vector for saving values of Nc/Nad
XAmaxima=[]; Vector for saving values of XA in the
maximum value of dv
incXA=0.001; % Increment of the conversion among two
consecutive points
```

```
for Tc=335:350 % Values of Tc
    dv=[];
    dv(1)=11400/Tc^2*(To-Tc); % Initial value of dv
    T(1)=((dv(1)*Tc^2)/11400)+Tc;
    XA=0;
    i=1;

    while XA<1
        i=i+1;

        Nad=11400*dTad/Tc^2;
        kc=(3.94*10^12)*exp(-11400/Tc);
        Nc=0.2/kc;
        dv(i)=dv(i-1)+(Nad-Nc*(dv(i-1)*exp(-dv(i-1))/(1-
XA)))*incXA;
        T(i)=((dv(i)*Tc^2)/11400)+Tc;
           XA=XA+incXA;
    end

   XAv=0:incXA:1;

        hold on
        figure(1)
        plot(XAv,T). % Temperature-conversion
        xlabel('Conversion')
        ylabel('Temperature (K)')
        figure (2)
        plot(XAv,dv). % dv-conversion
        xlabel ('Conversion')
        ylabel('dv')
       DVmax=[DVmax max(dv)]; % Vector saving the maximum
value of dv for each value of Tc
       NcdivNad=[NcdivNad Nc/Nad]; % Values of Nc/Nad
       XAmaxima=[XAmaxima 1-Nc/Nad*max(dv)*exp(-max(dv))];
% Values of XA corresponding to max(dv)
end


   figure (3)
   plot(XAmaxima,DVmax,'+')
   xlabel('Conversion')
   ylabel('DVmax')
   figure (4)
   plot(Thv, NcdivNad)
   xlabel('Cooling temperature (K)')
   ylabel('Nc/Nad')
   figure (5)
   plot(XAmaxima, DVmax, 'or')
   xlabel ('conversion')
   ylabel('DVmax')
```

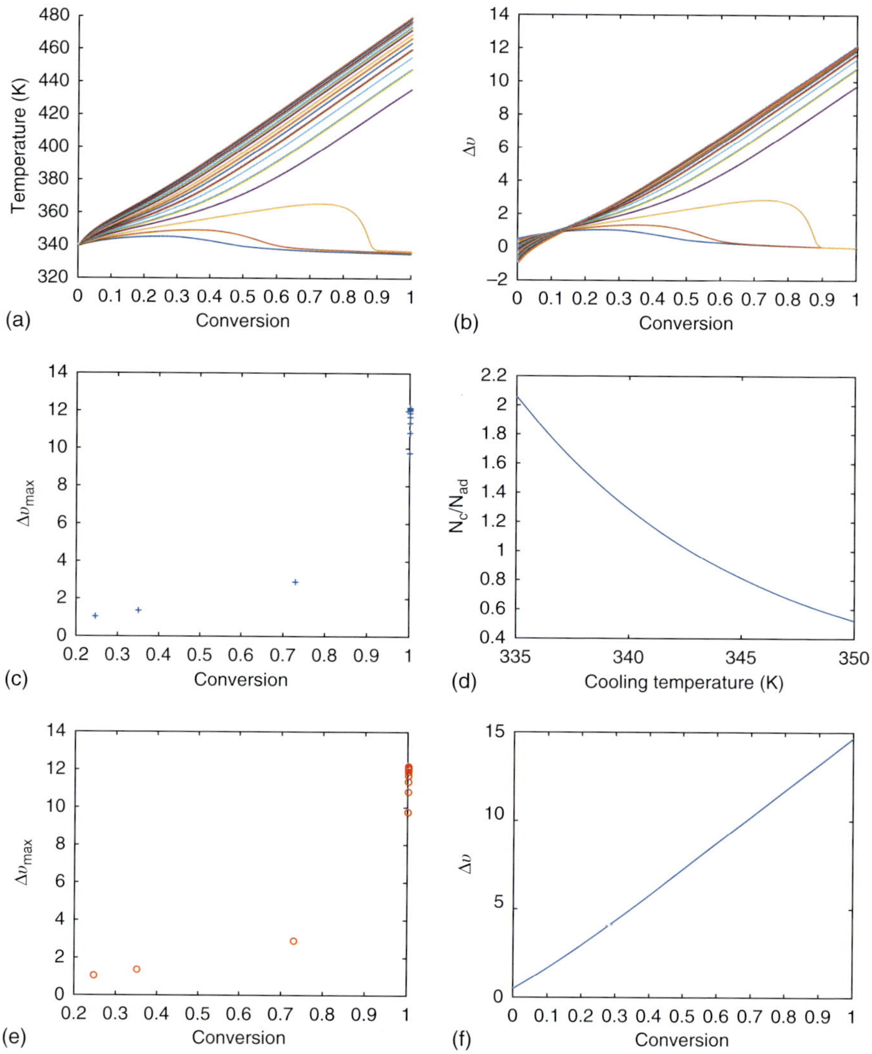

Figure 6.10 Solution of Example 6.4.

The results are shown in Figure 6.10a–e.

Two different situations can be distinguished in the previous plots. In the first case are those lines that represent controlled systems, in which by increasing the conversion, the system is able to keep the temperature low ($T_c \leq 337$ K). The second case is represented by those lines that for conversion values close to 1, a runaway is produced, since there is a loss of temperature control in the reactor, which continues to increase without control ($335 < T_c < 350$ K).

The maxima of the $\Delta\upsilon - \text{X}$ curve can be clearly observed (Figure 6.10c) until $T_c > 337$ K, where the curves have lost their maximum and the runaway is produced.

(e) If $T_c = 335$ K and diameter is four times higher, then we will have that relation of cooling area to volume "A/V" will decrease four times. In this situation, $N_c = 0.2/(4 \cdot k_c)$.

The Matlab program is very similar, using only $T_c = 335$ K. The corresponding plot can be observed in Figure 6.10f.

(f) The corresponding Matlab is presented, and the results are shown in the figure.

In the figure, the arrows represent the value of the slope $\frac{d\Delta\upsilon}{dX_A}$. As can be observed, if the reaction begins with $\Delta\upsilon_0 < 1$, the value of $\Delta\upsilon$ will not increase much, never exceeding the value of 1 and will decrease following the red line, that is, the reaction is easily controllable. The point $\Delta\upsilon = 1$ is the stability limit, and if the reaction starts with higher values of $\Delta\upsilon$, it will be out of control. This point is known as point "mm" and is the maximum of the curve of maximums.

If we are inside the curve (to the right), it is observed that the arrows indicate a decrease in the value of $\Delta\upsilon$, that is, if the reaction is between the red lines, the value of $\Delta\upsilon$ will decrease, following the trajectory of the lower red line.

Program Listing 6.5

```
% Stability of a tubular reactor
Nad=24;
Nc=exp(1)*Nad;
% Creating the mesh
v=linspace(0.1,5,20);
ta=linspace(0.05,0.8,20);
[V,Ta]=meshgrid(v,ta);

dta=ones(length(ta)).*0.1;
dv=(Nad-Nc.*V.*exp(-V)./(1-Ta)).*dta;
quiver(Ta,V,dta,dv);
axis([0,1,0,5])
xlabel('Conversion A')
ylabel('Dv')
text(0.02,1.2, 'Maximum of the curve')
text(0.02,1,'of maxima')
title('Nc/Nad=e')
zoom
hold on

% Curve of maxima
v=linspace(0,5,100);
ta=linspace(0,0.8,100);
tta=-Nc/Nad.*v.*exp(-v)+1;
plot(tta,v,'r')
hold off
```

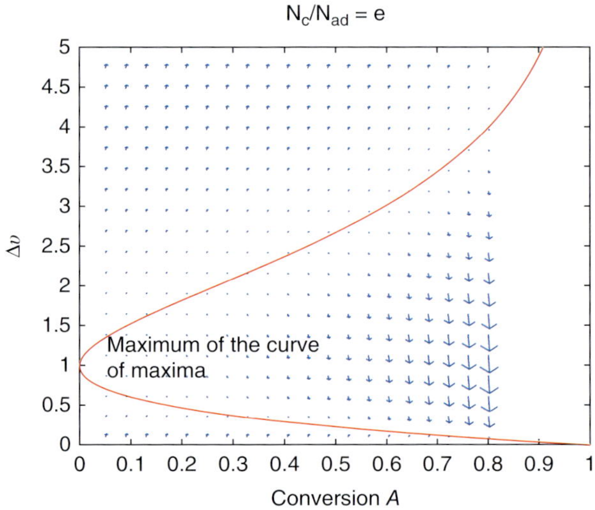

Example 6.4 ***Change in Initial Concentration***

The liquid-phase sulfuric acid catalyzed hydration of 2,3-epoxypropanol ($\Delta H_r = -88\ 200$ J/mol) has been studied in a batch reactor by Heiszwolf and Fortuin. The reaction takes place by a first-order kinetics reaction dependent on the concentration of the alcohol. The kinetic constant is defined by

$$k = 8.25 \cdot 10^9 \cdot \exp\left(-\frac{8809}{T}\right)\ \mathrm{s}^{-1}$$

The authors show that the system is stable when the coolant temperature is 306 K starting at a concentration of alcohol that produced a dimensionless adiabatic temperature rise of 13.1, but it presents a high increase in temperature when this temperature rise is 15.9. In all runs, the heat transfer capacity was maintained constant at a value of 23.7. The density of the fluid is 1000 kg/m^3. The overall heat transfer capacity is 50 W/K and the overall heat capacity of the mix and vessel is 3500 J/(kg K).

(a) Reproduce the system behavior and check if the limits in ΔT_{ad} produce any change according to the reasoning presented before.
(b) The system will be scaled up to a plug flow tubular reactor with a diameter of 5 cm. Study the behavior of the continuous system and predict one stable and one nonstable initial concentration of the alcohol.

Solution

(a) Let us focus on the heat balance in a batch reactor:

$$\rho \cdot C_P \cdot \frac{dT}{dt} = (-r_1) \cdot \Delta H_1 + U \cdot (T_c - T) \cdot \frac{A}{V}$$

That is equivalent to Eq. (6.9) representing the balance in a tubular reactor, substituting dt by dz/u.

If the balance is multiplied by the total volume, we have

$$m \cdot C_P \cdot \frac{dT}{dt} = (-r_1) \cdot V \cdot \Delta H_1 + U \cdot (T_c - T) \cdot A$$

For this system, we can define modified dimensionless values, N'_{ad} and N'_c, equivalent to those used in the plug flow reactor:

$$N'_c = \frac{U \cdot A}{k_c \cdot m \cdot C_P} \qquad N'_{ad} = \frac{E \cdot \Delta T_{ad}}{R \cdot T_c^2}$$

In the same way as in the plug flow reactor, we can check that Eq. (6.20) is valid in a batch reactor using these last dimensionless values:

$$\frac{d(\Delta\upsilon)}{dX_A} = N'_{ad} - N'_c \frac{\Delta\upsilon \cdot \exp(-\Delta\upsilon)}{(1 - X_A)}$$

The values of dimensionless adiabatic temperature rise given in the statement, refer to the N'_c and N'_{ad} we have just defined, so we have

$$N'_c = 23.7$$

$$N'_{ad1} = 13.1 \quad \frac{N'_c}{N'_{ad1}} = 1.81 \quad \text{(Experimentally System stable)}$$

$$N'_{ad2} = 15.9 \quad \frac{N'_c}{N'_{ad2}} = 1.49 \quad \text{(Experimentally Runaway)}$$

Both situations present a value of $\frac{N'_c}{N'_{ad}}$ much lower than "e," so a runaway situation is probably found, depending on the specific conditions. Let us graphically see how the system will perform. For that, a program very similar to that shown in Program Listing 6.6 will be used:

Program Listing 6.6

```
T0=306; % Input temperature (K)
Tc=306;
Nc=23.7;

for Nad=[13.1 15.9] % Values of Nad to be used
XA(1)=0; % Initial conversion
dv(1)=(T0-Tc)*11600/(Tc^2); % Initial value of the
dimensionless temperature difference
% Integration of the equation by using finite increments:
for i=2:99 % We will use 100 increments in XA
    D=Nad-Nc*dv(i-1)*exp(-dv(i-1))/(1-XA(i-1)); % Value of
the derivative d(Dv)/dXA
    XA(i)=XA(i-1)+1/100; % The increment of conversion is
1/100
    dv(i)=D*(XA(i)-XA(i-1))+dv(i-1); %Corresponding value of
dimensionless temperature difference
end
plot(XA,dv,'.')
hold on
```

```
end
axis([0 1 0 12])
legend('Situation 1','Situation 2')
xlabel('X_A')
ylabel('Dimensionless temperature difference')
```

As a result of this calculation, the following plot is obtained

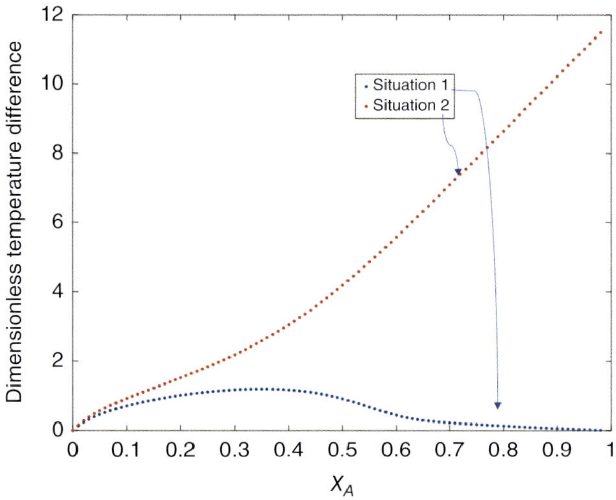

As we can check, situation 2 is much more conducive to a runaway situation. If the initial temperature of the mix is somewhat higher than that of the cooling media ($\Delta \upsilon > 0$), there will probably be a quick increase in temperature making a runaway.

(b) In the proposed tubular reactor, we have $A/V = 2/R = 2/(0.05/2) = 80\,\text{m}^{-1}$. Let us calculate the adiabatic and cooling numbers from the data:

$$\Delta T_{\text{ad}} = \frac{C_{A0} \cdot (-\Delta H_r)}{\rho \cdot C_P} = \frac{C_{A0}\,\text{mol/l} \cdot 88\,200\,\text{J/mol}}{1000\,\text{kg/m}^3 \cdot 3500\,\text{J/(m}^3\,\text{K)}}$$
$$= 0.0252\,C_{A0} \quad (\text{in K when } C_{A0} \text{ in mol/l})$$

Assuming that the global transfer coefficient is constant, the dimensionless numbers are

$$N_c = \frac{U \cdot \frac{A}{V}}{k_c \cdot \rho \cdot C_P} = \frac{\frac{50}{A}\,\text{J/(s m}^2\,\text{K)} \cdot 80\,\text{m}^{-1}}{2.56 \cdot 10^{-3}\,\text{s}^{-1} \cdot 3500\,\text{J/(kg K)} \cdot 1000\,\text{kg/m}^3}$$
$$= \frac{0.4464}{A} \quad \text{with } A \text{ in m2}$$

$$N_{\text{ad}} = \frac{E \cdot \Delta T_{\text{ad}}}{R \cdot T_c^2} = \frac{8809 \cdot 0.0252\,C_{A0}}{306^2} = 0.002\,37\,C_{A0} \quad \text{with } C_{A0} \text{ in mol/l}$$

In that situation, $\dfrac{N_c}{N_{\text{ad}}} = 188.3/(C_{A0} \cdot A) \quad \text{with } A \text{ in m}^2 \text{ and } C_{A0} \text{ in mol/l}$

In the limit $\frac{N_c}{N_{ad}} = e = 188.3/(C_{A0} \cdot A)$ we can deduct that the following must be fulfilled:

$$C_{A0} = 69.27/A$$

For example, if the total area of heat transfer is $A = 1\ m^2$, the maximum initial concentration will be 69.27 mol/l. If the initial concentration is higher, the adiabatic temperature increment will be higher than the limit with the given heat transfer characteristics. The result is somewhat surprising because a higher area will permit a lower concentration, this is because the total heat capacity is maintained constant, which is not a usual situation.

Bibliography

Barkelew, C.H. (1959). Stability of chemical reactors. *Chemical Engineering Progress* 55: 37–46.

Heiszwolf, J.J. and Fortuin, J.M.H. (1996). Runaway behaviour and parametric sensitivity of a batch reactor – an experimental study. *Chemical Engineering Science* 51 (11): 3095–3100.

Jiang, J., Jiang, J., Pan, Y. et al. (2011). Investigation on thermal runaway in batch reactors by parametric sensitivity analysis. *Chemical Engineering and Technology* 34 (9): 1521–1528.

Morbidelli, M. and Varma, A. (1982). Parametric sensitivity and runaway in tubular reactors. *AIChE Journal* 28: 705–713.

Moulijn, J.A., Makkee, M., and van Diepen, A. (2013). *Chemical Process Technology,* 2e. Wiley.

Van Welsenaere, R. and Froment, G.F. (1970). Parametric sensitivity and runaway in fixed bed catalytic reactors. *Chemical Engineering Science* 25: 1503–1516.

Vogel, H. (2000). Process development, Principles of Chemical Reaction Engineering and Plant Design. In: *Ullmann's Encyclopedia of Industrial Chemistry,* vol. B4 (eds. B. Elvers, S. Hawkins and G. Shulz), 437–475. Weinheim, Germany: Wiley-VCH Verlag GmbH & Co. KGaA.

Westerterp, K.R., Van Swaaij, W.P.M., and Beenackers, A. (1984). *Chemical Reaction Design and Operation,* 2e. Chichester: Wiley.

7

Forced Unsteady State Operation of Chemical Reactors

7.1 Introduction

This chapter tries to give an overview of the operation of chemical reactors in an unsteady regime, which is usually known by its acronym: FUSO (forced unsteady state operation). The four most used forms in the practice of carrying out an operation in an unsteady regime are analyzed: periodic variation of the input conditions, periodic flow inversion, operation with variable volume, and use of oscillating pressure. We review how to carry out this operation, giving their advantages and disadvantages.

It will be concluded that FUSO is not less efficient than the equivalent operation in steady regime and, on many occasions, its practical interest can be considerable. It is true that some of the possibilities discussed in this work are rarely used, some have only been tested on a laboratory scale, and others are at the pilot plant level, while only a few have been applied on an industrial scale. However, it is expected to see greater use in the future. An important exception is the system used for automotive emission control of NOx, called diesel oxidation catalyst (DOC, see Chapter 8), where a periodical switching between fuel-rich and fuel-lean combustion generates pulses that are more efficient.

On the other hand, the consideration of operations in an unsteady state is an interesting exercise to better understand the operation of chemical reactors.

When developing a chemical process, the production capacity marks whether to carry it out continuously or in batch systems. Once made, by the high production, the choice of a continuous process tends to consider the steady state as the best option and even as the only one possible. This is usually due to the fact that, in a continuous industrial process, a constant product output is sought, which maintains its characteristics over time, so that an elementary analysis leads to the conclusion that it is best to use a continuous process in steady state. And this choice is supported, in addition, by certain additional practical considerations, such as that the control of the process is much easier in steady state and it is easier for the operators to understand, and that a good prediction of the behavior can be achieved (more stable and safe), since the equations that govern the process are simpler, the control of the temperature is simpler, the assemblies can be considerably cheaper, etc.

Chemical Reactor Design: Mathematical Modeling and Applications,
First Edition. Juan A. Conesa.

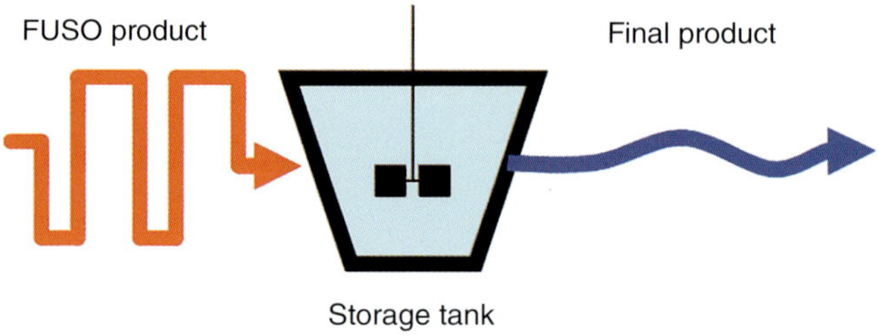

Figure 7.1 Effect of regulation in unsteady operations.

However, despite designing to work in a steady state, the chemical engineer must face situations in which the process is not stable over time, as, for example, in start-up processes and stops or regeneration of catalysts, which are usually problematic, by disturbing the work routine of the plant, so generally we tend to avoid them as much as possible. Why then consider work in an unsteady regime if it seems to be clearly disadvantageous? This chapter tries to answer this question, considering the most common ways of operating in a FUSO regime.

In the first place, as already indicated, it is necessary to specify that when speaking of operation in FUSO, it does not refer to a batch reactor, in which the design variable is precisely time, but to the intentional disturbance of the steady state in a continuous reactor. The regeneration of catalysts has points in common with this type of operation, but it is not considered in this study since, except in rare cases, the periods between regenerations take days at the least, and often months and even years, by what is the normal operation in the steady state. What we are dealing with here are continually transitory processes. In addition, the regeneration tries to return the initial performance to the process, while the objective of the FUSO is to improve that performance with respect to the steady regime.

However, we must not forget that we are still looking for a homogeneous production over time, so any type of steady-state disturbance cannot be used. It is intended as a periodic variation, so that the average production in time is constant, and that is why the FUSO is also known as periodic operation or cyclic operation. In addition, it will be necessary to introduce storage tanks that collect the different fractions and provide an approximately constant output, whether it is the final output of the process or whether it will pass to other units or devices (separation, purification, another reactor, etc.). Figure 7.1 schematizes the objective sought with the FUSO.

7.2 Objectives and Types of FUSO

Obviously, if it is proposed to use the FUSO as a way of working, some improvement in the process has to be produced with respect to the steady regime. For example, in catalytic systems, the phenomena of adsorption, desorption, diffusion, etc. always play a role. Essentially, for a catalytic reaction to take place, these phenomena are dynamic, and this aspect could be exploited in a more efficient way in a periodic reactor. From another point of view, the steady state is simply a particular case of the periodic operation, when the period tends to zero (or to infinity); therefore, it does not seem logical to think that the permanent regime will be the best option in all cases.

In fact, the studies carried out suggest that the optimal state of the catalyst surface to achieve a high conversion or selectivity occurs in an unsteady operation, in which the inertia of the catalyst is used. Increasing catalyst activity, expressed as an increased conversion or reaction rate, is often the reason for studying the periodic operation. However, as already indicated, FUSO is more complicated and more expensive than the operation in steady state, so this increase in conversion could be achieved in a more economical way by increasing the size of the reactor or the amount of catalyst used. Consequently, working with a periodic operation is only reasonable when the increase in catalyst activity is considerable.

Still, there are situations in which this conversion increase is a reasonable goal, for example, in reactions with extreme temperatures and/or pressures or with an enormously expensive catalyst. It is also interesting when it is necessary to increase the capacity of an existing reactor, and it could also be considered in processes whose conversion is limited, for example, by the equilibrium. The objective can be even more ambitious, focusing on improving selectivity and/or production, where operation in an unsteady regime proves to be perhaps the best technique to achieve it, except for the development of new catalysts. There is also evidence that the periodic operation can decrease the rate of deactivation of the catalyst, and this can also be an objective to be considered when considering the application of a FUSO.

Operation in an unsteady regime has been studied for more than 70 years, and during this time numerous studies have been published. They deal with ways to get a periodic operation, describing the improvements that have been achieved. According to these works, it can be seen that the alternatives to achieve a periodic operation can be classified into four types, which are described in the following sections:

1. Periodic variation of the input (concentration, flow or temperature),
2. Periodic flow inversion (periodic flow reversal),
3. Operation with variable volume (variable volume operation, "VVO"),
4. Operation with oscillating pressure (pressure swing).

7.3 Periodic Variation of the Input

The terms periodic operation, cyclic operation, forced periodicity, modulation, etc. are used interchangeably and refer to those operations in which one or more of the inputs to a chemical reactor vary with time, but in such a way that this variation is repeated causing a periodic signal.

Figure 7.2 shows a system in which there is the flow of reagents A and B periodically between two values, so that a pulse string representing the variation of the concentration of reactants in the stream is generated. This example corresponds to a well-studied system, but there are other variables or input conditions that can also be varied, such as reactor temperature, total flow velocity, and pressure or volume of the reactor.

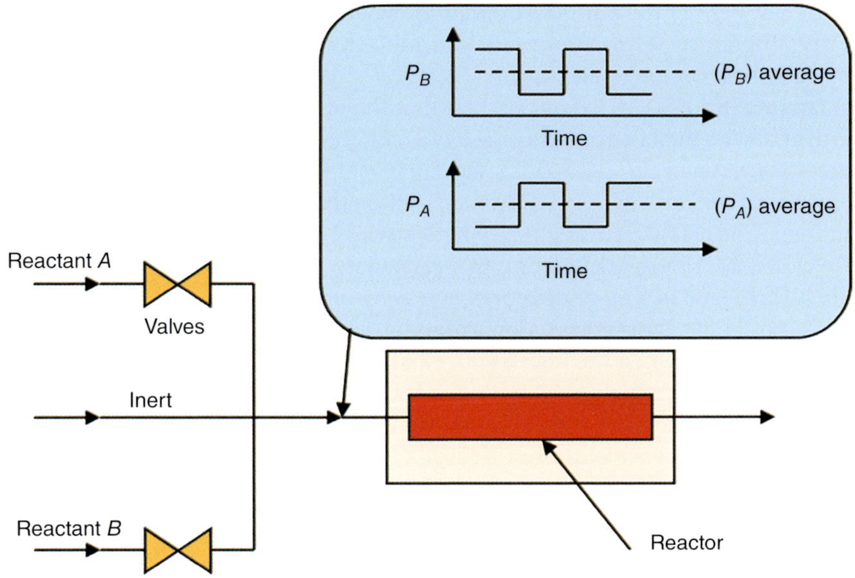

Figure 7.2 Flow variation of reagents A and B periodically between two values.

A series of terms are associated with this type of systems that operate in periodic conditions when attempting to describe their periodicity. These terms can be better understood by referring to Figure 7.3, in which the input signals and process rate are schematically shown.

The variables that appear are as follows:

(a) *Period* (τ): Refers to the time passing between the repetitions of the variation in the input condition.
(b) *Division (s) (split)*: The duration of one of the parts of the cycle related to the period. A value of $s = 0.5$ indicates that each of the two parts of the cycle has the same duration, and it will be a symmetrical cycle. The split measures the duration of a portion of the cycle in which one of the reactants is at its highest concentration; for this reason, when we speak of a split, we must refer to one of the reactants.
(c) *Amplitude (A)*: The change in the value of the entry condition over the average value. The amplitude will have a single value when there is symmetry, but if $s = 0.5$, two amplitudes should be given, one for each portion of the cycle.
(d) *Phase lag*: The variations of the composition in the figure are 180° or π radians out of phase, but another delay could have been used. A special case would be that of a pulse operation, in which the phase shift would be zero.

7.3.1 Modes of Operation

Many types of cycles can be carried out in periodic operations. On the one hand, there are systems with more or less equal cycle divisions (with $s = 0.5$, the most studied), and on the other are those where one of the parts of the cycle is very short, that is, the "pulse mode."

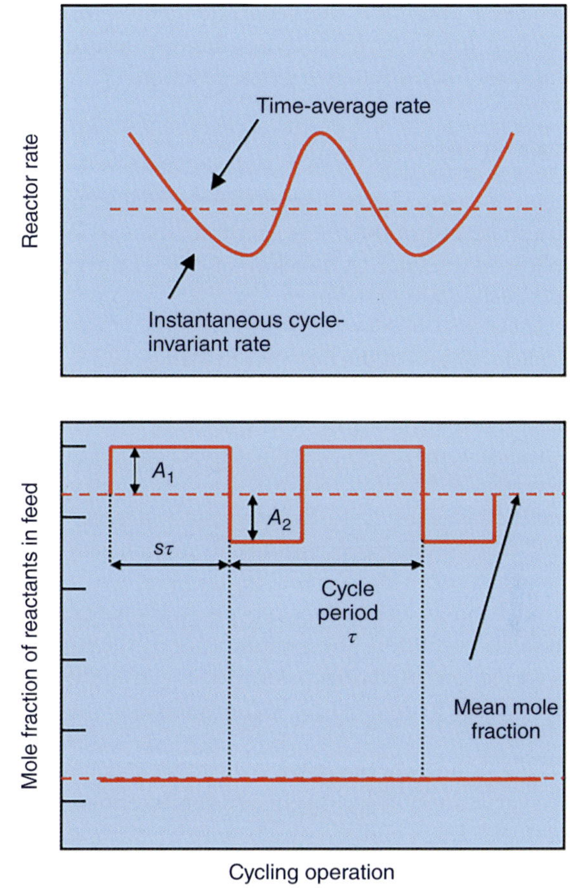

Figure 7.3 Comparison of the input signals of a steady- state operation with another carried out in an unsteady state. Source: Silveston et al. (1995). Adapted with permission of Elsevier.

The classification of the modes is usually done according to the characteristic response of the system or the relaxation time (t_c) of the catalytic reaction or reactor that is being forced periodically. At one extreme are the cycles with long periods ($\tau \gg t_c$), in this case the reactor behaves as if it were operating in a steady state, and, in this way, it is called "quasi-steady mode." At the other extreme, of short periods ($t_c \gg \tau$), two situations can appear:

(i) If there is a good and rapid mixture, then the system behaves as if it were in a steady state, and the average value of the forced variable is obtained at the input. This situation is reached in reactors with recirculation or in which the mixing level is very high (special tanks).
(ii) If the degree of mixing is negligible, or the section is small, the "relaxed steady-state mode" is reached.

Between both extremes, there is a region where $\tau \cong t_c$, in which two modes can also be given: on the one hand, there is the "pulse operation," when the offset at the inputs is zero, and on the other hand is the "standard cyclic mode," when the offset is 180° (while reagent A increases, B decreases). The "pulse mode" is interesting in practice, since it can cause a "chromatographic" effect to be achieved, directing the separation between the reagent and the product, and causing improvements in production and/or selectivity when limitations are imposed for the balance of the reaction.

7.3.2 Design Strategy

The choice of which entry is the one that interests to vary, the way to vary it, and the structure of the cycle that must be carried out are part of the strategy to be followed in order to develop an efficient periodic operation. The steps to get it established are listed here.

7.3.2.1 Choice of the Entrance to Manipulate

One input to a reactor is a variable that can be manipulated, and most inputs can be forced periodically. In the same system, one can force more than one entry simultaneously. The possible inputs to be manipulated are flow rate (it has only been studied by simulation; it does not seem very useful for single-phase systems, but it does seem useful for biphasic and three-phase systems); composition of the feed (they are the most studied systems); pressure (systems little studied because the effects are scarce); feeding temperature (they are also well-studied systems by simulation as well as by experimentation); and wall temperature (i.e. heat transmission condition, the temperature of the furnace, or a cooling or heating fluid is varied, but high thermal inertia makes it difficult that the periodic cycles affect the ordinary functioning of the system).

7.3.2.2 Choice of Handling

If a variation of the concentration is intended to be carried out, at least two components are required, which may be both reactants, or a reagent and an inert component. In this case, two options can be followed: that both components vary simultaneously, or that the concentration of one of them remains constant and that of the other varies periodically.

If three components are considered, the possible combinations are more numerous, and, therefore, there are more possibilities to establish the periodicity in the system: varying three, two, or only one component.

For the cases in which the periodic variations are in phase, as the composition of the feed varies, the input flow varies. To avoid this, a new component must be introduced, an inert one, whose flow will also vary periodically with time (thus, the number of options is also extended).

7.3.2.3 Choice of Mode

There will be different possibilities: choice between periodic mode or operation by pulses, relaxed or quasi-steady state, exchange between pure reactants or a mixture of them, or between an inert and a mixture of reagents, etc.

Some cases are distinguished from others according to the variables of periodicity that are modified. The difference between periodic and pulsed strategy lies in the value of the split(s), and in the same way, the difference in the exchange between mixtures and pure reactants is in the amplitude or that of the quasi-steady states and the relaxed state, which are distinguished by the period of the cycle. Therefore, the mode and the establishment of certain variables for the system are practically the same.

Among all the different signals that can be developed when establishing the periodic cycles of the inputs, three types stand out: rectangular signal (when the range of the input values oscillates instantaneously from the maximum to the minimum and vice versa) such as that shown in Figure 7.2; triangular signal (when the value of the input decreases with time from the maximum to the minimum and goes back up); and sinusoidal signal (when the function of the input variable oscillates between the maximum and minimum values describing a sinusoid).

Of all of them, it has been shown that the most significant effects in the unsteady state are produced by the application of a rectangular signal, being even greater at a greater amplitude of the cycle (greater difference between the maximum and minimum values of the entrance).

7.3.3 Periodic Variation of Concentration

This is undoubtedly the most studied FUSO together with the periodic reversal of the flow, although it has not yet reached the industrial level. In this mode of operation, the input that varies is the reactive concentration.

For example, if the catalytic oxidation of CO is carried out with O_2, in a steady-state operation the CO adsorbed on the catalyst surface predominates, with O_2 adsorption being the rate-limiting step due to the absence of empty pores in the catalyst. However, if a periodic operation is used by cyclically introducing CO and O_2, the distribution of the adsorbed species is substantially varied, and the reaction rate is increased, although the total concentration of adsorbed CO varies very little. The explanation for this phenomenon can be found using the model of the islands (Figure 7.4).

In the steady state, large sets of adsorbed CO (islands) are formed, and if the reaction rate is proportional to the perimeter of these islands, it would be convenient to have a smaller size, increasing the number of islands and the total perimeter. This effect is achieved thanks to the periodic variation of the concentration. In addition, another favorable situation is reached, and that is of a new species being formed on the most reactive surface (CO mixed with surface O), which leads to a higher reaction rate. This is summarized in Figure 7.4, where part (a) represents the steady-state situation and (b) the periodic operation.

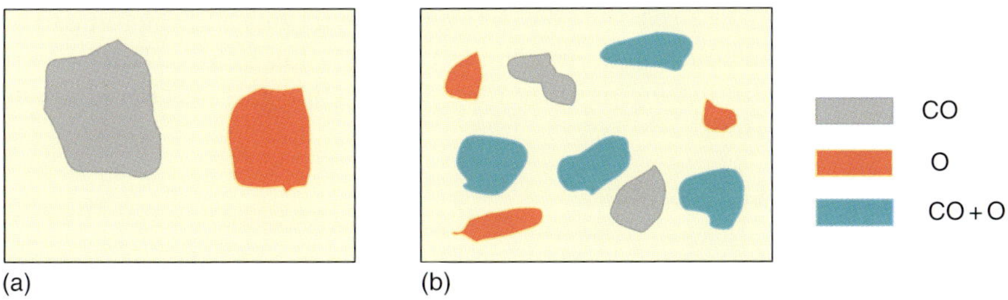

Figure 7.4 Island model. (a) Steady- state operation, (b) FUSO. Source: Zhou et al. (1986). Adapted with permission of Elsevier.

Significant increases in selectivity can also be achieved. For example, it is possible to achieve the partial oxidation of certain hydrocarbons in the absence of air using redox catalysts. The catalyst provides oxygen to the hydrocarbon by forming the products of partial oxidation, and in a second step and in the absence of hydrocarbon, the catalyst can be oxidized again by air to return the initial activity. If both reagents are not in direct contact, and one is introduced first and then the other, the selectivity of the process improves significantly. Even if the conversion decreases, production (which is what interests the engineer) may increase. Studies on the oxidative dehydrogenation of propane to obtain propene show increases in production of more than 30%.

7.3.4 Periodic Variation of the Flow

This way of causing the system to work in an unsteady state is applied to systems in which the presence of more than one phase is necessary, as is the case of three-phase fixed-bed reactors (trickle-bed), where coexists the gas, liquid, and solid catalyst.

Cyclically varying the flow rate of the fluid phase in a catalytic reactor can modify the hydrodynamics in the catalyst bed and improve its operation. This variation can cause an increase in the degree of wetting of the catalyst, since the film of liquid that surrounds each catalyst particle is much thinner than that which would result in a steady-state operation. Thus, an increase in the reaction rate can be achieved.

In addition to influencing the mass transfer from the system, the modulation of the fluid affects in a very direct way the contact time of the liquid, the catalyst particles, and the interfacial holes, so it alters the permanence of the liquid in the catalytic bed, thus influencing the residence time of the reagents.

The temperature of the catalyst will also be affected, since a permanent temperature oscillation will be created between the catalyst particles, which can cause an increase in the average reaction time.

Anyway, the most important effect of the periodic variation of the input flow to the reactor will be that produced on the mass transfer between the liquid and the solid, since they will be provided new contact channels previously unavailable.

The industrial implementation of this type of systems is difficult. The decrease of the liquid flow can cause a bad distribution of the temperature and the formation of hot spots in the system. The fact of periodically shutting off the flow of liquid supposes a strong risk for very exothermic reactions, although it can be applied for less exothermic reactions, such as the oxidation of SO_2. This process for the regenerative elimination of sulfur dioxide from flue gases is characterized by the following reaction:

$$SO_2 + \frac{1}{2}O_2 + H_2O \rightarrow H_2SO_4$$

where water is used both to remove the reaction products in the form of sulfuric acid and to absorb heat. The resistance to transport offered by the aqueous phase decreases the reaction rate, but this phase is essential for the development of the reaction, so it cannot be eliminated.

Steady operation is generally not economical, because the acid obtained has a low concentration. In the dynamic operation, the SO_2 and oxygen are introduced to the reactor continuously, and the oscillation is applied to the water. In this way, during the operation there is a period when the water flows, and then it does not flow in the next period. This causes an increase in the rate of mass transfer, on the order of 40%, and the production of concentrated sulfuric acid. Sulfuric acid accumulates in cavities during the period when there is no water flow. Because sulfuric acid is easily absorbed by water, it takes a long time for the catalyst to be deactivated by the accumulation of SO_3, and high concentrations of acid are obtained.

7.3.5 Periodic Variation in Temperature

Another input that can be changed periodically for FUSO is the feed temperature. The investigations in this sense are not very advanced, mainly due to the difficulty in the control. These studies are limited to the presentation of some satisfactory data that improve the steady process, but do not go further.

For reversible exothermic processes, oscillating temperatures in the catalytic bed open an opportunity to increase the conversion, without having to resort to elimination of heat in the bed. This option has been investigated for many processes, for example, in the synthesis of ammonia.

7.4 Periodic Flow Reversal

Let us have an exothermic reaction carried out in an adiabatic fixed-bed reactor, with the catalyst, preheated, initially at a high temperature. There are several possibilities of behavior, but one of them is as follows (Figure 7.5): when the relatively cold feed current is introduced, the temperature of the bed logically decreases while that of the feed increases; and during this period there is basically no reaction. As soon as the ignition point is reached, the temperature rises sharply and the reaction begins to elapse. The next fraction of food entering the reactor finds the cold catalyst in the first zone of the bed due to the previous cooling by the reactant stream, then it reaches the hot zone, which is cooled by this new current at low temperature until, a little more advanced in the reactor, the point of ignition is reached again. The result is a temperature front that moves as time passes along the reactor, which logically causes a reaction front; it is observed that as time passes, in more reactor length there is no appreciable reaction. This is observed in Figure 7.5.

This behavior has two important characteristics. The first is that the maximum temperature of the front is much higher than that predicted by the adiabatic behavior of the system and it does not depend only on the kinetics of the reaction, but on the transport processes in the bed as well. The second refers to the rate of propagation of the front, considerably lower than the rate of mass transfer, which carries reagents and products. These two effects are a consequence of the time scale and the inertia commented on earlier and make the carrying out of catalytic

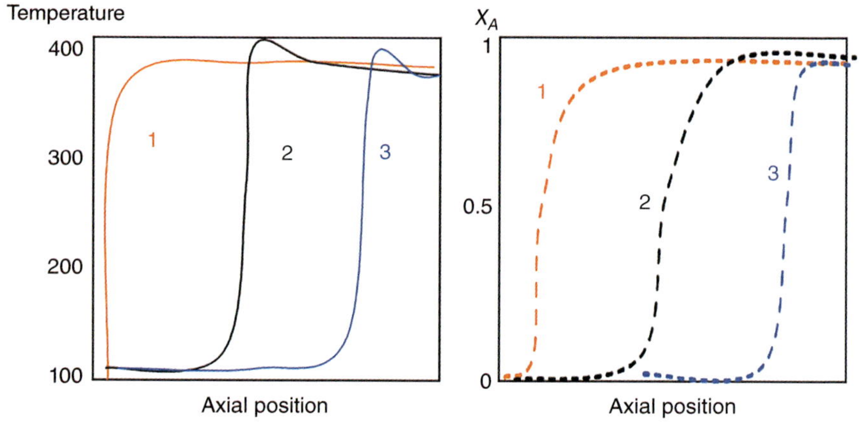

Figure 7.5 Operation diagram of an operation in adiabatic stages and one of "reverse flow."

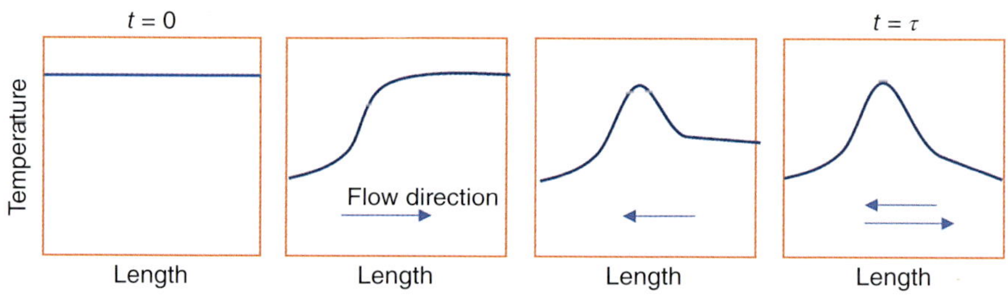

Figure 7.6 Development of a temperature profile during operation in "reverse flow."

processes in this regime of temperature fronts interesting (for example, in front of an adiabatic operation). However, after a certain time, the temperature front would "come out" of the reactor, and then the bed would have to be heated again, which reduces the possible interest in this technique.

A periodic inversion of flow (periodic flow reversal) has been proposed to avoid this phenomenon (Figure 7.6). The idea is that before the temperature front leaves the reactor, the food stream is inverted, that is, it is introduced through the other end of the reactor. With this, it is possible to generate a new temperature front that moves in the opposite direction. After a certain time, a typical temperature profile is developed, as can be seen in Figure 7.6, which can be experimentally achieved with a simple set of valves. In addition, with the bell-shaped profile obtained, a lower temperature at the outlet of the reactor is achieved, which favors the equilibrium of the exothermic reaction. It could be said that this way of operating forces the reaction to go "beyond the balance." These profiles, which are practically impossible to achieve continuously, are very close to the optimal behavior of the system.

Obviously, an analogous reasoning could be done with an endothermic reaction generating inverted temperature profiles compared to the previous ones, but this would no longer be an advantage to displace the equilibrium, but a problem.

In the example described, implemented industrially in some countries (or on which studies are being carried out in a large pilot plant, especially in Russia), what is achieved is to combine the reactor and the heat exchangers in one unit, in such a way that the heat of the reaction is used, saving on the use of additional equipment and passing the reactor from several stages to one, and saving also on necessary catalyst. In addition, the reaction occurs mainly in the center of the reactor, so it is not necessary that the entire bed is catalytic, being able to put at the beginning and at the end of the reactor an inert filler that fulfills its tasks as a heat transmitter, and yet is cheaper.

Figure 7.7 shows a simple way to get reverse flow in a reactor with a combination of two three-way valves. The flow direction is controlled by opening and closing simultaneously both valves, as indicated in the figure.

7.4.1 Operation Design

The design of a periodic operation of this type must be carried out through experimentation for each particular system. However, unlike other FUSOs, such as the periodic variation of concentration at the entrance, there are not many parameters to be determined, the main design parameter being the half-cycle time, which must simply be checked to be less than the time that it would take the generated front to leave the reactor. The main problem of the operation with periodic inversion of the flow is that at the beginning of each half cycle there are large variations in the reactor output concentrations. To solve this, several methods have been proposed, among which it is important to collect all the output fractions in a storage tank that provides a constant output (Figure 7.1); however, this would

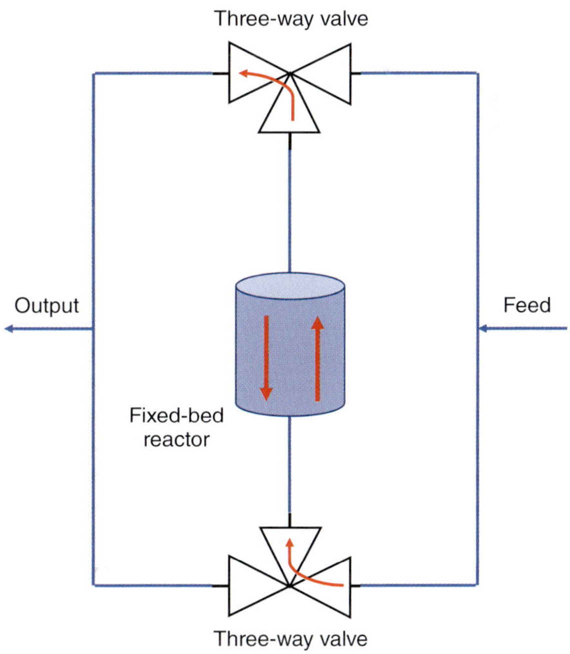

Figure 7.7 Reverse flow in a reactor with a simple combination of two three-way valves.

greatly lower the final concentration of the product. Vanden Bussche and Froment (1996) devised an association of three reactors in which each bed operates cyclically in three phases (inlet, recirculation, and exit), and in each cycle, each reactor goes through the three phases with the same time in each of them. The inversion of flow in the reactor occurs when passing from the recirculation phase to the exit phase.

7.5 Operation with Variable Volume (VVO)

From the point of view of the reaction mixture, a chemical process can operate at an intermediate point between the two ideal mixing ends, that is, between the total mixture (ideal perfect stirred) and minimum mixing (ideal piston flow). Operating in an unsteady regime can also lead to this intermediate mixing situation.

Manipulating the type of mixture is equivalent to varying the residence time, and there is no doubt that this is a very important parameter in all types of reactors; therefore, the possibility of acting on it is very interesting. This can be achieved by the so-called VVO, which is a convenient method for manipulating the residence time in a stirred tank reactor. This mode of operation can convert a batch process into a continuous one without losing many of the benefits of its batch nature while at the same time reducing its typical disadvantages.

The mode of operation of a VVO reactor is outlined in Figure 7.8. We can observe up to four different periods in each cycle: the feeding period, the "batch" period (in which the volume in the reactor is kept at the maximum value), the discharge period, and the "rest" period (in that the reactor volume is at its minimum value). The duration of the individual periods, as well as the amplitude of the oscillations, are parameters or conditions to be optimized.

VVO is widely applied in biotechnological processes, where microorganisms are used as catalysts. This specific application of VVO reactors is known as "fed-batch" reactors (see Chapter 10). In this equipment, the crop starts as it does in a discontinuous reactor, but at a later stage, when the period of exponential growth has been exceeded, more food is introduced into the reactor, without extracting material simultaneously (which is the difference of a continuous fermentation). The system continues feeding food with a certain flow rate, which is reduced with time in a predefined way.

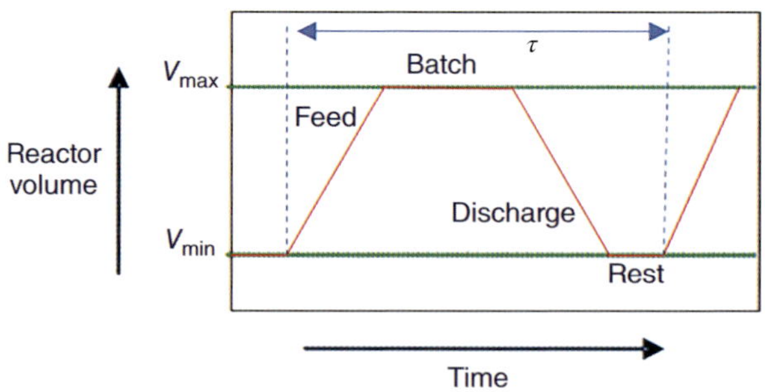

Figure 7.8 Mode of operation of a variable volume operation.

The reagent is not completely converted into products, as part of it is used for the growth and maintenance of microorganisms. In processes in steady state, it is often not possible to work with this type of catalysts, since the microorganisms mutate spontaneously, which results in a reactor not working as expected.

In some conventional batch reactors, a high initial concentration of the reactants inhibits the conversion carried out by the microorganisms. For this reason, in the VVO, the reagent is added to the reactor during the feeding period, and subsequently it is discharged in the "batch" period. The volume of the reactor that remains in the "rest" period contains enough microorganisms to keep the four cycles as short as possible: if almost all the microorganisms were eliminated in the discharge period, the feeding period would be too long. Usually, after a few cycles, the reactor is completely drained to get rid of all the microorganisms that no longer work satisfactorily.

So far, the VVO has also been applied to the preparation of fine chemical products, in order to prevent runaway reactions. In the fed-batch mode, the reagent is slowly added to the reactor so that the heat released by the exothermic reaction can be effectively dissipated, for example. Another application that the VVO operation could have is in plants where multiple products of fine chemistry are manufactured, in which different types of chemical reactions must be carried out in the same reactor.

7.6 Oscillating Pressure

In reactors that work with oscillating pressure, the pressure varies periodically, and the rest of the conditions and/or variables will do so accordingly. In these cases, the equilibrium of the system can be favored because the products of the reaction are eliminated, for example, in a reaction catalyzed by a solid, by desorption.

A simple system operating with oscillating pressure is seen in Figure 7.9. In this case, a tubular reactor contains a mixture of catalyst and adsorbent. The gaseous reactant is pumped into the packed bed at high pressure during a part of the operation cycle. An additional volume ("surge volume") is placed between the reaction bed and the outlet stream to allow reflux (purging) and perhaps higher product purities.

As can be supposed, there are several modes of variation of the pressure, in the form of sinusoidal, triangular, square, and "bang–bang" or "on–off" signal. The

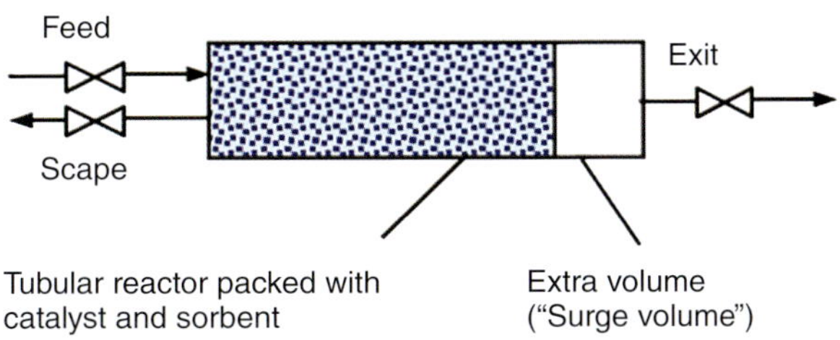

Figure 7.9 System for a periodic separation reactor.

easier to implement is the square variation, and since it has not been shown that another type is more effective is the most studied.

When changing the input pressure of a system with catalyst and adsorbent, the system passes through three different stages:

(i) Feed is introduced under pressure.
(ii) The catalytic reaction takes place. The adsorption of the product in the adsorbent material also occurs, which will be mixed in the bed with the catalyst.
(iii) The bed is depressurized and the reactants and the products are separated at low pressure. The desorption of the products occurs.

We can consider the whole process as a combination of two processes: the periodic operation of the reactor and the cyclic separation process of adsorption/desorption by oscillating pressure. Large pressure variations occur near the feed, but due to the low permeability of the bed, a current (which we will call output current) can be continuously drawn from the other end of the reactor.

In addition, in the feeding phase of each cycle, the concentration can be varied (as we have seen in a previous section), there being two limits: (i) introduce one reagent and then the other in stages (case of reaction with two reagents) and (ii) introduce a constant composition mixture. Although this second form is chosen, we do not avoid the cyclic variation of the concentration due to the separation of reactants because of the adsorbent.

The use of this method, combining chemical reaction and adsorption by oscillating pressure, is relatively new. Although a new type of periodic operation, it is more complicated to carry out than the periodic inversion of flow, and more expensive, so that the application on an industrial scale seems far away. However, it is theoretically interesting because the conversion can exceed what is predicted by equilibrium.

Bibliography

Creaser, D., Andersson, B., Hudgins, R.R., and Silveston, P.L. (1999). Cyclic operation of the oxidative dehydrogenation of propane. *Chemical Engineering Science* 54: 4437–4448.

Haure, P.M., Hudgins, R.R., and Silveston, P.L. (1989). Periodic operation of a trickle-bed reactor. *AIChE Journal* 35: 1437–1444.

Lange, R. and Hanika, J. (1994). Investigations of periodically operated trickle-bed reactors. *Chemical Engineering Science* 49: 5615–5621.

Matros, Y.S. (1989). *Catalytic Processes Under Unsteady State Conditions*. Amsterdam: Elsevier Science B.V.

Reshetnikov, S.I., Ivanov, E.A., Kiwi-Minsker, L., and Renken, A. (2003). Performance enhancement by unsteady-state reactor operation: theoretical analysis for two-sites kinetic model. *Chemical Engineering and Technology* 26: 751–758. https://doi.org/10.1002/ceat.200301640.

Silveston, P.L., Hudgins, R.R., and Renken, A. (1995). Periodic operation of catalytic reactors-introduction and overview. *Catalysis Today* 25: 91–112.

Stankiewicz, A. and Kuczynski, M. (1995). An industrial view on the dynamic operation of chemical converters. *Chemical Engineering and Processing: Process Intensification* 34: 367–377.

Sterman, L.E. and Ydstie, B.E. (1991). Periodic forcing of the CSTR. An application of the generalized II-criterion. *AIChE Journal* 37: 986–997.

Vanden Bussche, K.M. and Froment, G.F. (1996). The STAR configuration for methanol synthesis in reversed flow reactors. *Canadian Journal of Chemical Engineering* 74: 729–734.

Xiao, W.D., Wang, H., and Yuan, W.K. (1999). Practical studies of the commercial flow-reversed SO_2 converter. *Chemical Engineering Science* 54: 4645–4652.

Zhou, X., Barshad, Y., and Gulari, E. (1986). Transient response and rate enhancement through forced concentration cycling. *Chemical Engineering Science* 41: 1277–1284.

Zwijnenburg, A., Stankiewicz, A., and Mouljin, J.A. (1998). Dynamic operation of chemical reactors: friend or foe? *Chemical Engineering Progress* 94: 39–47.

Part II

Catalytic, Multiphase and Biochemical Reactor Design

8
Industrial Catalysis

It is important for the chemical engineer who must design a catalytic reactor to know not only the design equations but also the chemical fundaments of the catalysts that will be used in the industry. Thus, the chemical bases of catalysis are discussed in this chapter, with special reference to the most important industrial catalytic processes. The structure of the catalysts is detailed (with special interest those that are supported on ceramic materials). With this information, we proceed to the design of the reactors, dealt with specifically in Chapter 9.

8.1 Introduction

8.1.1 Reactors for Solid-Catalyzed Reactions

Reactions catalyzed by solids usually involve two or three phases. Solid catalyst and gaseous and liquid reactants are brought in contact to achieve the desired conversion. Some of the reactor types used are briefly presented here for background information, with generalized remarks on their advantages and disadvantages.

Gas–solid reactors are the most well-known two-phase catalytic reactor types operated continuously. The major reactors for solid-catalyzed gas-phase reactions are the fixed-bed, the fluidized-bed, and the entrained-flow reactors.

The fixed-bed reactor is either operated adiabatically or isothermally. In the latter case, a multitubular reactor system is often chosen where the parallel reactor tubes are surrounded by a heat-exchanging fluid. The fixed character of the catalyst bed necessitates a long catalyst life. If deactivation occurs, it should be possible to regenerate the catalyst inside the reactor.

Advantages of the adiabatic fixed-bed reactor are its simplicity, the high catalyst load per unit volume, little catalyst attrition, and little back mixing. The disadvantages are the high pressure drop, difficult temperature control, and the long diffusion distances.

A reactor system similar to the fixed-bed reactor is the moving-bed reactor, where the deactivation rate is relatively low, but too high for pure fixed-bed operation, e.g. in some catalytic reforming processes. Catalyst particles should be spherical and of uniform size for smooth flow through the bed.

Chemical Reactor Design: Mathematical Modeling and Applications,
First Edition. Juan A. Conesa.

The circulation of gases through fixed beds approximates the plug flow and is completely different from the circulation in fluidized beds, where the gas flow is not easily defined. The gas flow in a fluidized bed is very different from the plug flow in that it can present a large bypass, because part of the gas passes in the form of bubbles. This makes the contact not effective, needing a high amount of catalyst if high conversions are desired. If, in a particular case, the effectiveness of the contact is very important, it is better to use a fixed-bed reactor.

The fluidized-bed reactor is mainly chosen because of its good heat exchange properties and the uniform temperature distribution. Furthermore, due to the fluid-like behavior of the bed, the catalyst can be added and withdrawn during operation for regeneration or replacement in case of fast deactivation. In addition, due to the relatively small particle size, the diffusion distances are short. On the other hand, the disadvantages are the occurrence of catalyst attrition and back mixing, and the fact that fluid-bed reactors are difficult to scale-up.

It is difficult to control the temperature in the fixed beds, especially if they are large since their heat conductivity is low; therefore, in very exothermic reactions, hot zones or mobile hot fronts will form, which can damage the catalyst and even deactivate it. This does not happen in fluidized beds where the temperature can be easily homogenized when there is a rapid mixture, and the bed can be considered practically isothermal.

If the catalyst used in the process is easily deactivated and it has to be regenerated very often, it is better to use a fluidized bed since the behavior of the solids as a fluid permits to handle them much more easily.

The entrained flow is used when very short contact times are required, for example, in case of highly active catalysts that deactivate fast. In fluid catalytic cracking (FCC), the recirculating catalyst also supplies part of the heat for the endothermal reaction. Depending on the catalyst loading, one distinguishes dilute and dense phase "risers." Advantages of entrained-flow reactors are the high mass transfer rates, short contact times, and the possibility of continuous catalyst replacement. The disadvantages are the occurrence of catalyst attrition, reactor erosion, and the requirement to separate the catalyst from the product.

Intermediate reactor types between fluidized-bed and entrained-flow reactors also exist, like internally circulating bed reactors with an internal riser section and an external annular section of a fluid/moving-bed type.

Relatively new are the monolithic reactor types, mainly used in environmental applications. Examples include the three-way catalyst for the exhaust gas cleaning of gasoline engines, the selective catalytic reduction (SCR) for NOx reduction with ammonia, the oxidation of volatile organic compounds, etc. All have extremely low pressure drops due to the structured character of the catalyst and can be used at high flow rates.

Usually, no reactors of the fluid bed or slurry type are used for liquid–solid reactions. In most cases, a gas phase is also involved, resulting in a three-phase ("multiphase") operation. Many applications of gas/liquid/solid reactors are found in industry. Selective oxidations and hydrogenations, hydrotreating, hydrations, and aminations of a liquid reactant (mixtures) catalyzed by a solid are important industrial examples. Two extremes exist in three-phase operations: the trickle-bed and the slurry reactor (Figure 8.1).

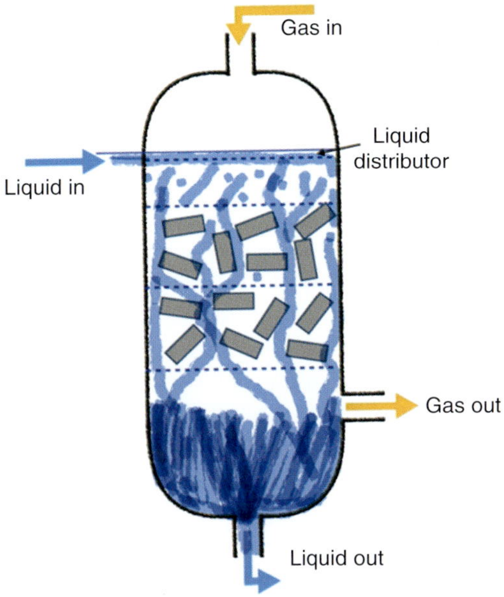

Figure 8.1 Trickle-bed reactor.

In the trickle-bed reactor, the gaseous and liquid reactants flow cocurrently downward over a packed bed of catalyst. The liquid phase must be distributed such that a good wetting of the catalyst is obtained. Uneven wetting may lead to local hot spots and runaways. The gas phase is the continuous phase. High gas flow rates are used to remove heat. Since the liquid covers the catalyst particles, reactant diffusion is much slower than that in the gas-phase operation. The nature of the packed bed requires catalyst bodies of a few millimeters, so the catalyst effectiveness is restricted.

Advantages of trickle-bed reactors are high catalyst load, no need for catalyst separation, no attrition, and very limited back mixing. The disadvantages are reduced catalyst effectiveness and selectivity, cocurrent operation, flow maldistributions, and large pressure drop.

In slurry reactors, small catalyst particles (10–100 nm) and gas bubbles are present suspended in a liquid phase by mechanical mixing. These reactors can be operated in batch or semi-batch mode, with the gas phase usually supplied continuously. Also, the continuous mode is possible.

Upflow reactors are sometimes used for small-scale testing. Here, the liquid phase is the continuous phase through which the gas bubbles rise, and wetting problems are absent.

Countercurrent operation may have certain advantages like removal of intermediate products to avoid side reactions, to overcome equilibrium limitations (hydrogenation of aromatics), or to avoid product inhibition. Also, temperature profiles along the reactor length differ, which may be used advantageously. The major problem is flooding, i.e. entrainment of the liquid by the gas phase. Larger catalyst particles must be used to allow sufficiently high flow rates, thereby further lowering the catalyst efficiency.

8.1.2 Solid Catalysts (Supports)

One important aspect in the catalytic reactor design is the nature and properties of the catalyst itself. It is important then to know the most important required properties of the solid catalysts and the industrial way to get these properties. This would give the engineer a background knowledge that will improve his/her ability to do a good design.

Generally speaking, the chemical nature of the material used for catalyst development determines whether a catalyst's applicability may be conditioned by its physical properties.

Figure 8.2 shows a scheme of the main requirements for an industrial catalyst, which is to get a material with high activity and selectivity for the reaction of interest. This material can exist as a simple catalyst, but usually it is necessary to support the active catalyst on other materials that can have or not have catalytic activity.

The external surface of a catalyst is generally too small for the required properties of surface-to-volume ratio, so the catalyst should be porous. It is difficult to achieve at the same time high porosity and high mechanical strength, so the necessity of using catalyst supports is justified.

Supports are usually essential to prepare porous and nonporous catalysts. Support is generally an inert material where the active component is dispersed. Materials used for supports must have a high mechanical strength, high surface area, high pore volume, etc. and the selection of the material is done as a function of the operation parameters and the reaction system. Besides the mechanical strength, the supports provide the following other properties:

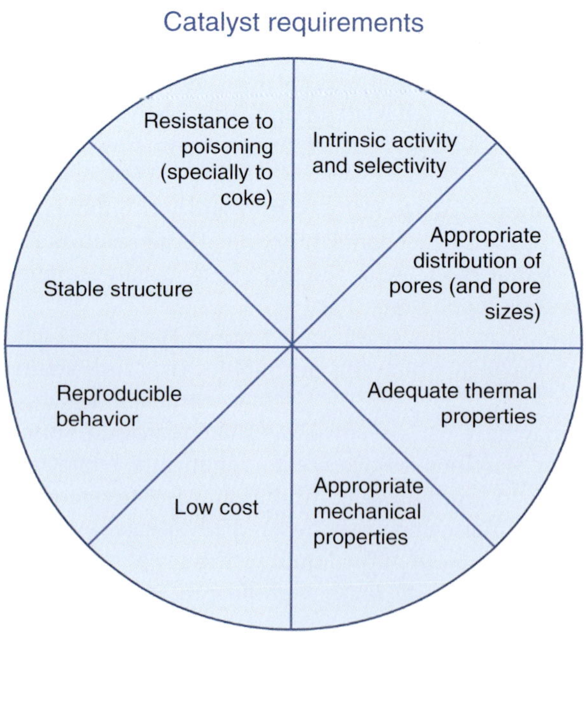

Figure 8.2 Catalyst key points.

- They can improve the stability by separating the small crystals of the catalyst, avoiding sintering.
- If a chemical that reacts with the catalyst is used as a support, a chemical with higher catalytic activity can be formed.
- Increasing the surface area of the support reduces the sensitivity of poisons.
- If the support has a high thermal capacity and thermal conductivity, it can dissipate the heat, avoiding local overheating that can cause sintering.

The most commonly used supports in the chemical industry are silica, alumina, zeolites, and active carbon.

Preparation of these supports is very diverse, and depending on the necessary properties, the preparation method can vary. Taking as support alumina, the texture will depend on many factors. The major phase in the material will have an important effect on the surface characteristics and the porosity; the area can vary from 500 m^2/g for γ-alumina type to 2 m^2/g for α-alumina.

The chemical structure of alumina is complicated due to the existence of different phases. When alumina is generated by precipitating an aluminum solution, the nature of the precipitate can be either a hydrogel of ammonium or a crystalline structure; and in the latter case, the particular structure is dependent on the temperature, pH, and agitation of the initial solution.

Rarely, an industrial catalyst consists only in the active phase and a support material, as mentioned before. The final catalyst will have different additives, mainly promoters, substances with no catalytic properties by themselves, but increasing the rate or the selectivity of the reactions, or both.

The catalytic stabilizers may help usually to prevent the sintering of the active phase. For example, in the metallic catalysts, these compounds act as a barrier among the different metallic points, in order to avoid sintering. Other additives are used in the preparation of the catalysts, as is the case of those used to ease the extrusion of the pellets or the cementation of the support granules.

During the preparation of the catalysts, the incorporation of the active phase on the support can be done in different ways: precipitation, adsorption, impregnation, ionic interchange, etc. But the important point is the arrangement of the active phase on the support. The objective is to obtain an optimum profile of the active phase over the support to be applied to a particular reaction system. Figure 8.3. shows typical distribution profiles of the active phase for a spherical pellet along its diameter.

In the uniform distribution, the material is dispersed along the diameter. This is the most favorable case when the chemical step is controlling the reaction rate, and no diffusion control exists. Nevertheless, if the reaction is limited by mass transfer phenomena, it is better to use a catalyst where the active phase is on the exterior surface of the sphere (second drawing in Figure 8.3 "Outer egg shell"). On the contrary, in systems where the catalyst is exposed to different poisons or extreme conditions, it may be better to use a geometry where the active phase is on the center of the pellets ("inner egg yolk"), or a situation among them ("middle egg white"). In order to get these active profiles in the support, usually a competitive adsorption with other ions is used, which can later be eliminated or minimized in any way.

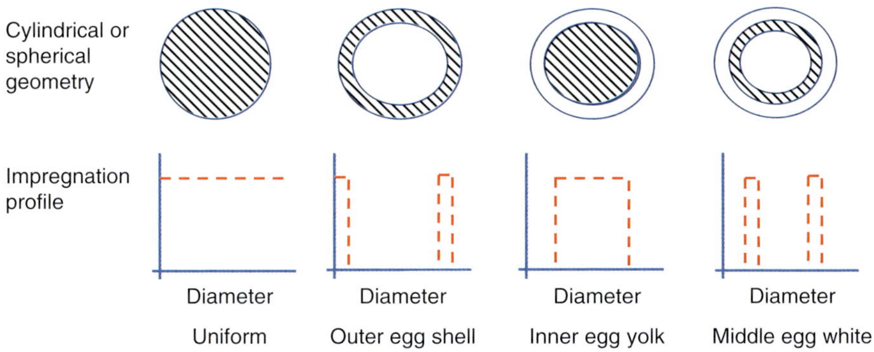

Figure 8.3 Typical distribution profiles of active phase in a spherical support.

Catalysts with the same kind of global distribution of active phase can present differences in the distribution at a smaller scale. In this way, the activity and selectivity of the catalysts will be a function of the size of the metallic particle and the morphology of the support.

8.1.2.1 Choice of Catalyst Support

The choice of the support depends on several factors. The first requirement is that it must be stable under the reaction conditions. Many materials are not resistant to high temperatures. An example is the transformation of highly porous γ-alumina into α-alumina at temperatures exceeding 1400 K. Activated carbon has a high thermal stability, but it is oxidized in an oxygen-containing environment. Acidic or basic solutions can cause the dissolution of oxidic supports. In addition, catalysts requiring regeneration must be based on supports that are stable under regeneration conditions. For example, hydrocracking and FCC catalysts must be stable in an oxidative environment at high temperature.

The texture of the catalyst support (average diameter of the pores, shape, and connection among pores, etc.) influences the mass and heat transfer characteristics. In the case of pores of very low diameter (too narrow) or very long ones, the selectivity will be reduced in partial oxidation processes, because reactions completing the oxidation are favored. In this respect, zeolites are a special case due to their pores of molecular dimensions.

The interaction between the support and the active phase also is critical. One of the functions of the support is to stabilize the active phase against sintering, as mentioned, and a strong interaction favors this stability, but care must be taken because the chemical characteristics of the active phase could be also modified.

Obviously, the cost of the support must be as low as possible. Other properties of interest are density, heat capacity, and size/shape.

8.1.2.2 Comparison and Uses of Supports

In Table 8.1 we find the main characteristics of some materials used as catalyst supports. Silica and alumina are the most common supports. Alumina is available in many forms, of which a high surface area (γ-Al_2O_3) and a low surface area (α-Al_2O_3) are shown.

Table 8.1 Characteristics of some support materials.

Material	Surface area (m^2/g)	Pore volume (cm^3/g)	Mean pore size (nm)
γ-Al_2O_3	180–200	0.6	3.5
α-Al_2O_3	3–5	—	—
SiO_2	150–400	0.5–1.2	Bimodal (5–50 and >50)
SiO_2–Al_2O_3	150–160	0.2–0.7	10
TiO_2 (anatase)	50	0.34	9
C (activated carbon)	200–1500	0.1–2	1 (often bimodal)
Zeolites	400–700	0.2–0.6	0.4–1

The inorganic oxides are dissolved in acidic or basic media, so activated carbon is used in these cases. In addition, the acidity of supports such as silica and alumina is not always desired, and more inert materials like anatase or activated carbon are used. Furthermore, silica cannot be used under conditions of high temperature and steam atmosphere, because evaporation takes place.

8.1.2.3 Silica, Alumina, and Mixtures

Silica and alumina are both prepared by a method that starts with precipitation from a concentrated solution of the metal salt. The resulting hydrogel, after purification, can be formulated into different shapes, such as powders, pellets, extrudates, or simple spheres.

Silica has no catalytic activity of its own. It is used as a support for Ni, Pd, or Pt in gas-phase hydrogenation reactions, and also as a support for vanadium pentoxide (V_2O_5) in the oxidation of SO_2.

γ-Alumina is used as support in the catalysts for exhaust-gas treatment, hydrotreating catalysts, and some hydrogenation catalysts. This material itself has catalytic activity, and is used as catalyst for alcohol dehydration and in the conversion of H_2S into the elements.

Silica-alumina is a mixed oxide that can be produced by impregnation of porous silica with an Al^{2+} solution. This support contains Lewis and Brønsted acid sites (this is explained later), constituting a remarkable strong acid, more acidic than silica or alumina alone.

8.1.2.4 Zeolites

The majority of silica-aluminates are amorphous materials, that is, in which there is no ordering of the atoms in a large distance forming a crystallographic phase. However, zeolites are silica-alumina materials presenting a very interesting crystallinity. Zeolites are crystalline and have a well-defined pore structure with pores of molecular dimensions.

Zeolites are formed by assembling tetrahedral units of SiO^4 or $(AlO^4)^-$. These units are first grouped in a building block composed of 24 tetrahedra, called sodalite, and then in units of sodalite, giving rise to the superstructure of the

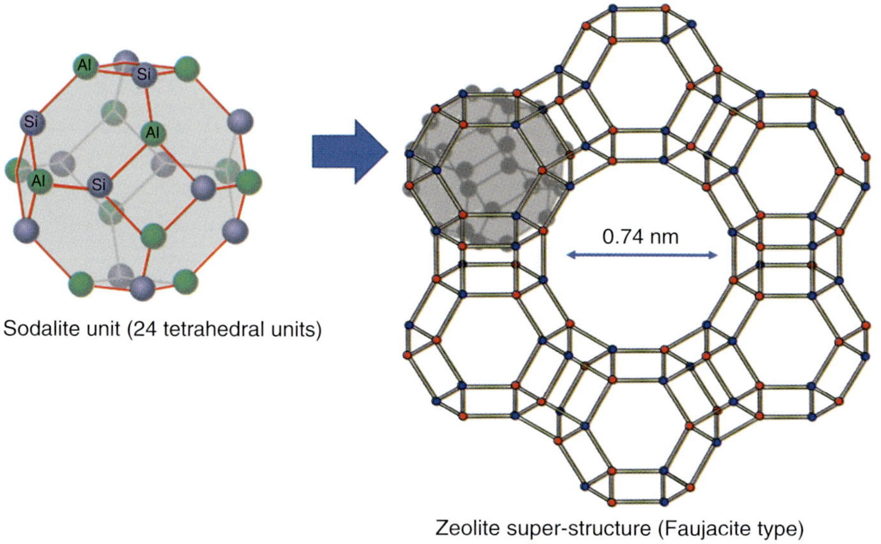

Figure 8.4 Structure of sodalite unit and superstructure of the zeolites. Source: Adapted from https://commons.wikimedia.org/wiki/File:Faujasite_structure.svg.

zeolites. Figure 8.4 shows the structure called faujasite, in which the Si or Al atoms are in the intersecting lines and the O atoms are between them.

In the inner part of the structure of the zeolite, we find cavities with large holes, whose size depends on the way in which the sodalite units are assembled. The size of these cavities determines the size of the molecules that can be adsorbed inside and that can react within the "box" of zeolite. Other ways to assemble the sodalite units include the following:

- Sodalite, formed by units of sodalite directly above each other, with a diameter between the cavities of only 0.26 nm, where only small molecules such as water or methanol can be adsorbed;
- Zeolite A, in which the units are separated by an oxygen layer forming cavities of 0.42 nm in diameter that can adsorb somewhat larger molecules;
- Faujasite, with cavities of 0.74 nm (Figure 8.4);
- Mordenite, with cavity size of 0.80 nm;
- ZSM-5, with cavity of 0.60 nm.

More than a hundred zeolite structures are well defined on an atomic scale. Thus, bond length and angles of the atoms comprising the catalyst material are known in detail. Depending on the type of zeolite, the microporosity shows a sharp, almost unique pore size distribution in the range of the molecular dimensions, enabling molecular shape selectivity.

Based on the chemical and physical compatibility of alumina and silica on an atomic scale, relatively strong Brønsted acidity is achieved by substitution of aluminum that is counterbalanced by protons. The catalytically active sites induced by a post-synthesis procedure of the zeolites are accessible for reactant molecules, since all the atoms in the zeolite framework are exposed to the pore volume.

Table 8.2 Use of zeolite in a modern oil refinery.

Oil fraction	Catalytic process
Gas	Aromatization and alkylation
Light ends	Isomerization
Naphta	Hydrotreating and reforming
Middle distillates	Hydrotreating and dewaxing
Luboils	Dewaxing
Flashed distillates	Catalytic cracking and hydrocracking
Residue	Hydrodesulfurization and hydroconversion

Many reactions in the oil refining and petrochemical industry are accompanied by the formation of carbonaceous by-products (as coke) that usually deactivate the catalysts. Zeolites are thermally stable and can, therefore, be regenerated at high temperature by oxidation of these carbonaceous deposits.

The contribution of zeolites in the industry has been mainly developed in the past 40 years, and it can be considered essential for the development of many industrial processes. Table 8.2 shows the main processes where zeolites are present in refineries.

8.1.2.5 Activated Carbons (ACs)

ACs are carbons with high specific surface area. Carbon is extensively used in industrial adsorption applications, but also in catalytic chemistry, due to the exceptional properties it presents, as they possess the following:

- High stability both in the presence of acids and bases,
- Very high adsorption capacity of a variety of molecules, especially organic molecules,
- Wide variety of textures properties,
- Extraordinary resistance to high temperatures,
- Not much erosion on pumps.

Other properties of the ACs make difficult their application in oxidizing environments, as they are easily burnt. Also, the AC can present a weak mechanical strength, making it difficult to avoid the formation of fines, which can make separation difficult.

Many biomass feedstocks can be used for preparing ACs. Wood, coal, and lignite are the most common. Usually, the manufacturing process consists of pyrolysis (heating in inert atmosphere) followed by activation (with steam or oxygen) to increase the porosity and surface area.

8.2 Industrial Preparation of Catalysts

The industrial manufacture of the catalytic converters is characterized by its secret end, not only by the manufacturer but also by the user. The catalyst market is large and fragmented in numerous segments, each with its own characteristics. These characteristics are generally determined by the chemical nature of the catalyst and/or the market in which, normally, the segments are not competitive with each other.

8.2.1 Structure of Commercial Catalysts

As mentioned, during the design of the catalysts, special attention is required to understand the factors that can limit the rate of the process. In the design of environmental catalysts, the diffusion phenomena are minimized by synthetizing them in a process divided into three steps.

In this way, many catalysts have a monolithic ceramic cylindrical structure (cordierite) that has a three-dimensional scale. The cordierite comprising the monolith cannot be used to support the catalysts due to its low surface ratio since it is not capable of efficiently dispersing the metal particles. In this sense, a stable material of high surface (coating or "washcoat") is adsorbed in the cordierite structure. Then, the metal active particles are dispersed in the coating (Figure 8.5, see also Figure 9.8). Some examples of materials used as coating are Al_2O_3, SiO_2, TiO_2, CeO_2, ZrO_2, V_2O_5, La_2O_3, and zeolites.

8.2.1.1 Synthesis of Zeolites

Zeolites are produced in the "hydrothermal process." This is the common name of the process, which involves water and heating of different sodium solutions. Usually, sodium hydroxide, sodium aluminate, and sodium silicate are used, and an "aqueous glass" is formed. This forms a gel that is then converted into the crystalline zeolite. The structure and shape of the zeolite will depend on the composition, temperature, reaction time, and pH of the medium. A scheme of the process is given in Figure 8.6.

By the type of reagent used in the synthesis, the sodium salts of the zeolites are actually obtained, not the acid form. To convert them into the active form, they

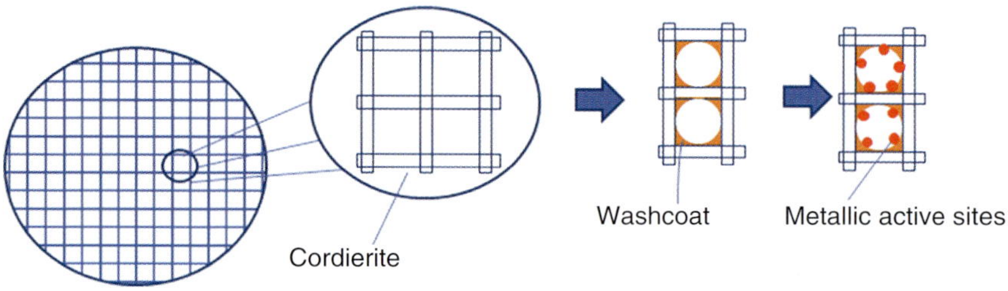

Figure 8.5 Catalyst structure and active components.

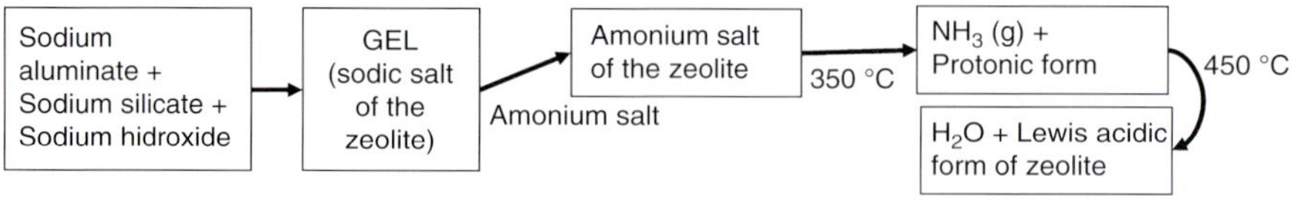

Figure 8.6 Scheme of the production of zeolites.

must be treated with ammonium salts, so that the ammonium salt of the zeolite is produced, ending with a 350 °C heating that releases the ammonia and leaves the proton form of the zeolite (Brønsted acid). Subsequent heating to 450 °C will liberate the water from the structure forming the Lewis acid sites.

8.2.1.2 Manufacturing of the Catalytic Support

The conventional supports of the catalysts for refining of petroleum have as raw material the aluminum oxide, Al_2O_3, usually in the γ type. A plant for preparing alumina supports has the following basic parts:

(1) *Wet mixer*: In this unit, the mix of the alumina powder with water is done, and other ligands, with the aim of getting a mixture with some particular rheological properties. It is obvious that one of the main steps to obtain a good catalytic support is to get a good mixture in this step.
(2) *Extruder*: The mix, when reaching the desired rheological properties, is dosed to an extruder. This equipment can be assembled with a double or a simple spindle, and dispose of a system for evacuating the air that can be occluded in the mix. Also, the extruder will have a system for heating or cooling the mix, in order to regulate the working temperature. Also, the extruder head should be modified in accordance with the desired physical form of the final product. The geometry of the support is of special importance, modulating the surface-to-volume ratio.
(3) *Pre-drying oven*: Usually, the extruded product exits with a high humidity, and it is necessary to eliminate it before the calcination step.
(4) *Calcination oven*: Composed of a conveyer belt with programmable velocity, in order to vary the time the support is in the range of temperature 300–800 °C. Heat is obtained by infrared resistances or with burners that use natural gas as fuel.
(5) *Sifting machine*: The product leaves the calcination oven with different particle sizes, and in the sifting machine they are separated into three parts (fines, thick material, and the desired product).

8.2.1.3 Impregnation with Active Metals

Active metals for catalytic activity are obtained from different salts that are dispersed in the surface of the supports. From the metals most used in the manufacturing of catalysts, the following can be highlighted.

Nickel The industrial manufacture of catalysts that have nickel as active element generally uses nickel nitrate ($Ni(NO_3)_2$) as mineral salt, because it presents an excellent solubility in water and has a very low cost. Contrarily, if this salt is used, a system to eliminate NOx will be needed, and also a loss of activity as this is not the salt giving the best activity.

Another possibility is to use nickel formate ($C_2H_2NiO_4$), which presents some advantages such as higher activity and that a system for eliminating contamination will not be necessary. Nevertheless, the salt has low solubility and implies use of ammonia solutions.

Molybdenum The salt that is used in most industrial preparations is ammonium heptamolybdate ($(NH_4)_6Mo_7O_{24}$), which is soluble in water, although it can be expensive.

Cobalt It is similar to nickel, that is, the main salt used is cobalt nitrate ($Co(NiO_3)_2$), but cobalt acetate ($Co(CH_3CO_2)_2$) can also be used. Currently, the drawback of these salts is the high cost.

8.2.2 Key Definitions in Catalysts Performance

It is important to be familiar with some characteristics of a catalyst, related to the ability to increase the rate of a reaction. First of all, the activity of a catalyst refers to its ability to accelerate the overall conversion of a reactant to products. Mathematically, activity can be considered as the quotient between the reaction rate in a precise moment and the reaction rate with a fresh catalyst (usually, $t = 0$):

$$\boldsymbol{a} = \frac{(-r_A)_t}{(-r_A)_{t=0}} \tag{8.1a}$$

In the studies arising from the deactivation of a catalyst, i.e. its loss of activity, the reaction rate at any moment is given by

$$(-r_A)_t = (-r_A)_{t=0} \cdot \boldsymbol{a} \tag{8.1b}$$

while the loss of activity is modeled through equations of the type:

$$\frac{d\boldsymbol{a}}{dt} = f(C, \boldsymbol{a}) \tag{8.2}$$

where C is be the concentration of the species producing the deactivation.

On the other hand, the residence time in a catalytic reactor can be calculated as the volume of the catalyst bed divided by the reagent flow rate, that is

$$\text{Residence time} = \frac{\text{catalyst volume}}{(\text{volume reagents/time})} \tag{8.3}$$

While this concept is equivalent to the reaction time in a batch reactor, it does not give a measure of the ability of the catalyst to convert reactants into products. For example, if we have 100 mL of a catalyst A capable of converting 100% ethane flowing at 100 ml/min, it will have the same residence time as a catalyst B that can convert 100% of a current that enters only 10% ethane in the catalyst bed. However, catalyst B is only able to convert 1/10 of the amount that catalyst A converts.

A better measure of the conversion of reactants in a particular catalyzed reaction is the "space velocity." In that concept, the conversion is measured per unit time and catalyst, so it indicates the rate at which a given catalyst is able to convert reactants into products per unit of time. This spatial velocity can be expressed with a volumetric base (which could be "liquid hourly space velocity," LHSV, or "gaseous hourly space velocity," [GHSV] that will be given in vol/vol-h) or in a mass basis (weight/weight-h, WWH). Both parameters are usually expressed in inverse units of time.

Table 8.3 Examples of space velocity.

Process	WWH or LHSV (h^{-1})	GHSV (h^{-1})
Chemicals products manufacturing	0.01–1	10–1 000
Refinery	1–10	1 000–10 000
Exhaust gas elimination	10–100	10 000–100 000

Table 8.3 shows typical spatial velocity values for the three most important types of catalytic applications, which are the production of chemical products, operation in refinery, and exhaust gas conversion.

For instance, a typical catalytic reactor in a chemical plant may contain 500 000 kg of catalyst, which has a capacity of 500 000 kg per day. This represents a weight/weight-h velocity of

$$\text{WWH} = 500\,000\ \text{kg product}/(500\,000\ \text{kg cat} - \text{day}) = 1\ \text{day}^{-1} = 0.04\ \text{h}^{-1}$$

This mass of catalyst may be also necessary to convert 100 000 barrels ($1.4 \cdot 10^8$ kg) of oil into gasoline and diesel oil in a refinery, which will be equivalent to a space velocity of $1.2\ \text{h}^{-1}$.

The catalysts used in the exhaust gas cleaning typically fill a volume of 5 l and need to convert flow rates of approximately 250 000 l/h. These amounts results in a required gas hourly space velocity of $\text{GHSV} = 50\,000\ \text{h}^{-1}$

Another usual way to measure the ability of a catalyst is the "turnover rate." The abovementioned parameters do not give an idea of how a particular catalyst is functioning on a molecular scale. This type of information could be needed to understand if a catalyst is working correctly from the chemical point of view. It is also helpful to compare the performance of different catalysts of diverse physical properties with different structures.

This type of measurement is carried out with the so-called turnover rate or the turnover number, which is the number of catalytic cycles per active site of the catalyst and per unit of time:

$$\begin{aligned}\text{Turnover rate (TR)} &= \frac{\text{moles of product}}{[(\text{moles of active sites}) \cdot (\text{time})]} \\ &= \frac{\frac{\text{grams of reactive}}{\text{MW reactive}}}{\left[(\text{g cat}) \cdot \left(\frac{\text{sites}}{\text{g cat}}\right) \cdot \text{h}\right]} \\ &= \frac{\frac{\text{grams of reactive}}{(\text{g cat})\cdot \text{h}}}{\left[\text{MW reactive} \cdot \left(\frac{\text{sites}}{\text{g cat}}\right)\right]} = \frac{\text{WWH}}{\text{MW} \cdot \text{SD}}\end{aligned} \tag{8.4}$$

where SD is the density of sites, which is measured by spectroscopic or chemical methods.

TR is a valid definition of reaction rate. Indeed, it is the most fundamental one, as it reflects the frequency at which molecules are reacting on the catalyst. Turnover rates, as other reaction rates, must be defined at a definite temperature and concentration of reactants. To be acceptable, TRs should be measured under conditions where only chemical effects are observed, i.e. in the absence of heat and mass transport limitations. This, as will be seen in Chapter 9, is generally got at low temperature and low conversion.

TRs have units of molecules/site-second or s^{-1}. Values of TRs for commercial catalysts are typically in the range of about 10^{-3}–10^{1} s^{-1}, while TR values for enzymes are in the range of 10^{2}–10^{4} s^{-1}.

8.3 Main Catalytic Processes in Industry

Some important processes catalyzed in industry are summarized in Table 8.4, with special interest in the oxidation/reduction reactions. In this section, we batch them into four groups of reactions.

Table 8.4 Important industrial catalyzed reactions with the corresponding catalyst.

Reactants	Main products/coproducts	Type of catalyst	Type of reaction[a)]
Ammonia/air	NO	Pt–Rh gauze	SO
Ammonia/air	N_2	Pd-V_2O_5-WO_3/ TiO_2–SiO_2	SO
H_2S/air	S_8	Fe_2O_3/SiO_2	SO
Methane/O_2/NH_3	HCN	Pt–Rh gauze	SO
CH_4 or $(CH_2)x/O_2$	Syn gas (CO/H_2)	Supported Rh or Ni	SO
Methanol/air	Formaldehyde	Ag, Fe–Mo–oxide	SO
Ethylene/O_2	Ethylene oxide	Ag(K,CI)/a-AI_2O_3	SO
Ethylene/air or O_2/HCl	1,2-Dichoroethane	Cu/CI/O/K-yAl_2O_3	SO
Ethanol/O_2	Acetaldehyde	Ag, Cu	SO
Ethylene/acetic acid/air	Vinyl acetate	Pd/Au/K-aAI_2O_3	SO
Ethylene/O_2	Acetaldehyde	Pd/CI/acetate supported	SO
Propene/air	Acrolein	Bi/Mo/Fe/Co/ K-oxide	SO
Propene/air/NH_3	Acrylonitrile	Bi/Mo/Fe/Co/ K-oxide	SO
Propene/ hydroperoxide	Epoxide/alcohol	Mo-complexes or silica supported Ti	SO
Acrolein/air	Acrylic acid	V/Mo/W-oxide	SO
n-Butane/air	Maleic anhydride	V-P-oxide	SO
n-Butane/O_2 or air	Acetic acid	Co, Mn salts	SO
n-Butane/air	Butenes/butadiene	Bi-Mo-P-oxide	SO
F-butyl alcohol	Methacrolein	Bi-Mo-Fe-Co- K-oxide	SO

Table 8.4 (Continued)

Reactants	Main products/coproducts	Type of catalyst	Type of reaction[a)]
Isobutene/air	Methacrolein	Bi-Mo-Fe-Co-K-oxide	SO
Methacrolein/air	Methacrylic acid	V/Mo/W-oxide	SO
Cyclohexane/air	Cyclohexanone	Co salts	SO
Cyclohexanone/ HNO_3	Adipic acid	Co salts	SO
Cyclohexanone/ H_2O_2/NH_3	Cyclohexanone oxime	Ti-silicalite	SO
Benzene/air	Maleic anhydride	V-Mo-oxide	SO
o-Xylene/air	Phthalic anhydride	V-W-P-Cs-TiO_2	SO
p-Xylene/air	Terephthalic acid	Co/Mn/Br	SO
Cumene/O_2 or air	Phenol/acetone	Co-Mn salts	SO
Naphthalene/air	Phthalic anhydride	V-K-oxide/SiO_2	SO
Alkanes	Alkenes	Cr/Al_2O_3, Cr/ZrO_2, Pt/Sn/Al_2O_3, Mo/Al_2O_3	DH
n-Butane	Isobutene	Cr/AI_2O_3, Pl/Sn/Al_2O_3, alkali promoted Pt/Sn/Al_2O_3	DH
Ethylbenzene	Styrene	Fe_2O,/K_2O/Cr_2O_3	DH
Cyclohexanes	Benzenes	Ag, Cu gauze, Pt/C, Pd/C	DH
Alcohols/ aldehydes	Ketones	Cu, Cu chromite. Cu/K/SiO_2	DH
Cyclohexanol	Cyclohexanone	Cu-ZnO/AI_2O_j, Cu/MgO, Cu/K/SiO_2, Ru/C	DH
Acids	Alcohols	Re, Ru	H
Aliphatic aldehydes	Alcohols	Ru	H
Aromatic aldehydes	Alcohols	Pd	H
Aromatic aldehydes	Hydrocarbons	Pd	H
Unsaturated aldehydes	Unsaturated alcohols	Os, Re, Pt	H
Mono-alkenes	Alkanes	Pd	H
Alkadienes	Alkenes	Pd	H
Cyclohexenes	Cyclohexanes	Pt, Rh	H
Cyclohexenes	Cyclohexane + benzene	Ru, Rh	H

(Continued)

Table 8.4 (Continued)

Reactants	Main products/coproducts	Type of catalyst	Type of reaction[a)]
Alkynes	*cis*-Alkenes	Pd	H
Alkynes	Alkanes	Pd	H
Anilines	Cyclohexylamines	Rh, Pd	H
Anilines	Cyclohexanones	Pd	H
Epoxides	Alcohols	Pd	H
Hydrazones	Hydrazines	Pt	H
Imines	Amines	Pt	H
Aliphatic ketones	Alcohols	Ru	H
Aromatic ketones	Aromatic alcohols	Pd	H
Benzyl compounds	Aromatic hydrocarbons	Pd	H
Benzyl compounds	Cyclohexyl derivatives	Rh, Ru	H
Aliphatic nitriles	Primary amines	Pd, Pt, Rh	H
Aliphatic nitriles	Secondary amines	Rh (solvent)	H
Aliphatic nitriles	Tertiary amines	Pd, Pt (solvent)	H
Aromatic nitriles	Benzylamines	Pd	H
Aromatic nitriles	Dibenzylamines	Pt	H
Aromatic nitriles	Aldehydes	Pd	H
Nitriles	Secondary amines	Rh	H
Nitriles	Tertiary amines	Pd, Rh	H
Phenols	Cyclohexanones	Pd	H
Phenols	Cyclohexanols	Rh, Ru	H
Phenols	Cyclohexanes	Pt, Ir	H

a) Type of reactions: SO, Selective oxidation; DH, dehydrogenation; H, hydrogenation.

8.3.1 Acid Catalysis

In the classic Brønsted definition, acids are sources of protons, but acids can be also contemplated as substances accepting electrons (in the definition of Lewis acids), i.e. acids are electron-poor materials. The elements of the periodic table used in acid catalysis are those showing acid characteristics, as is the case of hydrogen (in the form of H^+), transition metals with high oxidation states (Si, Al) , and elements of the upper right part of the periodic table. Also, halogens are used due to their very high electronegativity and the high strength of the conjugate acid. The elements that are the most used in industrial acid catalysts are H and Si/Al/O such as silica, alumina, and a wide variety of materials containing these compounds in the form of zeolites or clays.

Brønsted and Lewis acid catalysts are used. Protonic acids (Brønsted) react with substances by donation of a proton, while Lewis acids do not possess protons, but can accept electrons from the reactants. Commonly, the Brønsted acids can be converted into the corresponding Lewis acid by loss of a molecule of water, and the hydration reaction will produce the Lewis acid, in reactions of the type (usually reversibly):

$$\text{Lewis acid} \underset{-H_2O}{\overset{+H_2O}{\rightleftarrows}} \text{Brønsted acid}$$

Some of the most important industrial processes related to the acid catalyst are those related to the refinery plants. In these, the main processes in which the acid catalysts intervene are the following:

- Catalytic cracking (FCC);
- Disproportionation of aromatics (reaction in which multi-alkylated aromatics are generated without substitutions, which is used to increase or decrease the octane level of gasoline);
- Alkylation of isobutane/isobutylene in which a C_8 molecule is produced from two C_4, and super acids are required as catalysts;
- Synthesis of methyl *tert*-butyl ether (MTBE);
- Methanol-to-gasoline (MTG) process, in which a zeolite-type ZSM5 catalyzes the formation of $(CH_2)_n$ and H_2O starting from methanol.

Acid catalysis is also present in the chemical industry, in general. Some of the production processes of products in the chemical industry, which require acid catalysts, are the following:

- Benzene ethylation, producing ethylbenzene;
- Production of cumene by reaction of benzene and propylene;
- Isomerization Beckman producing caprolactam by reacting a cyclohexanone oxime with a strong acid;
- Esterification, forming ethyl acetate mainly.

8.3.1.1 Fluid Catalytic Cracking

Let us see in more detail the FCC process. During the FCC, the heavier fractions of petrol are converted into products of lower boiling point and higher added value. The cracking reaction takes place in a reactor ("riser") at 500 °C approximately and a residence time of two to four seconds. The gasoil is preheated and contacts the catalyst coming from the regenerator (see Figure 8.7) in the bottom part of the riser. The products and the catalyst are separated in the unit called "stripper"; the deactivated catalyst is then passed to the regenerator. In that equipment, the combustion of the coke deposited in the surface of the catalyst is produced. This process is done at 700 °C approximately in the presence of air and steam. Later, this catalyst is turned back to the riser, continuing the cycle.

The catalyst used in the FCC process has a quite complicated formulation. Among others, the catalyst contains alumina, clays, and silica in the support, and platinum, antimony, and sodium in the active phase, although the main component is a zeolite. Each of these active points intervenes in different reactions. For example, zeolites are active in the cracking of heavy fractions, and platinum is

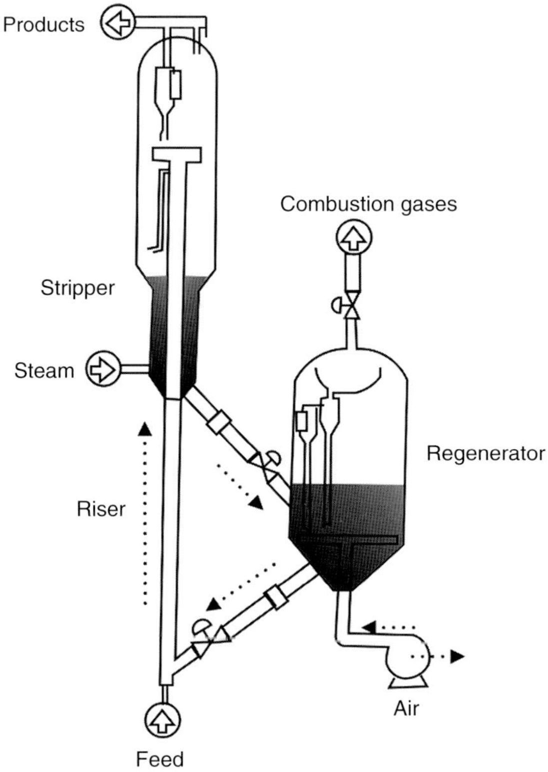

Figure 8.7 Scheme of the FCC process.

active in the combustion of the coke. Zeolite is the component determining the selectivity of the reaction to the different products.

The present FCC catalysts incorporate zeolite of the type Y as the main active component. In this, not all the active centers have the same acid strength, and there exists a distribution depending on the amount of Al in the net, i.e. on the ratio Si/Al.

Due to the high temperatures to which the catalyst is subjected, especially in the regenerator and in the presence of steam, the zeolite in the fresh catalyst suffers a de-aluminization process, whereby Al is extracted from the net and is deposited on the surface of the zeolite.

Another characteristic of zeolite Y affecting its catalytic behavior is the size of the particles. In fact, the smaller the particle size, the higher the amount of acid sites in the surface and they are accessible to gasoil molecules. Also, the diffusion of the cracking products is favored with small particle sizes, and the possibility of producing secondary reactions is reduced.

Zeolite ZSM-5 has also been used in the FCC process. This is a zeolite with smaller channels than type Y, and can be useful to obtain products of low molecular weight (gas fraction). Also, ZSM-5 is more resistant to the deactivation by coke. ZSM-5 is usually mixed in a low proportion (2–5%) with Y-zeolite, thus optimizing the process.

8.3.1.2 Ethylbenzene Production

Ethylbenzene ($C_6H_5CH_2CH_3$) is an alkyl aromatic compound with a single ring. It is also known as phenyl ethane and ethylbenzol. This compound is mainly (>90%) used as an intermediate for the manufacture of styrene ($C_6H_5CH{=}CH_2$). It is one of the most important large-volume chemicals and is the monomer of polystyrene.

The main method of obtaining ethylbenzene is from the alkylation of benzene and ethylene. The technology and the advances in catalysts have allowed developing a method respectful of the environment and with safe conditions. It is based on an alkylation reactor, where the reaction takes place, a transalkylation reactor, to recover the aromatic compounds with more than one ethyl radical, and a purification unit to separate the desired product (Figure 8.8).

The alkylation unit consists of the supply lines of the two necessary reagents and the auxiliary equipment needed. The main element of this zone is a multitubular reactor, which is where the reaction takes place.

Benzene enters via line 1 (Figure 8.8) and pump B-1 compresses it to the reactor pressure. Exchangers E-1 and E-2 heat it up to a temperature of 235 °C. Line 2 is that of ethylene and works in a similar way. Compressor C-1 brings ethylene up to 40 bars and exchanger E-1 decreases its temperature until it is the same as that of benzene to enter the reactor. Lines 9 and 10 run through the cooling water of the multitube reactor to control the temperature.

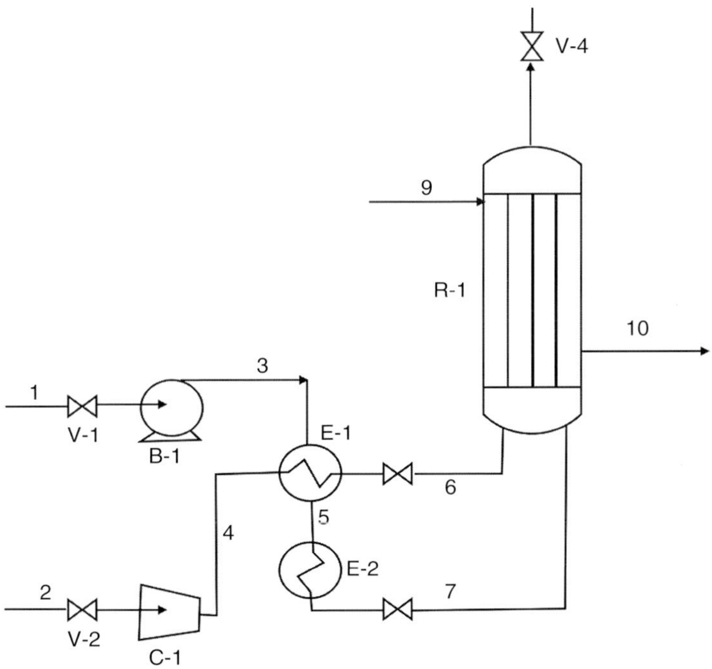

Figure 8.8 Flowchart of the ethylbenzene production process. V-1, V-2, and V-4 – valves; C-1 – compressor; B-1 – pump; E-1 and E-2 – heat exchangers; R-1 – reactor.

The most important factor for modeling the reactor is the kinetics of the reactions that occur in it. Two reactions occur, the main reaction, catalyzed by zeolite, and the secondary reaction, where an undesired product is produced:

$$C_6H_6 + C_2H_4 \xrightarrow{\text{Zeolite Y}} C_6H_5C_2H_5$$

$$C_6H_5C_2H_5 + C_2H_4 \rightarrow C_6H_4(C_2H_5)_2$$

This reaction is carried out at 190 °C in the presence of a zeolite catalyst.

8.3.2 Oxidation Catalysis

The formation of partial oxidation products in a preferential manner was not commercialized until selective oxidation was done in a realistic manner, in the late 1960s, when appropriate catalysts were discovered. These catalyst are constituted by elements that should present a cycle between several oxidation states, offering a path of low activation energy to form partial oxidation products. In this way, the elements present in the oxidation catalysts must present multiple oxidation states, among which the transition metals stand out, and of course the oxygen. The most important will be V, Mn, Cr, Co, Fe, Ni, Nb, Cu, Mo, Ru, Ru, Pd, Rh, Ag (transition metals) and O, Sb, S, Te, Sn, Bi as other active elements.

The design of a catalyst for selective oxidation is based on many parameters. Among them, the selection of the metal that will be part of the active center is obviously one of the most important. The isolation of active sites is another key concept in the design of the catalyst. A strong oxidant such as Cu_2O_3 is very active, but not very selective for the oxidation of olefins. Nevertheless, the selectivity can be increased if some part of the copper oxide is reduced using a reductant such as H_2. Also to be borne in mind is that the surface of an oxidation catalyst is usually chemically different from the internal zone.

Table 8.4 shows the most important processes in this industrial oxidation catalysis. We describe in detail two of these processes.

8.3.2.1 Ethylene Oxide Production from Ethylene

Ethylene oxide (C_2H_4O) is an organic compound that belongs to the family of cyclic ethers. At room temperature, ethylene oxide is a flammable, toxic, pale gas that readily dissolves in water.

This compound is an important intermediate in obtaining ethylene glycol, this being its main use. A small amount (less than 1% of production) is used in hospitals for sterilization of medical equipment and supplies that cannot be sterilized by thermal processes. The world production of ethylene oxide accounted to approximately 20 million tons in the past years.

For a long time, the only method to obtain ethylene oxide was the so-called Würtz synthesis, which consisted of a two-stage process (with ethylene chlorohydrin as an intermediate product). In 1931, the French chemist Theodore Lefort managed to develop a silver catalyst able to directly oxidize the ethylene. This process was much cheaper on an industrial scale than previously used; therefore, from 1975 onwards, all the production of ethylene oxide in the United States was exclusively by direct oxidation of ethylene.

The main causes that motivated the boom of the direct oxidation process were the large number of chemical products required by the traditional method of obtaining it, as well as the high cost of chlorine, which had a marked influence on the total cost. Also, the considerable contamination of the water could be a problem.

The direct oxidation process is carried out using oxygen or air on a silver catalyst, as already mentioned. This reaction takes place in the gas phase, unlike the traditional chlorohydrin process that was carried out in the liquid phase. The main reactions that take place in obtaining ethylene oxide by direct oxidation are shown here:

$$O_2 + 2C_2H_4 \rightarrow 2C_2H_4O$$
$$O_2 + 1/3\,C_2H_4 \rightarrow 2/3\,CO_2 + 2/3\,H_2O$$

The process is usually carried out in fluidized-bed reactors. The catalyst used in the reactor contains a wt. 5% of silver, using a support composed of alumina and silica (8% by mass of Al_2O_3 and 92% of SiO_2). Spherical particles of very small diameter are used, of a density close to 1800 kg/m^3. The efficiency factor of such a catalyst under the conditions of the reaction is practically unity. The process is done at 550 K and 10 atm. of pressure, and the input mix to the reactor must have less than 6 vol% of oxygen, to avoid the total combustion of the ethylene. Under these conditions, the total conversion of the ethylene is close to 0.40, so it is convenient to improve the global conversion by recirculating the unreacted ethylene. For this reason, a unit for the separation is needed, as shown in Figure 8.9. On the other hand, as the gas velocity is usually high, the catalyst particles are dragged and a cyclone at the exit separates the solid from the gas, returning the valuable catalyst to the bed.

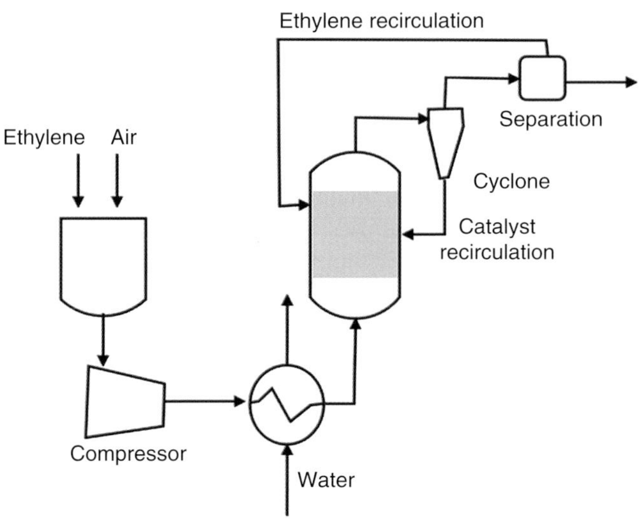

Figure 8.9 Ethylene recirculation diagram.

8.3.2.2 Acrylic Acid from Acrolein

Acrylic acid production in the industrial scale has been studied for many years. Different processes have been tested in the past, involving the use of acetylene (HC=CH) or ketene (CH_2=C=O) as reactants. Also, the hydrolysis of acrylonitrile was introduced as a valid method, but usually it occurs as an excessive polymerization, producing big amounts of ammonium bisulfate as by-product, causing a disposal problem.

A promising route to produce acrylic acid is the partial oxidation of propylene in vapor phase. It is generally carried out in a two-step process, implicating acrolein as an intermediate. The general reaction scheme is the following one:

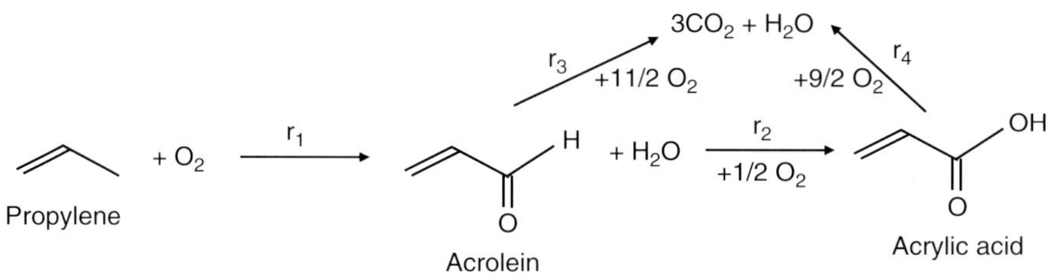

This is a complex reaction scheme, in which there is a precursor, acrolein, formed from the reaction R1, and by-products of carbon oxide and water by the combustion of both acrolein and acrylic acid. Acetic acid, formaldehyde, and acetaldehyde are also produced in the process, although the amount produced depends to a large extent on the conditions under which the reaction is carried out. Both the oxidation of propylene to acrolein and the oxidation of this intermediate to acrylic acid are partial oxidation reactions, carried out in catalytic reactors.

The second reaction of the previous scheme, the production of acrylic acid by oxidation of acrolein, is usually carried out using a catalyst containing antimony, nickel, and molybdenum in the form of oxides.

The commercial applications of acrylic acid and its esters are numerous. For example, they are used as binders for paints, and as adhesives. The process of production of paper, leather, and textiles also has a final treatment with these chemicals.

The temperature of the process is usually in the range 285–360 °C, with a clear increase of the conversion (0.15–0.45) with temperature. Nevertheless, the selectivity of the reaction decreases as temperature increases, so a low-medium temperature is adopted. At 285 °C, the conversion is 0.2 approx. for a space time of two seconds, and the selectivity is increased at low catalyst particle size, with an optimum of 92% of selectivity with a particle diameter of 0.8 mm.

8.3.3 Reduction Catalysis

Fuels and other chemicals are produced using catalysts that operate in a reducing atmosphere. Hydrogen, the simplest and cheapest reducing species, is included as a reactant or product in these reactions.

Table 8.5 Catalytic processes in the syngas production.

Reaction	Catalyst used
Reforming with methane vapor: conversion of CH_4 and H_2O to CO/H_2	NiO/SiO_2
Methanation: conversion of CO/H_2 to methane	NiO/SiO_2
Methanol synthesis: conversion of CO/H_2 to methanol	$Zn/Co/SiO_2$
Methanol decomposition: conversion of CH_3OH to CO/H_2	Co/SiO_2
Fischer–Tropsch reaction: conversion of CO/H_2 to hydrocarbons in the range of gasolines	Co/SiO_2 Fe/SiO_2
Steam reforming of hydrocarbons: conversion of hydrocarbons and water into CO/H_2	Co/SiO_2
Oxygenated synthesis: conversion of CO/H_2 into oxygenated hydrocarbons of higher molecular weight	Rh/SiO_2

The catalysts employed in hydrogenation catalysis or reduction are typically transition metals in their low oxidation states. Usually, noble metals such as platinum, palladium, and rhodium are used. Also, basic metals (Fe, Co, Ni, Zn) are active in some cases. In some processes, water is used as reactant and the H of this molecule is used as reducing agent.

The synthesis gas, a mixture of H_2 and CO, is an important raw material for the synthesis of numerous chemicals and fuels. Some of the products are as important as methanol, methane, higher hydrocarbons, and oxygenates. The synthesis gas can be obtained by coal gasification, and then represents a source of chemical products apart from petroleum.

The proportion of CO in the synthesis gas can be modified by converting it to H_2 through the water-shift reaction, so that the reactions involved in the synthesis gas would be those indicated in Table 8.5, together with the most commonly used catalysts. Some of them are commented on in detail.

8.3.3.1 Steam Reforming of Alcohols

The complete process for energy production using a biomass source may be divided into four stages, namely: bioethanol production, hydrogen production, hydrogen purification, and power generation. Usually, these stages are treated separately, and scarce information is available on the process integration. Obviously, the first stage (Figure 8.10) depends largely on the type of biomass used and also on the purification steps. The main products are ethanol and water. In the second stage (H_2 production from bioethanol), usually the steam reforming of ethanol (SRE) is run in order to increase the hydrogen yield.

The SRE process is an endothermic process, mainly represented by the reaction:

$$CH_3CH_2OH + 3H_2O\ (+\ \text{heat}) \rightarrow 2CO_2 + 6H_2$$

This equation actually represents a very wide network of reactions that will depend on the catalyst used, and also obviously on the reaction conditions. The final product is a complex mix of by-products. Among them, the main are carbon

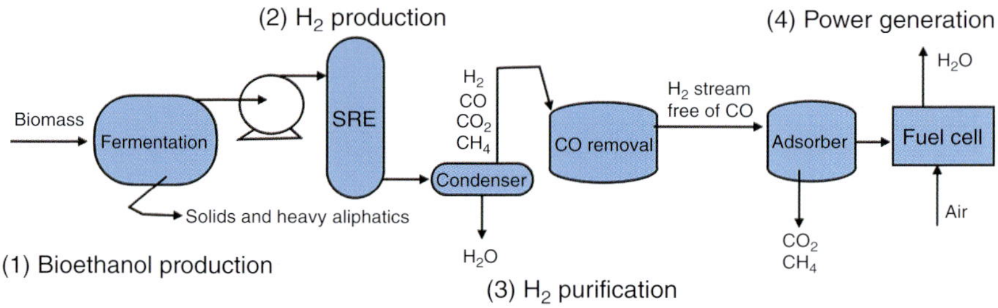

Figure 8.10 Complete process for energy production from biomass.

monoxide, ethane, methane, and ethylene, and, in some cases, acetaldehyde (CH_3CHO).

One of the most important reactions being produced in this process is the water–gas shift reaction (WGSR):

$$CO + H_2O \rightleftharpoons CO_2 + H_2 \text{ (+ small amount of heat)}$$

This WGSR favors CO_2 formation with increasing H_2 and decreasing CO concentration in the outlet stream.

In the power generation stage (Figure 8.10), fuel cells (FCs) are usually used, most of them demanding a very low (<100 ppm) CO concentration. If this is the case, the by-product formation should be minimized, and a previous H_2 purification stage is usually needed. Actually, CO removal from post-reformed streams is currently one of the biggest problems to connect H_2-producing systems to commercially available FCs devices.

SRE catalysts are usually constituted by a support where base metals such as Ni, Cu, Al, Zn, and Co, and also noble metals such as Pt, Rh, Pd, Ru, and Ir are dispersed. The ethanol decomposition route depends on the metal present, but the catalyst support can also influence the reaction. Furthermore, a low metal loading is usually used (typically, less than 1%), due to the high costs. This promotes the role of the support on the reaction mechanism. A support avoiding a high oxygen storage is usually more effective, properties that are conferred by supports that can be reduced, such as CeO_2, ZrO_2, and La_2O_3.

Conventionally, CO removal by conversion through WGSR produces an outlet CO content close to 2000–5000 ppm, too high for FC use. A second stage may be needed, which ensures a low CO (<100 ppm).

For reducing the amount of CO in the mix, two different processes are common: (i) preferential oxidation of CO (CO-PROX, Eq. (8.5)) and (ii) CO methanation (Eq. (8.6)):

$$2CO + O_2 \rightarrow 2CO_2 \quad \text{Preferential oxidation of CO, CO} - \text{PROX} \tag{8.5}$$

$$CO + 3H_2 \rightarrow CH_4 + H_2O \quad \text{CO methanation} \tag{8.6}$$

Nevertheless, high H_2 loss is common during the final stage because of the presence of secondary reactions such as

$$CO_2 + 4H_2 \rightarrow CH_4 + 2H_2O \quad CO_2 \text{ methanation} \tag{8.7}$$

$$2H_2 + O_2 \rightarrow 2H_2O \quad H_2 \text{ oxidation} \tag{8.8}$$

8.3.3.2 Steam Reforming of Hydrocarbons (Methane)

Actually, bioethanol reforming is not the main source of hydrogen. Nowadays, more than 95% of the hydrogen produced in the world is produced by reforming of natural gas (methane) in large central plants. In the industrial H_2 production, natural gas containing methane (CH_4) is used as a precursor of the hydrogen, by treating it with steam at high temperature (700–1000 °C). Methane (and other hydrocarbons) reacts under high pressure in the presence of a catalyst to produce the hydrogen and CO in the WGSR.

A final process can be included in the reaction network, called "pressure-swing adsorption," where CO_2 and other by-products are removed from the gas stream, producing hydrogen as pure as possible. It is also possible to obtain a CO/H_2 mix by partial oxidation of hydrocarbons, with a limited amount of oxygen. This is an exothermic process that is usually much faster than WGSR, but produces less hydrogen per unit of the input fuel than is obtained by steam reforming of the same fuel.

8.3.3.3 Methanation: CO/H_2 to Methane

Methanation is the chemical process where carbon monoxide and/or carbon dioxide are converted to methane. The reaction is the opposite to that of the WGSR, and is based on the following exothermic reaction:

$$CO_2 + 4H_2 \rightarrow CH_4 + 2H_2O \quad \Delta H < 0$$

Carbon monoxide conversion to methane is also possible (CO methanation), by

$$CO + 3H_2 \leftrightarrow CH_4 + H_2O \quad \Delta H < 0$$

Generally, CO_2 is used for methane production, and few plants use CO instead. The former reactions are exothermic, so they are favored at low temperatures. The temperature control (efficient heat removal) is one of the greatest challenges involved in methanation, closely linked to reactor design. Also, the equilibrium is displaced to the products under conditions of high pressures.

The simplest option for a catalytic reactor is probably the adiabatic fixed bed. The reactor consists of a bed of catalytic pellets, and its temperature is increased along the reactor due to the heat production. The reaction rate is quite high and a low amount of catalyst is usually needed. As temperature is increased, the CO_2 conversion decreases, and the catalyst can be deactivated via sintering. For that reason, the reactor feed gas is usually diluted, to limit in any way the temperature increment.

Nickel-based catalysts are considered the best option for methanation. Nickel oxide (NiO) exhibits a very high activity and selectivity for producing methane. This is mainly due to the ability of NiO to undergo a reduction process. Another reason for the high activity of NiO is that it usually presents a bimodal pore structure, i.e. two sizes of pores, which is optimum for the adsorption of both reactants (CO_2 and H_2). Other more expensive metals such as Rh, Ru, Pd, and Pt also present high activity in the methanation of CO or CO_2, but usually their high prices do not make them competitive.

Nevertheless, it has been proved that the addition of Pd and Pt considerably enhanced the behavior of Ni/Mg(Al)O catalysts, allowing decreasing the temperature and increasing the hydrogen consumption.

Also, the selection of support is important. As already commented on, it may influence both the activity and selectivity of the reactions. Particularly important is the fact that the addition of alumina eliminates the interference that introduces the presence of H_2S, thus increasing methanation activity. Therefore, Al_2O_3 is generally an inexpensive and convenient support.

8.3.4 Environmental Catalysis

The reduction of environmental pollutants is one of the most important objectives achieved by the chemistry of catalysis. The United States, Europe, and Japan have required in the past years a drastic reduction of pollutants emission. Especially important are the depletion of emissions of particulates (mainly carbonaceous particulates), NOx, hydrocarbons, and CO. This applies to both stationary and mobile sources, but the most important problems are in nonstationary sources (vehicles), which will be treated with greater importance.

For example, the simultaneous reduction of carbon and NOx in diesel combustion gas was a particularly complicated technical problem since the required reduction levels were more than 90% compared to historical levels. Almost all the techniques that produce this reduction call for the presence of catalysts.

Catalysts for the conversion of exhaust gases fall into one of these four categories:

Noble metals	Basic metals	Alkaline earth metals	Rare earth metals
Pt, Pd, Rh (alumina supported)	V, Fe, Co, Ni, Cu (silica supported)	Mg, Ca, Sr, Ba	La, Ce

The composition of the exhaust gases varies widely depending on the conditions of the engine and its type. Table 8.6 shows a set of typical values for gasoline and diesel engines.

Table 8.6 Typical composition of the exhaust gases of diesel and gasoline engines.

Component	Gasoline	Diesel
NOx	1 050 ppm	270 ppm
Soot/PM	60 mg/km	1 300 mg/km
CO	6 800 ppm	350 ppm
Hydrocarbons (HC)	750 ppm	350 ppm
H_2	2 300 ppm	120 ppm
O_2	5 100 ppm	60 000 ppm

The main reason for these differences is that typical temperatures are much higher in gasoline engines. This produces a high emission of NOx (that is a product of excessive combustion) and CO, but less soot (carbonaceous material, the main component of particulate matter), consistent with the higher level of combustion. The much greater presence of O_2 in the exhaust of a diesel engine makes the simultaneous elimination of coal, CO, HC, and NOx very complicated. In addition, the high carbon content causes the catalysts to become poisoned, which will require treatment.

The gases could be divided into oxidants and reducers. In the first category would be O_2, NOx, and SOx, while the reducers would be HC, carbon, CO, and H_2 (that is presented usually in very small proportion). In addition, ammonia (NH_3) produced by decomposition of urea ($CO(NH_2)_2$) can be added to the exhaust gases as a reducing agent. This will permit elimination of NOx in a process called "selective catalytic reduction," or SCR.

8.3.4.1 Catalytic Reactions for the Removal of Pollutants in the Exhaust Gases

NOx reduction can occur through various chemical processes:

$$2NO + 2CO \rightarrow N_2 + 2CO_2 \quad (3 - \text{way catalyst})$$

$$C_mH_n + (2m + n/2)\,NO \rightarrow (m + n/4)\,N_2 + mCO_2 + n/H_2O \quad (\text{lean NOx catalyst})$$

$$C_mH_n + NOx + O_2 \rightarrow N_2 + CO_2 + H_2O \quad (\text{lean NOx catalyst} - \text{oxygen rich})$$

$$NO + NH_3 \rightarrow N_2 + H_2O \quad (\text{selective catalytic reduction})$$

In the three-way catalysts used in gasoline engines, the first reaction is involved, NO being removed by the presence CO at low concentrations of O_2, producing N_2 and CO_2. Hydrocarbons (HCs) can also react with NOx to form N_2, in the lean NOx catalyst that is described later. Also, in the presence of O_2, some HCs will be oxidized to form extra CO_2 and water. In this case, all HCs may be eliminated and do not react with NO. Finally, urea can be added to produce NH_3, which reduces NOx by SCR. Since most NOx (>90% in diesel) are present as NO (with 10% NO_2), the main reaction will be that shown earlier.

On the other hand, the main process for the elimination of CO will be the oxidation to CO_2, which occurs in the three-way catalysts in gasoline engines, and also in a "diesel oxidation catalyst" (DOC):

$$2CO + O_2 \rightarrow 2CO_2 \quad (3 - \text{way catalyst, DOC})$$

$$CO + H_2O \rightarrow H_2 + CO_2 \quad (3 - \text{way catalyst, DOC})$$

A minority reaction that also follows in these catalysts is the WGSR forming N_2 and CO_2 from CO and water, discussed in the previous topics. Hydrogen is a very potent reductant that can be used to reduce NOx to N_2 and water. The HCs are also removed by oxidation to CO_2 with the three-way catalyst.

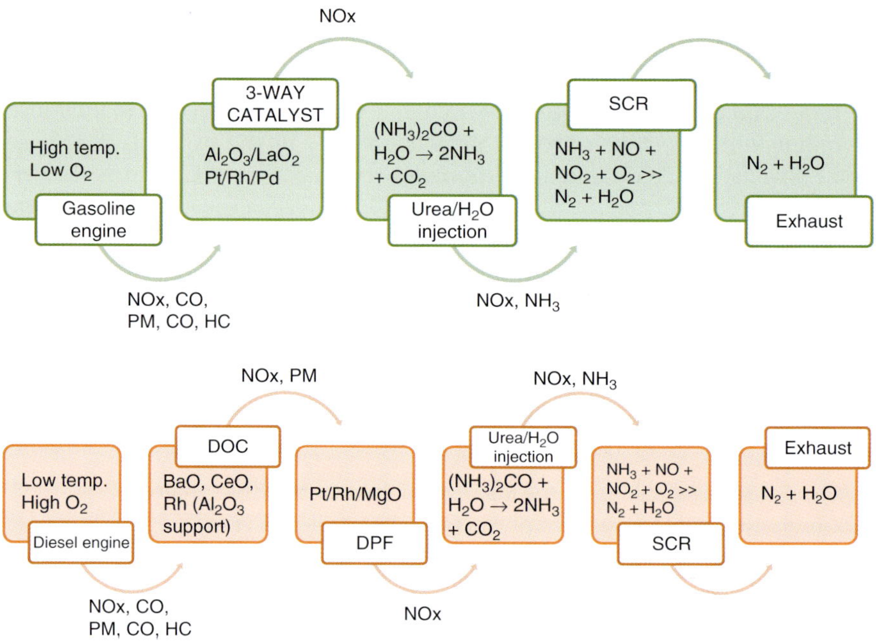

Figure 8.11 Some catalytic processes present in the gas exhaust of diesel and gasoline engines.

8.3.4.2 Components of the Cleaning Systems

An outline is shown in Figure 8.11.

8.3.4.3 Three-Way Catalyst

The catalyst used for the simultaneous elimination of NO, CO, and HC has the following main components:

* Alumina of the γ type (γ-Al_2O_3) with high surface. This component acts as a support of the noble catalytic metals such as Cerium. The noble metals act as oxygen buffers, storing O_2 when it reaches a certain critical level.
* LaO_2 (lanthanum oxide), which stabilizes the γ-Al_2O_3 and prevents Rh from being eliminated from the support.
* Platinum (Pt), which is the oxidation catalyst of CO, HC, and NO_2.
* Rhodium (Rh), which is the catalyst for the decomposition of NOx.

In some compositions, Pa (palladium) is used for replacing a part of the Rh or Pt. The emission of CO, HC, and NOx in the vehicle engines mainly depend on the ratio air:fuel, and also on other factors as the precise moment of the ignition. The general behavior of the emissions with the ratio air:fuel is shown in Figure 8.12. For obtaining a high conversion of all pollutants simultaneously, a stoichiometric air:fuel ratio (marked in figure by S*, close to 14.6) should be maintained.

The main reactions being produced in the system are as follows:

$$2NO + 2CO \rightarrow N_2 + 2CO_2 \tag{R1}$$

$$C_mH_n + (m + n/4)O_2 \rightarrow mCO_2 + n/2\,H_2O \tag{R2}$$

$$2CO + O_2 \rightarrow 2CO_2 \tag{R3}$$

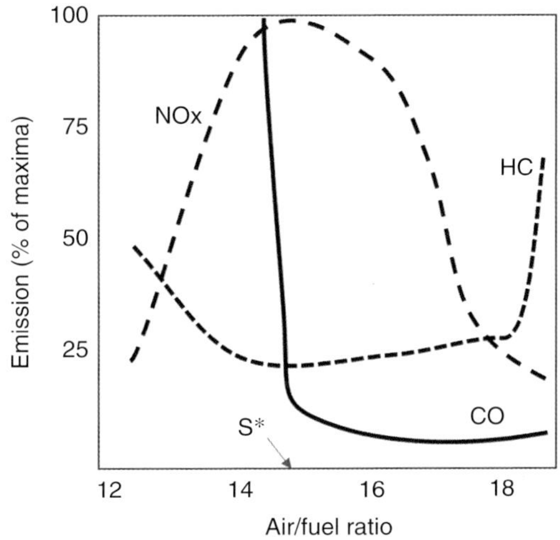

Figure 8.12 Variation of the conversion of the three-way catalyst reactions as a function of air/fuel ratio.

In the case of a low amount or air (small air/fuel ratio), there is an excess of fuel and the conversion of CO and HC does not occur because there is not enough oxygen. Under these conditions, the elimination of NOx is favored, and the power of the engine is high, but the fuel consumption is very high. Under these conditions, there is much more CO than NO, so the NO is not enough to eliminate all the CO present.

On the other hand, when the ratio air/fuel is very high (very oxidant conditions), there is a drastic increase of the emission of HC and a loss of engine power. At values of ratio higher than 24, the ignition cannot be produced. In usual driving conditions, engines usually run at a ratio somewhat poor, minimizing fuel consumption.

In gasoline engines, this is achieved by a careful control of the oxygen content by means of sensors ("lambda probe"), which keeps the oxygen level very close to the stoichiometric. Since there is no excess oxygen, it does not compete with NO for the oxidation of CO or HC and therefore it cannot be effectively reduced to N_2 in commercial three-way catalysts.

8.3.4.4 SCR Catalyst

SCR is the process in which ammonia is introduced as a reducer for the conversion of NO to N_2. As mentioned before, the ammonia is produced in situ by hydrolysis of urea. Three types of catalysts are used in this process:

- *Supported basic metal*: V/TiO_2 or W/TiO_2, or mixtures of both metals. Vanadium catalyzes the reduction, anatase (titanium oxide) provides high surface area, and tungsten oxide stabilizes anatase, which is necessary to have a high conversion. A good control of the ratio between ammonia and NOx is always required. If the ratio is too high, ammonia, which is also a pollutant, will be emitted and this will increase the cost of the process. Contrarily, at low values of the ratio, NOx conversion will not be sufficient. The optimal molar ratio is $1:1 = NH_3:NO$.
- *Supported noble metal*: Pt/Al_2O_3. Platinum acts as a catalyst for the reduction of NO with ammonia. Alumina provide high surface area.
- A zeolite, which catalyzes the reduction of NO by NH_3. It may contain Fe and Cu.

The last two systems provide NOx reduction at a lower and higher temperature than V_2O_5. Pt-based systems are better for heavy diesel engines that have a lower exhaust temperature.

The SCR process is also present in many industries, in processes where NH_3 or urea is injected in the gas before emission, in order to decrease the NOx emission. An example of the design is discussed in Chapter 9.

8.3.4.5 Diesel Oxidation Catalyst (DOC)

The DOC catalyst uses a Pt system supported in alumina to convert CO into CO_2. The oxidation of HC and soot also occurs. The DOC also has a NOx absorber system to reduce NOx in exhaust gases. This process combines components that catalyze the reduction of NOx with the capture of NO. Barium oxide (BaO) is used as a storage medium. In poor fuel conditions (excess air), NO is initially oxidized to NO_2, followed by reaction with barium oxide to form barium nitrate:

$$NO + \tfrac{1}{2}\,O_2 \rightarrow NO_2$$

$$BaO + NO_2 + \tfrac{1}{2}\,O_2 \rightarrow Ba(NO_3)_2$$

For a diesel engine, the exhaust contains an excess of air so that most of the time the system is in this state, until all the available BaO is transformed into $Ba(NO_3)_2$. At this stage, BaO must be regenerated in fuel-rich conditions by decomposition of the nitrate, forming NOx and BaO:

$$Ba(NO_3)_2 \rightarrow BaO + 2NO + 3/2\,O_2$$

$$Ba(NO_3)_2 \rightarrow BaO + 2NO_2 + 1/2\,O_2$$

These two reactions are promoted by the presence of ceria in the catalyst (in the form of oxide CeO) and also of rhodium. Also, in fuel-rich conditions, the NOx will be reduced with CO to form N_2 and CO_2:

$$NO + CO \rightarrow \tfrac{1}{2}\,N_2 + CO_2$$

This last reaction is produced in the Al_2O_3 support. The fuel-rich condition only happens during a very small fraction of time and returns to fuel-poor conditions, working this system in a forced unsteady state operation (FUSO). It will change between controlled conditions “lean” and “rich”; it is known as modulation lean-rich “poor-rich” and is necessary for the system to work.

These catalysts have a structure similar to that of the diesel particulate filter (DPF), but the channels are all open.

8.3.4.6 Diesel Particulate Filter (DPF) Catalyst

The DPF system catches particulate matter (PM), formed mainly by carbonaceous particles, in a ceramic filter where a Pt/Rh/MgO catalyst is supported. The presence of platinum catalyzes the oxidation of the trapped carbon (soot) to CO_2, in addition to other hydrocarbons or CO present. Rh inhibits the oxidation of SO_2 to sulfate salts at the outlet, which would contribute to PM. Magnesium works as a promoter for the processes catalyzed by the platinum.

In this device, the end of every two adjacent cells in the monolith is closed, which forces the gas to pass through the walls of each cell. In the process, the walls

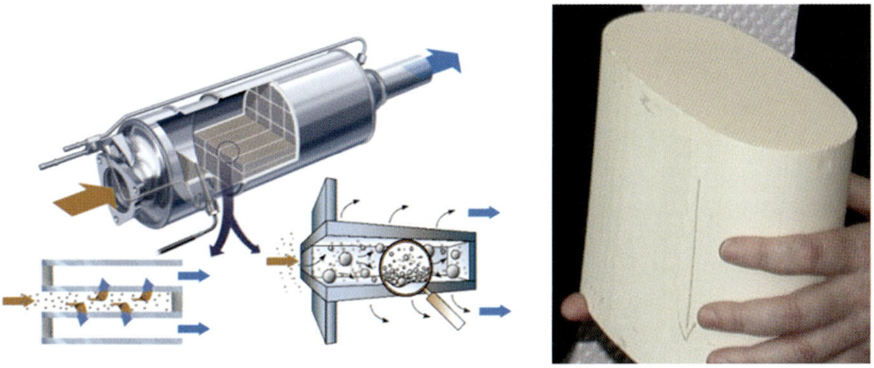

Figure 8.13 Flow dynamics in a DPF and general aspect. Source: Adapted from https://commons.wikimedia.org/wiki/File:Dpf_filter.png and https://commons.wikimedia.org/wiki/File:DPF_Cordierite_Alumunium-Titanat.jpg.

act as a filter, eliminating the carbon formed (and other PM), which is deposited on the walls. These walls are coated with a compound that disperses the noble metal (Pt/Rh/MgO) or the basic metal (V_2O_5). Figure 8.13 shows a scheme of the flow dynamics.

The appearance of these filters is the one seen on the right side. The filter acts in two periods. The first period is passive, and the particles are accumulating while they are slowly oxidized by the air. Later, when the pressure loss is too important, the temperature of the catalyst increases due to the heat produced by the combustion of the particles. The second period occurs when the temperature exceeds a certain threshold at which the rate of combustion of the soot is high and, in contact with the catalyst, CO_2 is produced. Thus, begins a new cycle. A typical temperature of the first period is 350 °C, while at 500 °C combustion will start ("soot light off temperature").

Systems based on noble metals (Pt and Pd) are the most used in the DPF. This is related to the high activity presenting at the low temperatures of the diesel systems. However, Pt and Pd are very sensitive to the presence of sulfur in the exhaust gases. One of the unwanted processes in DPF and DOC is the formation of sulfate salts from SO_2 and SO_3. In this way, the efficiency of a DPF dramatically decreases in the presence of small amounts of S in the fuel, from about 50 ppm.

Bibliography

Afandizadeh, S. and Foumeny, E. (2001). Design of packed bed reactors: guides to catalyst shape, size, and loading selection. *Applied Thermal Engineering* 21 (6): 669–682.

Anderson, J.R. and Boudart, M. (eds.) (1981–1985). *Catalysis: Science and Technology*, vol. 8. Berlin: Springer-Verlag.

Burrington, J.D. (2016). *Industrial Catalysis: Chemistry and Mechanism*. Imperial College Press.

Cifuentes, B., Valero, M., Conesa, J., and Cobo, M. (2015). Hydrogen production by steam reforming of ethanol on Rh-Pt catalysts: influence of CeO_2, ZrO_2, and La_2O_3 as supports. *Catalysts* 5: 1872–1896. https://doi.org/10.3390/catal5041872.

Clark, J.H. (2001). Catalysts for Green Chemistry. *Pure Appl. Chem.* 73: 103–111.

Cobo, M., Becerra, J., Castelblanco, M. et al. (2015). Catalytic hydrodechlorination of trichloroethylene in a novel NaOH/2-propanol/methanol/water system on ceria-supported Pd and Rh catalysts. *J. Environ. Manage.* 158: 1–10. https://doi.org/10.1016/j.jenvman.2015.04.035.

Gates, B.C. (1992). *Catalytic Chemistry*, 1992. Wiley.

González Velasco, J.R. and Gutiérrez Ortiz, M.A. (1997). *Catálisis: una ciencia multidisciplinar con presente & futuro*. Universidad del País Vasco.

Greenwood, N.N. and Earnshaw, A. (1997). *Chemistry of the Elements*. Butterworth-Heinemann.

Kent, J.A. (2012). *Handbook of Industrial Chemistry and Biotechnology*. London: Springer https://doi.org/10.1007/978-3-319-52287-6.

Reshetnikov, S.I., Ivanov, E.A., Kiwi-Minsker, L., and Renken, A. (2003). Performance enhancement by unsteady-state reactor operation: Theoretical analysis for two-sites kinetic model. *Chemical Engineering and Technology* 26 (7): 751–758. https://doi.org/10.1002/ceat.200301640.

Satterfield, C.N. (1991). *Heterogeneous Catalysis in Industrial Practice*, 2e. McGraw-Hill.

White, M.G. (1990). *Heterogeneous Catalysis*. Upper Saddle River, NJ: Prentice Hall.

9

Catalytic and Multiphase Reactor Design

9.1 Introduction

As mentioned in Chapter 8, the rate of a reaction can be modified by the presence of certain substances, which are usually neither initial reactants nor products. These substances are called catalysts, and they produce an increase in the reaction rate.

A catalyst can vary the reaction rate of a process in thousands or millions of times, so its use in industrial processes is usually very interesting. The characteristics of a catalyst can be summarized as

1. *Selectivity* of the catalysts. It refers to the ability of these substances to act in certain reactions and not to do so in different ones.
2. A catalyst increases the rate of reaction or favors a reaction vs. another, but *in no case determines the equilibrium or the end point thereof*, which will always be conditioned by thermodynamic issues.
3. The activity of a catalyst *depends not only on its chemical constitution but also on its physical or crystalline structure,* since the catalyst can lose its activity above certain temperatures. Therefore, an exhaustive study of a catalyst must include an investigation of its surface. In this way, it is necessary to ensure that the catalysts have a large active solid surface per unit volume.
4. In the presence of a catalyst, *the reactant molecules undergo a weakening of their bonds* or form intermediates in the vicinity of the surface of the solid. Several models that try to explain this behavior have been proposed:
   (a) The intermediate product is an association of the reactive molecule with a region of the surface of the solid.
   (b) Reagent molecules are under the influence of surface forces, moving in an environment close to it.
   (c) Free radicals form on the surface of the catalyst, which move toward the gas causing a series of reactions, until finally they are destroyed; in this last theory, the reaction will take place within the gas, the surface of the catalyst being a generator of free radicals.
5. According to the transition state theory, *the catalyst reduces the potential energy barrier that must be exceeded in order for the reactants to pass into products.* This will, in turn, cause a decrease in the activation energy, and therefore, an increase in the reaction rate, as can be seen in Figure 9.1.

Chemical Reactor Design: Mathematical Modeling and Applications,
First Edition. Juan A. Conesa.

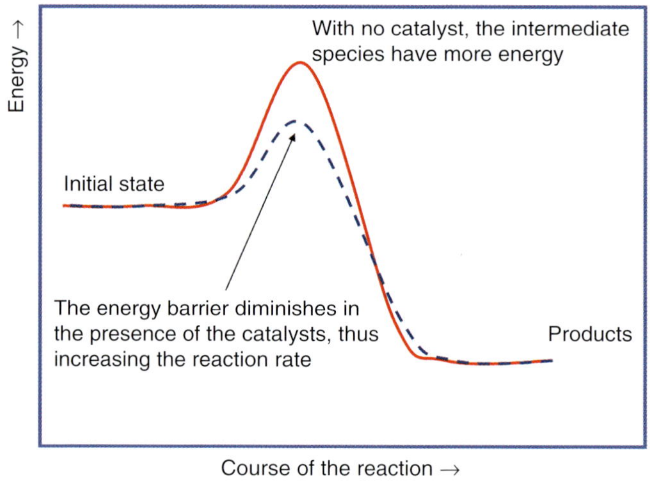

Figure 9.1 Representation of the action of a catalyst.

To be able to carry out a design of reactors for this type of reactions, it will be necessary to study in detail the rate equation in these processes from the study of the kinetics of the reaction.

9.2 Rate Equation in Catalytic Systems

The rate of a chemical process is defined by the following expression:

$$r_A = \frac{g_A}{\text{(extensive property)}} \tag{9.1}$$

where g_A is the generation term (mol/time) of a mole balance in the system. The extensive property is additive for subsystems, and usually for homogeneous systems, the total volume is used in such a way that the reaction rate is given, for example, in mol/(s m_R^3), the subscript "R" referring to the whole reactor.

For heterogeneous systems, the total volume is not always important. For example, imagine that we have a very big reactor, but a very small amount of catalyst inside it. If the reaction needs the catalyst to occur, the reaction rate will be quite low. In fact, for a catalytic reaction, the reaction rate expressed in many ways can be found. Table 9.1 shows some of the usual forms of expressing the reaction rate in heterogeneous systems. In the table, "S" is the external surface of the catalyst, "W" is the weight of the catalyst particles, "V" is the volume of the reactive mixture, and "V_p" is the volume of the catalyst particles.

Indeed, from the expressions in Table 9.1, by clearing g_A, the following equivalence can be demonstrated:

$$g_A = r_A S = r'_A W = r''_A V = r'''_A V_p \tag{9.6}$$

Table 9.1 Different properties used for defining reaction rate and their corresponding valid units.

Extensive property	Rate equation symbol (valid units)	Kinetic constant (valid units) assuming first-order kinetics	Relationship with generation term	
External surface of catalyst	$(-r_A)$, kmol/(s $m^2{}_{cat}$)	k, $m^3{}_L/(m^2{}_{cat}$ s)	$(-r_A) = g_A/S$	(9.2)
Weight of catalyst	$(-r'_A)$, kmol/(s kg_{cat})	k', $m^3{}_L/(kg_{cat}$ s)	$(-r'_A) = g_A/W$	(9.3)
Volume of reactor mix	$(-r''_A)$, kmol/(s $m^3{}_R$)	k'', $m^3{}_L/(m^3{}_R$ s)	$(-r''_A) = g_A/V$	(9.4)
Volume of catalyst	$(-r'''_A)$, kmol/(s $m^3{}_{cat}$)	k''', $m^3{}_L/(m^3{}_{cat}$ s)	$(-r'''_A) = g_A/V_p$	(9.5)

Note that the units of the kinetic constants will depend on the expression used to define the reaction rate, even for first-order kinetics. The table shows valid units for first-order kinetics in the different rate expressions; Eq. (9.6) could serve for changing units of kinetic constants or the reaction rate, if necessary.

Example 9.1 ***Second-Order Catalytic Reaction***

The second-order decomposition reaction: $A \rightarrow B + 2C$ is being carried out in a tubular reactor packed with a catalyst. Component A enters pure at a surface velocity of 3 m/s, at a temperature of 250 °C, and a pressure of 500 kPa. The rate constant of the chemical reaction at this temperature is

$$k' = 0.05\ m^6/(mol\ kg_{cat}\ s)\ \text{(second order reaction)}$$

Calculate the variation of the conversion of A with the length of the reactor for a total length of 0.1 m in:

(a) Isothermal conditions at 250 °C.
(b) Adiabatic conditions (initial temperature 250 °C)

Data:

$\Delta C_p = 0$
k' (at 400 °C) = 2 $m^6/(mol\ kg_{cat}\ s)$
$\Delta H_r = -800$ J/mol
Catalyst density = 2000 kg/m^3
Porosity of the bed = 0.4
Reagent heat capacity A = 15 J/(mol K)

Solution

First of all, let us calculate the pre-exponential factor and activation energy from the given data. We can do it using the Arrhenius equation:

$$k' = k'_0 \cdot \exp\left(-\frac{E}{RT}\right)$$

Knowing k'(at 250 °C) = 0.050 and k'(at 400 °C) = 2 m^6/(mol kg$_{cat}$ s), we calculate:

$$k'_0 = 7.71 \cdot 10^5 \frac{\text{m}^6}{\text{mol kg s}} \quad \frac{E}{R} = 8656.1\ \text{K}$$

(a) The mass balance in isothermal conditions in a plug flow reactor with catalyst is

$$dn_A = dg_A = (-r'_A) \cdot dW$$

where "dW" is the weight of the catalyst in a differential volume. From the data we know that $(-r'_A) = k' \cdot C_A^2$

In this case, the concentration depends on the molar conversion of the reactant (X_A) and also on the possible change in volume (or flow rate) due to the change in moles of the reaction. In this way:

$$C_A = \frac{n_A}{Q} = \frac{n_{A0} \cdot (1 - X_A)}{Q_O \cdot (1 + \varepsilon \cdot X_A)}$$

where ε is the expansion factor of the reaction, i.e. the number of moles produced by mol of reactants, in this case $\varepsilon = 2$; therefore:

$$C_A = C_{A0} \frac{(1 - X_A)}{(1 + 2 \cdot X_A)}$$

$$C_{A0} = \left(\frac{P_0}{RT_0}\right) = \frac{\left(\frac{500}{1013}\right)\text{atm}}{0.082 \frac{\text{atm l}}{\text{mol K}} \cdot (250 + 273)\ \text{K}} = 0.115 \left(\frac{\text{mol}}{\text{l}}\right)$$

Therefore, under isothermal conditions:

$$-n_{A0} dX_A = -\frac{k' C_{A0}^2 (1 - X_A)^2}{(1 + 2X_A)^2} dW$$

and, reorganizing:

$$\frac{dX_A}{dW} = \frac{k C_{A0}^2 (1 - X_A)^2}{n_{A0}(1 + 2X_A)^2}$$

The mass of the catalyst is related to the volume of reactor by

$$W\ (\text{kg}_{cat}) = V_{total}\ (\text{m}^3_{total}) \cdot \rho_{cat} \left(\frac{\text{kg}_{cat}}{\text{m}^3_{cat}}\right) \times (1 - \text{Porosity of the bed}) \left(\frac{\text{m}^3_{cat}}{\text{m}^3_{total}}\right)$$

And the volume relates to the length of the reactor (x) by $V = S \cdot x$, S being the section of the bed. On the other hand, $Q_0 = u_0 \cdot S$ and $n_{A0} = C_{A0} \cdot Q_0$. Finally, we obtain

$$\frac{dX_A}{dx} = \frac{\rho_{cat} \cdot (1 - \text{Porosity of the bed}) \cdot k' \cdot C_{A0}(1 - X_A)^2}{u_0 \cdot (1 + 2X_A)^2}$$

It is easy to apply a finite difference approximation to solve this differential equation, and we will get

$$\frac{X_A^{i+1} - X_A^i}{\Delta x} = \frac{\rho_{\text{cat}} \cdot (1 - \text{Porosity of the bed}) \cdot k' \cdot C_{A0}(1 - X_A^i)^2}{u_0 \cdot (1 + 2X_A^i)^2}$$

With initial condition $X_A = 0$, we will use Matlab® to solve the equation, as shown in Program listing 9.1:

Program Listing 9.1

```
clear all

u0=3;% m/s
porosity=0.4;
dcat=2000;% kg/m3 catalyst
k0=7.707e5;% m6/mol·kg·s

E_R=8656.1; % K
P=500000; % Pa
Ca0=P/(8.314*523); % mol/m3

T=250+273; % K
k=k0*exp(-E_R/T); % m6/mol·kg·s

XA(1)=0; % Initial value of conversion
L(1)=0;; % Initial value of length
incx=0.0001; % Increment of length
n=0.1/incx; % Number of points


for i=2:n
    L(i)=L(i-1)+incx;
    XA(i)=XA(i-1)+(dcat*(1-porosity)*k*Ca0*(1-XA(i-
1))^2/(u0*(1+2*XA(i-1))^2))*incx;
end

figure(1)
plot(L,XA)
xlabel('Length (m)')
ylabel('Molar conversion')
```

In the program, the different values of the variables are given, and the length increment is defined. The number of points is then calculated, and the differential equation is solved numerically using the finite differences method. The values of length and conversion are saved in two vectors that are plotted in the last part of the program. The following figure shows the result.

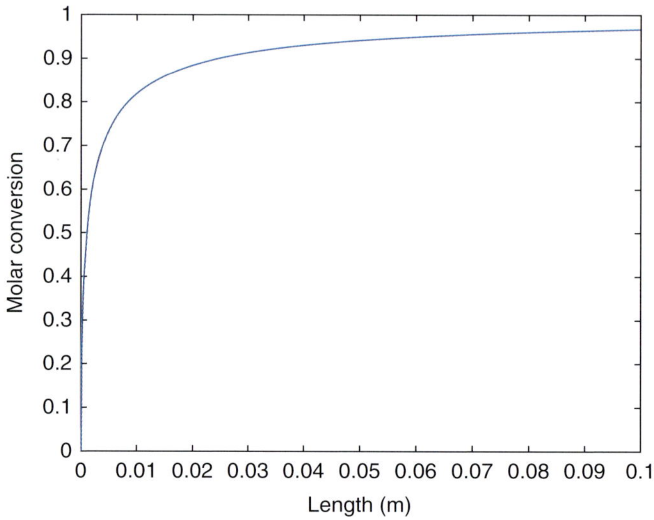

(b) The mass balance in adiabatic conditions is similar to that of the isothermal case, but the temperature change must be borne in mind. The energy balance in a PFR, with only one reaction and without transfer abroad (adiabatic), is

$$dT \cdot \sum_{j=1}^{s} n_{j0} C_{pj} + \Delta H_R \cdot \frac{dX_A}{n_{A0}} = 0$$

Bearing in mind that only reactant A is fed, and rearranging:

$$\frac{dT}{dX_A} = -\frac{\Delta H_R}{C_{pA}}$$

In the isothermal case, we checked that $Q = Q_O \cdot (1 + 2 \cdot X_A)$ but now a correction of the flowrate with temperature is needed, and the following results:

$$Q = Q_O \cdot \frac{T}{T_0}(1 + 2 \cdot X_A)$$

and then:

$$\frac{dT}{dx} = \frac{dT}{dX_A} \cdot \frac{dX_A}{dx}$$
$$= -\frac{\Delta H_R}{C_{pA}} \cdot \frac{\rho_{\text{cat}} \cdot (1 - \text{Porosity of the bed}) \cdot k \cdot C_{A0}(1 - X_A)^2}{u_0 \cdot \left(\frac{T}{T_0}\right)^2 \cdot (1 + 2X_A)^2}$$

In the previous equation, k should be calculated at the temperature of the differential of length. Applying the finite differences method:

$$\frac{T^{i+1} - T^i}{\Delta x} = -\frac{\Delta H_R}{C_{pA}} \cdot \frac{\rho_{\text{cat}} \cdot (1 - \text{Porosity of the bed}) \cdot k(T^i) \cdot C_{A0}(1 - X_A^i)^2}{u_0 \cdot \left(\frac{T^i}{T_0}\right)^2 \cdot (1 + 2X_A^i)^2}$$

Some small changes are needed in the program listing used in the previous isothermal case. A valid program is given, as shown in Program listing 9.2.

Program Listing 9.2

```
clear all

u0=3;% m/s
porosity=0.4;
dcat=2000;% kg/m3 catalyst
k0=7.707e5;% m6/mol·kg·s
CpA=15; %J /mol·K
DHR=-800;% J/mol
E_R=8656.1; % K
P=500000; % Pa
Ca0=P/(8.314*523); % mol/m3

T(1)=250+273; % Initial temperature, K
XA(1)=0; % Initial value of conversion
L(1)=0; % Initial value of length, m
incx=0.0001; % Increment of length, m
n=0.1/incx; % Number of points
k=k0*exp(-E_R/T(1)); % m6/mol·kg·s
ftemp=(T(1)/523)^2;

for i=2:n
    L(i)=L(i-1)+incx;
    dXAdt=(dcat*(1-porosity)*k*Ca0*(1-XA(i-
1))^2/(u0*ftemp*(1+2*XA(i-1))^2));
    XA(i)=XA(i-1)+dXAdt*incx;
    T(i)=T(i-1)+(-DHR/CpA)*dXAdt*incx;
    k=k0*exp(-E_R/T(i)); % m6/mol·kg·s
    ftemp=(T(i)/523)^2;
end

figure(1)
plot(L,XA)
xlabel('Length (m)')
ylabel('Molar conversion')

figure(2)
plot(L,T)
xlabel('Length (m)')
ylabel('Temperature')
```

The results are shown in the following two figures (conversion and temperature evolution vs. length of the reactor).

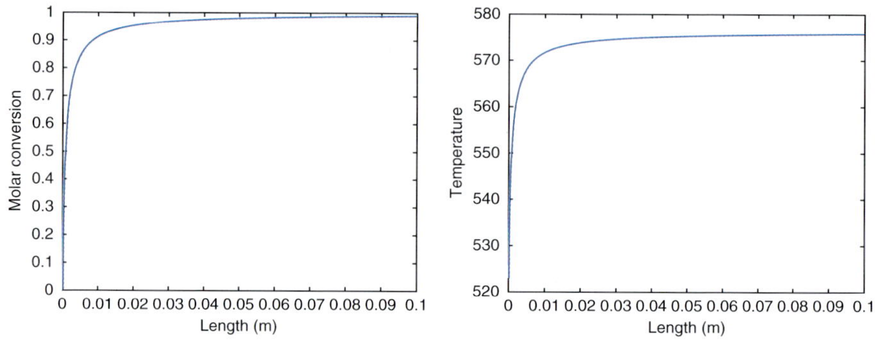

9.2.1 Steps in the Catalytic Reaction

Porosity, like density, is an important property of solid catalysts that is used in calculating and modeling pore diffusional and film mass transfer resistances. Figure 9.2 shows that the volume of a typical catalytic solid is partitioned into solids, pores, and voids. The pore volume is more typically distributed among pores of different sizes, with most of the volume residing in the largest pores.

In porous catalysts, the reaction occurs throughout the solid–fluid interface both on the external surface and inside the pores of the particle. In order to deduce the expressions of reaction rates, it is necessary to take into account the different steps (both physical and chemical) that take place in the reaction process and that may offer resistance to it, since one or the other of these predominates steps will have different expressions of rate. Figure 9.2 shows a diagram of the steps that a reagent A must undergo in order to reach the surface.

The step or steps with the lowest rate that limit the value of the reaction rate are known as controlling or limiting steps. The different steps are the following:

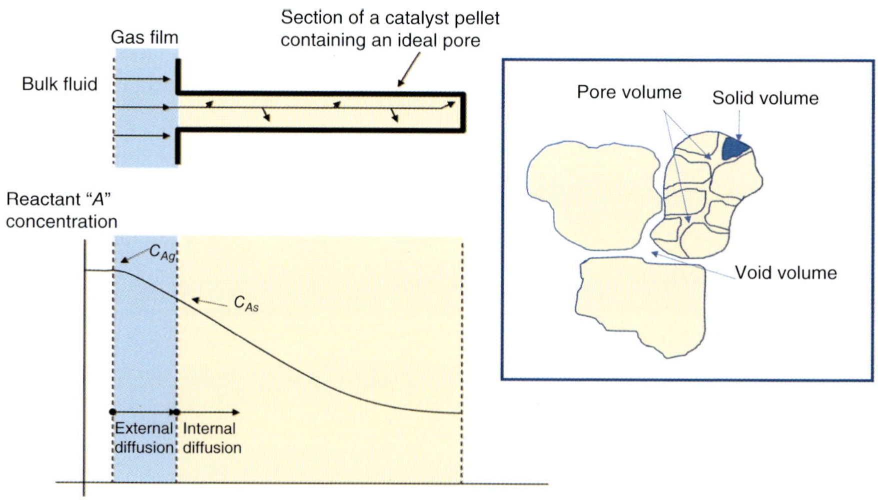

Figure 9.2 Porosity and void volumes in catalyst. Reaction steps in catalytic systems.

1. *External diffusion (or film resistance)*: Diffusion of the reactants toward the outer surface of the catalyst;
2. *Internal diffusion (or resistance to diffusion in the pores)*: Diffusion of the reactants to the interior of the catalyst, where the reaction will mainly occur;
3. *Chemical adsorption (or physical in certain occasions) on the surface*: During the union of the reagent molecules somewhere on the surface of the catalyst;
4. *Surface chemical reaction*: Reaction on the surface of the catalyst particle;
5. *Chemical desorption on the surface*: During the separation of the product molecules from the surface of the catalyst;
6. *Internal diffusion (or resistance to diffusion in the pores)*: Diffusion of products toward the exterior of the catalyst;
7. *External diffusion (or film resistance)*: During the exit of the products toward the gas phase.

Of all these reaction steps, that of internal diffusion has some particular characteristics. This step is considered as a controlling step, in the sense that if it is very slow the rate will be affected by the internal diffusion, since the reaction could also take place on the surface of the solid. Therefore, it is said that the resistance to internal diffusion is "in parallel" with the other resistances, all of them "in series."

9.3 Rate Equation When Chemical Step Is Limiting Reaction Rate

9.3.1 Mechanisms of Catalysis

The exact nature of the reasons why the intermediate complex, adsorbed on the surface of the catalyst, is easily formed is a research topic. Structural requirements can be foreseen for the formation and breaking of links, and this has led to the formulation of an important series of theories. As it has been proved, the transfer of electrons is involved in the formation of the bonds of the intermediate compound, and therefore the electronic nature of the catalysts is undoubtedly important. What we give is not a satisfactory explanation, since it does not exist, of the different mechanisms, but some examples are proposed with the corresponding explanations presented by the researchers in the bibliography.

For example, acidic catalysts, such as silica and alumina, can apparently act as Lewis acids (electron acceptor) or Brönsted acids (proton donor). The reaction mechanism for cracking a hydrocarbon would be as follows:

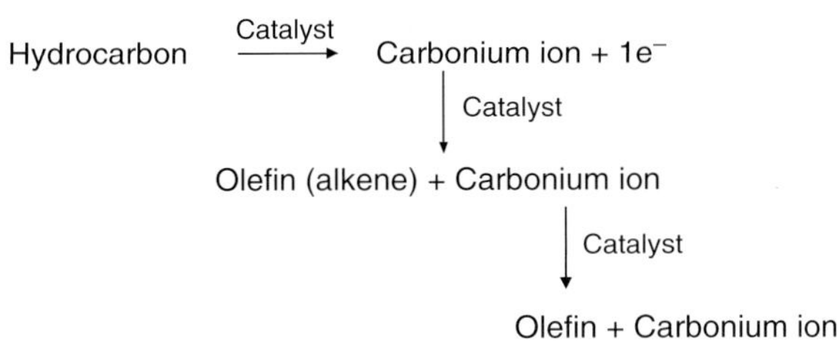

At the end, the carbonium ion can react with another initial hydrocarbon molecule to give a carbonium ion with a greater number of carbon atoms and start the breaking steps again. Among the currently used cracking catalysts are zeolites, as shown in Chapter 8. These compounds are crystalline aluminosilicates containing "boxes," often of molecular dimensions, which can block branched chains of the molecules and thereby favor breakage by a given bond, increasing the selectivity in a given component.

Metal catalysts are frequently used in dehydrogenations and hydrogenations. An example can be the synthesis of NH_3, which would follow the following scheme:

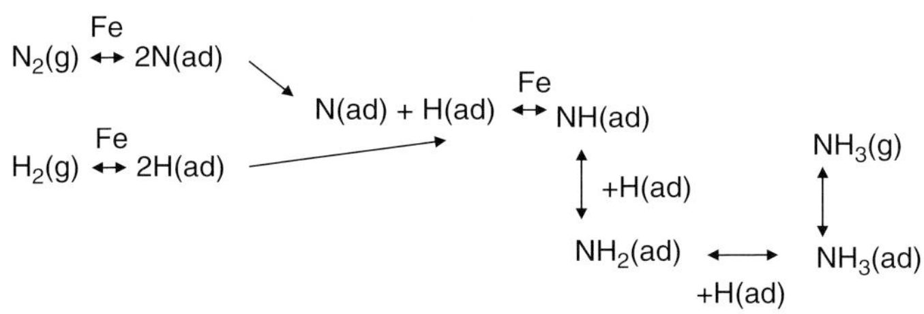

For the oxidation reactions of hydrocarbons, the active components of the catalyst are metal oxides (molybdenum, vanadium, chromium, etc.), capable of having several valences. In this case, it seems that the following reactions take place:

Hydrocarbon + Oxidized Catalyst

→ Reaction products + Partially reduced catalyst

Partially reduced catalyst + O_2 → Oxidized catalyst

It has been possible to verify these phases, marking the oxygen atoms (isotopes) of the catalyst, and subsequently checking the presence of these atoms among the products of the reaction.

The catalytic activity has been related to the electrical conductive character of the catalyst. It has been found that when the catalyst was a little conductive, no reaction took place; if it was very conductive, the most developed reaction was to total the carbon oxides and water, and for an intermediate situation the desired oxidized products were frequently formed (benzoic acid and benzaldehyde from toluene).

For many reactions, the reaction mechanism, which includes the adsorption and desorption of the active centers, can be represented by schemes such as the following (where L is the active center):

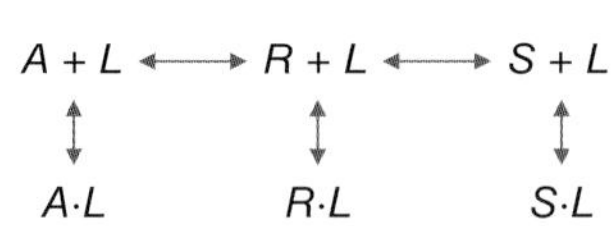

This scheme could correspond to the reaction $A \rightarrow S$, where the concentration of R in the fluid would be very small, since the equilibrium of adsorption is very displaced toward species $R{\cdot}L$.

9.3.2 Theories About Adsorption

The classical Langmuir theory of adsorption, which has justified a multitude of experimental data, explains adsorption as a phenomenon in equilibrium:

$$A + L \leftrightarrow A \cdot L$$

If $C_{A \cdot L}$ is the concentration of $A \cdot L$ (kmol/m^2), C_A is the gas-phase concentration (kmol/m^3), C_L is the concentration of active centers (kmol/m^2), when the equilibrium is reached, it will be fulfilled that:

$$K_A = \text{equilibrium constant of chemical adsoprtion} = \frac{C_{A \cdot L}}{C_A \cdot C_L} \tag{9.7}$$

The values of the rates of the direct and inverse process are given by

$$r_a(\text{adsorption rate}) = k_a C_A C_L \tag{9.8}$$

where

$$k_a = k_{a0} \cdot \exp(-E_a/RT) \tag{9.9}$$

$$r_d(\text{desorption rate}) = k_d C_{A \cdot L} \tag{9.10}$$

where

$$k_d = k_{d0} \cdot \exp(-E_d/RT) \tag{9.11}$$

In equilibrium, it will be fulfilled that:

$$k_a C_A C_L = k_d C_{A.L}$$

$$\frac{C_{A \cdot L}}{C_A \cdot C_L} = \frac{k_a}{k_d} = K_A \tag{9.12}$$

where K_A is the equilibrium constant.

We can relate C_{AL} to C_t in the following way:

$$C_t = C_L + C_{A \cdot L} = C_L + K_A C_A C_L = C_L(1 + K_A\, C_A) \tag{9.13}$$

where C_t is the concentration of active centers (active centers/m^2). Then, it is

$$C_L = \frac{C_t}{1 + K_A\, C_A} \tag{9.14}$$

and the portion of occupied active centers will be

$$\frac{C_{A \cdot L}}{C_t} = \frac{K_A C_A}{1 + K_A C_A} \tag{9.15}$$

The representation of $C_{A \cdot L}/C_t$ as a function of C_A has the form shown in Figure 9.3.

From the Langmuir model, expressions of the reaction rate can be obtained as a function of the concentration of reactants (and products, in reversible reactions). This can be seen in a simple example: In many catalytic hydrocarbon oxidation reactions, it has been found that the slow steps, which control the process, are the chemisorption of the O_2 and/or the superficial chemical reaction between the chemisorbed O_2 with the nonadsorbed hydrocarbon in the gas phase. We will deduce the kinetic equation for both cases. The mechanism of the reaction is as follows:

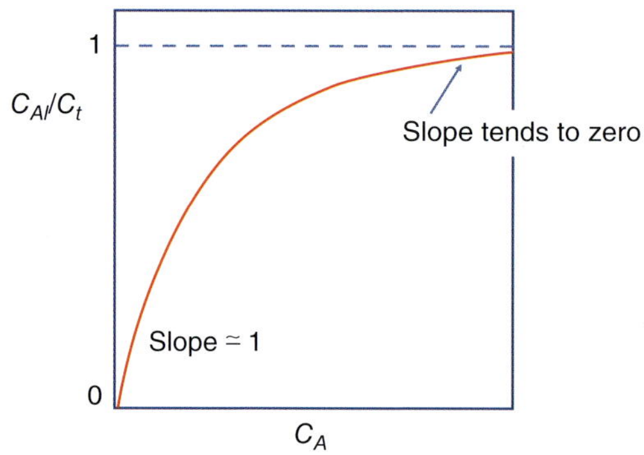

Figure 9.3 Variation of the proportion of active centers bonding to the ligand.

1. Bonding of oxygen to the surface (chemisorption):

$$O_2 + L \rightarrow O_{2-L}$$

2. Surface chemical reaction:

$$O_{2-L} + RH \rightarrow \text{Compound} - L$$

3. Desorption reaction:

$$\text{Compound} - L \rightarrow \text{Compound} + L$$

If the controlling process is the chemisorption of O_2, the reaction rate will be given by step 1 (the others will be very fast), and will be

$$r = k\, C_{O_2}\, C_L \tag{9.16}$$

On the other hand, it is true that the total "concentration" of the active sites is

$$C_t = C_L + C_{L\text{-}O_2} \tag{9.17}$$

As the surface chemical reaction is fast, $C_{L\text{-}O_2} = 0$ and then $C_L = C_t$. Therefore:

$$r = kC_tC_{O_2} = k'C_{O_2} \tag{9.18}$$

that is, the reaction rate is directly proportional to the concentration of oxygen and we will observe a pseudo first-order reaction.

If the controlling step is the superficial chemical reaction, step 1 will be in equilibrium (constant K_{ad}), step 2 will be slow and irreversible, and step 3 will be fast and irreversible. In this case, the rate will be given by

$$r = k\, C_{O_2 \cdot L} C_{RH} \tag{9.19}$$

and

$$C_t = C_L + C_{O_2 \cdot L} \tag{9.20}$$

Remembering that:

$$K_{ad} = \frac{C_{O_2 \cdot L}}{C_{O_2} \cdot C_L} \tag{9.21}$$

and therefore

$$C_{O_2 \cdot L} = K_{ad} C_{O_2} C_L \tag{9.22}$$

since the adsorption step is the slowest one, the reactants and the products will be in equilibrium. It follows that:

$$C_t = C_L + K_{ad} C_{O_2} C_L = C_L(1 + K_{ad} C_{O_2}) \tag{9.23}$$

So:

$$r = k\, C_{O_2 \cdot L} C_{RH} = k\,(K_{ad} C_{O2} C_L) C_{RH} = \frac{k\, K_{ad} C_t C_{RH} C_{O_2}}{1 + K_{ad} C_{O_2}} = \frac{k' C_{RH} C_{O_2}}{1 + K_{ad} C_{O_2}} \tag{9.24}$$

The two deducted expressions have been checked experimentally. For any other mechanism, the corresponding equations can be obtained. The rate expressions will coincide with those obtained experimentally, provided that the accepted hypotheses are correct (Langmuir mechanism, controlling step, etc.), in addition to the fact that the remaining steps of the reaction are faster. The value of rate of the given equations can be matched with any of the rate expressions, based on the catalyst mass, reactor volume, etc.). The units of the constant or constants will depend on it.

From the equations obtained, it can be predicted, if the model is correct, how the reaction rate will vary with the total pressure of the system, or with the concentrations of the reactants.

9.4 Rate Expression for External Diffusion as a Limiting Step

When the resistance to external diffusion is higher than all other resistances, the diffusion of the reactant from the bulk of the fluid to the surface of the catalyst is the limiting step of the reaction rate. The reagent flow that reaches the catalyst surface is governed by the material transport coefficient k_c between the fluid and the solid. The rate in this case would be given by

$$(-r_A) = -\frac{g_A}{S_{ex}} = k_C \cdot (C_{Ag} - C_{As}) \quad (\text{kmol/(s m}^2)) \tag{9.25}$$

where k_C = individual mass transfer coefficient (kmol/(s m^2)/kmol/m^3), S_{ex} = external surface of the particle (m^2), C_{Ag} = concentration of reagent A in the fluid (kmol/m^3), C_{AS} = concentration on the surface (kmol/m^3).

For cases where all other steps are rapid, including the external diffusion of the products, C_{AS} is the equilibrium concentration of reagent A with the reagents; if the other steps are very fast, including the chemical reaction, and if the reaction is irreversible, then C_{AS} is zero.

The step of external diffusion of a reagent is usually the dominant in the reactions that take place at high temperature and with very small catalyst particles. In these cases, the rate of the overall reaction will be the same for all solid–fluid reactions, both for catalyzed reactions and for those that are not.

The problem, when external diffusion is the controlling step, is usually the calculation of the mass transfer coefficient, although it can be estimated by semiempirical equations where dimensionless modules are used. We can relate k_C to the Sherwood module, as follows:

$$Sh = \frac{(k_C d_P)}{D_A} \tag{9.26}$$

where D_A = coefficient of molecular diffusion, which must be known or estimated by semiempirical equations, d_p = effective diameter of the particle, that is, the diameter of a sphere that has the same outer area.

The Sherwood module is related to the Reynolds and Schmidt numbers, by means of the parameter j_D:

$$j_D = \frac{Sh}{Sc^{\frac{1}{3}} \cdot Re} \tag{9.27}$$

where

$$Re = \frac{d_p \cdot \rho \cdot u}{\mu} \tag{9.28}$$

and

$$Sc = \frac{\mu}{\rho \cdot D_A} \tag{9.29}$$

where "μ" is the superficial rate, calculated as the flow/total section of the reactor.

For fixed-bed and for type $A \rightarrow B$ reactions in the absence of inert, the value of j_D depends mainly on the Reynolds modulus. There are numerical expressions that relate j_D to the Reynolds module, such as the Thodos equation:

$$j_D = \frac{0.725}{Re^{0.41} - 0.15} \tag{9.30}$$

With all these relations, we can estimate the value of the coefficient k_C.

Example 9.2 ***Cracking of Cumene***

Cumene is being cracked into benzene and propylene in a fixed bed of zeolite particles at 362 °C and atmospheric pressure, in the presence of a large excess of nitrogen. In a section of the reactor, where the partial pressure of the cumene is 0.0689 atm, the reaction rate observed is 0.153 kmol/(kg_{cat} h).

Calculate the % difference in concentration of the reagent between the medium and the surface of the particle.

Some properties of the reactive mixture and operating conditions are as follows:

- Average molecular weight of 34.37 kg/kmol,
- Density 0.66 kg/m^3,
- Viscosity 0.094 kg/(m h),
- Reynolds number 0.052,
- Diffusivity of the reagent through the mixture 0.096 m^2/h,
- Ratio external surface of catalyst/mass of catalyst = 45 $m^2{}_{cat}/kg_{cat,}$
- Mass flow rate of the mixture 56.47 kg/(m^2 h).

Solution

The basic equation of this system for calculating the reaction rate is given in Eq. (9.25).

Let us estimate the mass transfer coefficient k_C. For that, let us use the correlations in Eqs. (9.27) and (9.30).

We know the value of the Reynolds number, so we can estimate $j_D = 4.913$ using Eq. (9.30). The value of Sc can also be calculated using Eq. (9.29):

$$Sc = \frac{0.094}{0.66 \cdot 0.096} = 1.483$$

Using Eq. (9.27), we can calculate $Sh = 0.2913$.

From the value of the Reynolds number, we can estimate the particle diameter:

$$Re = \frac{d_p \cdot \rho \cdot u}{\mu} \quad \rightarrow \quad 0.052 = \frac{d_p \cdot 56.47 \frac{\text{kg}}{\text{m}^2\ \text{h}}}{0.094 \frac{\text{kg}}{\text{m h}}}$$

Therefore, $d_p = 8.6 \cdot 10^{-5}$ m. The definition of Sh (Eq. (9.26)) can be used to calculate the value of the mass transfer coefficient, obtaining a value of $k_C = 323.0$ m/h.

The value of the difference in concentrations ($C_{Ag} - C_{AS}$) can be then calculated:

$$(-r_A) = -\frac{g_A}{S_{ex}} = k_C \cdot (C_{Ag} - C_{As})$$

$$0.153 \frac{\text{kmol}}{\text{kg}_{cat}\ \text{h}} \cdot \frac{1}{45} \frac{\text{kg}_{cat}}{\text{m}^2_{cat}} = 323.0 \frac{\text{m}}{\text{h}} \cdot (C_{Ag} - C_{As}) \frac{\text{kmol}}{\text{m}^3}$$

So, $(C_{Ag} - C_{As}) = 1.05 \cdot 10^{-5}\ \text{kmol/m}^3 = 1.05 \cdot 10^{-2}\ \text{mol/m}^3$

The value of C_{Ag} can be also calculated, using the ideal gas equation: $C_{Ag} = P_A/RT = 1.322\ \text{mol/m}^3$, so the relative difference is $(C_{Ag} - C_{As}) \cdot 100/C_{Ag} = 0.79\%$.

9.5 Reaction Rate When Internal Diffusion Is Slow

As mentioned before, the internal diffusion is a step with some special characteristics. First of all, if there exists a high resistance to the diffusion in the pores, the reaction can still be produced in the exterior surface of the catalyst, so the effect of the internal diffusion is a kind of a correction of the rate. Mathematically, this correction is complicated, since the diffusion of the reactant through the solid is not related in a simple way with the other steps. The general expression of diffusion of a component A through a gaseous mixture $A + B$ is given by Fick's law:

$$n_A = x_A\,(n_A + n_B) - c\,D_{AB}\,\nabla x_A\ (\text{kmol}\,A/(\text{m}^2\ \text{s})) \tag{9.31}$$

In this equation, X_A is the mass fraction and the molar flux density, n_A, depends on two terms, the first of which is a consequence of the general movement and the second of diffusion. When there is no global movement, that is, if there is counterdiffusion and this only takes place in the X direction, it is fulfilled that $n_A = -n_B$ and the above equation becomes:

$$n_A = -c \cdot D_{AB} \frac{dx_A}{dX} \tag{9.32}$$

and if the concentration is constant:

$$n_A = -D_{AB} \frac{dC_A}{dX} \tag{9.33}$$

In multicomponent systems, the treatment is complicated because the term of global displacement is not negligible, and the diffusivities may depend on the composition. In gaseous systems, in the presence of a large amount of inert at rest, and in liquid systems, Eqs. (9.32) and (9.33) are applicable. Note that the units of the diffusion coefficient, by definition, are $m^3{}_{fluid}/(m_{cat}\ s)$, which are usually indicated as m^2/s.

The value of D_{AB} varies with temperature and pressure in different ways depending on the type of diffusion. In porous catalysts, and for a process $A \leftrightarrow B$, or in systems with a large proportion of inert, the diffusion law is applied as

$$n_A = -D_e \frac{dC_A}{dX} \tag{9.34}$$

n_A being the density of flow, defined as the kmol of A/s that are diffused per unit of total normal area, including that occupied by the solids. D_e is the effective diffusion coefficient that takes into account the free passage section of the holes. Usually, D_e is related to the coefficient D_{AB} by relations of the type:

$$D_e = \frac{D_{AB} \cdot \varepsilon}{\varphi}$$

where ε is the fraction of holes of the catalyst (porosity, $m^3{}_{holes}/m^3{}_{cat}$) and φ is the tortuosity factor (frequent value 3 or 4 or $1/\varepsilon$).

9.5.1 First-order Kinetics in Flat Particles

First of all, we will consider the case of a molecule of reagent A, in a reaction system $A \rightarrow B$, which is diffusing through a flat porous pellet through both sides (Figure 9.4). The chemical reaction step will be assumed of first order.

As it was treated in Chapter 4, we consider the diffusion of a species toward the center of a catalyst particle. Doing a mass balance in stationary regime for reagent A in the $dV = S\,dx$ of the considered pellet, we obtain

$$\text{Output} + \text{accumulation} = \text{generation} + \text{input (kmol } A/\text{s)}$$

$$[n_A]_{x+dx} \cdot S + 0 = r_A''' \cdot S \cdot dx + [n_A]_x \cdot S \tag{9.35}$$

n_A being the flow density of A (kmol $A/(\mathrm{s\,m^2})$) and r_A''' the reaction rate of A (kmol $A/(\mathrm{s\,m^3{}_{cat}})$), which is to be assumed to be of first order; so it is deduced that:

$$\frac{dn_A}{dx} = r_A''' = -k''' \cdot C_A \tag{9.36}$$

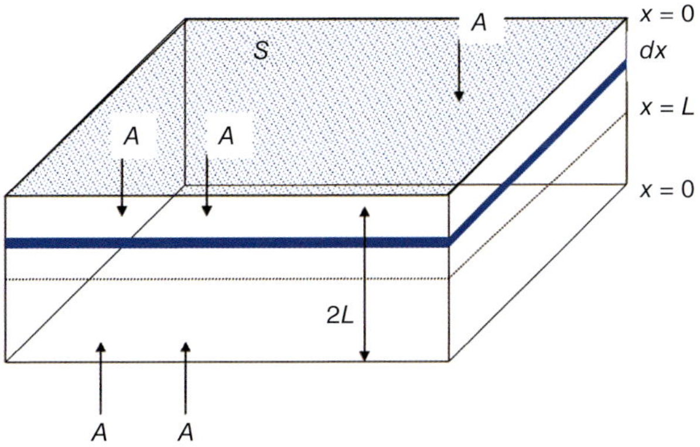

Figure 9.4 Porous catalyst pellet with reagent A entry on both sides.

and from here:

$$\frac{d\left(-D_e\frac{dC_A}{dx}\right)}{dx} = -k''' \cdot C_A \tag{9.37}$$

where k''' is the constant of the first-order reaction ($m^3{}_{fluid}/(m^3{}_{cat}\ s)$) (sometimes abbreviated as s^{-1}) and D_e is the diffusivity through the gas mixture (as already indicated, with units of $m^3{}_{fluid}/(m_{cat}\ s) \approx m^2/s$).

Therefore, in this particular case:

$$-D_e\frac{d^2C_A}{dx^2} = -k''' \cdot C_A \tag{9.38}$$

or

$$D_e\frac{d^2C_A}{dx^2} + k''' \cdot C_A = 0 \tag{9.39}$$

The general solution of this differential equation is

$$C_A = M_1 \cdot e^{mx} + M_2 \cdot e^{-mx} \tag{9.40}$$

where the value of m is given by

$$m = \left(\frac{k'''}{D_e}\right)^{0.5} \tag{9.41}$$

This module "m," as explained earlier, has the units of $1/m_{cat}$. The values M_1 and M_2 are constants that can be determined by the boundary conditions. One of these conditions is that on the surface of the pellet, the concentration of A is equal to that which exists in the fluid. The other assumes that the variation of the concentration in the middle of the porous pellet is zero (as treated in Chapter 4).

$$(1)\quad C_A = C_{As} \quad \text{for } x = 0 \tag{9.42}$$

$$(2)\quad \frac{dC_A}{dx} = 0 \quad \text{for } x = L \tag{9.43}$$

If we take into account the symmetry of the pellet, condition 2 is nothing more than indicating that in the center the concentration of A will present a minimum. Condition 1 is also logical.

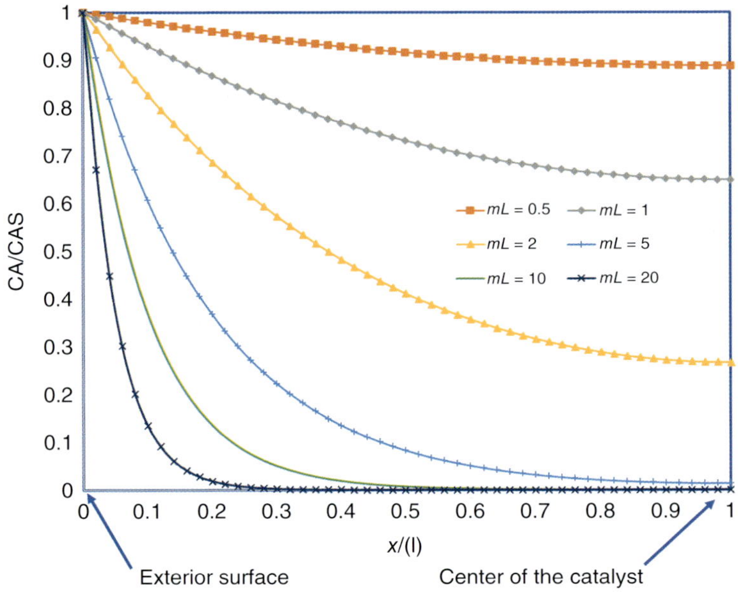

Figure 9.5 Variation of the concentration of *A* vs. distance *x*.

From here, we get

$$M_1 = \frac{C_{As}e^{-mL}}{e^{mL} + e^{-mL}} \quad M_2 = \frac{C_{As}e^{mL}}{e^{mL} + e^{-mL}} \tag{9.44}$$

and, therefore:

$$\frac{C_A}{C_{As}} = \frac{e^{m(L-x)} + e^{-m(L-x)}}{e^{mL} + e^{-mL}} = \frac{\cosh[m(L-x)]}{\cosh[mL]} \tag{9.45}$$

In view of this equation, it can be deduced that there is a decrease in the concentration of reagent as it progresses toward the interior of the pellet. This decrease depends on a dimensionless parameter "mL," which is defined as Thiele's module ϕ. Figure 9.5 shows different graphs C_A/C_{As} vs. x/L for different values of the Thiele's module ϕ.

It can be observed that the greater the Thiele module, that is, the faster the reaction to diffusion, the steeper the decrease in the concentration of A. However, for small values of the Thiele module, the diffusion through the pellet is very fast and the concentration of A does not vary appreciably.

To determine the decrease in the rate of the reaction in the case in which the diffusion through the porous pellet is important, a new magnitude is introduced, the efficiency factor, which is expressed by

$$\eta = \frac{\text{actual average generation (kmol } A/\text{s)}}{\text{average generation without internal diffusion effect (kmol } A/\text{s)}}$$
$$= \frac{\text{actual average reaction rate (kmol } A/(\text{s m}^3\text{ cat}))}{\begin{array}{c}\text{average reaction rate without internal}\\ \text{diffusion effect (kmol } A/(\text{s m}^3\text{ cat}))\end{array}}$$

$$\eta = \frac{\overline{(-r_A''')}}{(-r_A''')_{\text{no diffusion problems}}} = \frac{\frac{\int_0^V (-r_A''')dV}{V}}{k'''C_{As}} = \frac{\int_0^L (-r_A''')dL}{k'''C_{As}L}$$
$$= \frac{\int_0^L \left(-D_e \frac{dC_A}{dx}\right) dL}{k'''C_{As}L} = \frac{D_e}{k'''C_{As}L}\left[\frac{dC_A}{dx}\right]_{x=0} \tag{9.46}$$

where $(-r_A''')$and $\overline{(-r_A''')}$ are expressed in (kmoles of $A/(\text{m}^3{}_{\text{cat}}\ \text{s})$), V is the volume of the catalytic particle, and S_{total} represents the total surface through which there is input of reagent A (in the case considered of a laminar flat catalyst with entry of reagent by both sides, $S_{\text{total}} = 2S$). Note the difference between the average real rate $(-\bar{r}_A''')$, which can be determined experimentally, and the rate $(-r_A''')$ in each layer of the porous particle.

Then:

$$(-\bar{r}_A''') = k'''C_{As}\eta \tag{9.47}$$

which is a very comfortable expression to work with, since the efficiency factor modifies the reaction rate according to the effect of diffusion.

Taking the equation of the concentration profile deduced before (Eq. (9.45)) we obtain

$$\frac{dC_A}{dx} = \frac{C_{As}}{\cosh(mL)}(-m)\ \sinh(m(L-x)) \tag{9.48}$$

which for $x = 0$ provides

$$\left.\frac{dC_A}{dx}\right]_{x=0} = \frac{-m\ C_{As}}{\cosh(mL)}\ \sinh(mL) = -m\ C_{As}\tanh(mL) \tag{9.49}$$

Introducing this equation in (9.46), we obtain

$$\eta = \frac{\tanh(mL)}{mL} \tag{9.50}$$

This function can be approximated in two different regimes:

- When there is no resistance to diffusion in the pellet, $(\Phi = mL < 0.4)$, $\eta \cong 1$, and

$$(-\bar{r}_A''') = k'''C_{As} \tag{9.51}$$

- When resistance to diffusion is important, $(\Phi = mL > 4)$, $\eta \cong 1/mL$, and

$$(-\bar{r}_A''') = \frac{k'''C_{As}}{mL}\eta = \frac{C_{As}(k'''\cdot D_e)^{\frac{1}{2}}}{L} \tag{9.52}$$

At first sight, it could seem that the diffusion in the pellet can be considered as a corrective multiplicative factor of any of the previous cases, using η. But this is not the case, since this factor considers both the reaction term on the surface and the diffusion term.

9.5.2 First-order Kinetics in Other Geometries

The problem posed for the porous flat pellet with transport of A through both sides can be raised in other cases: porous pellet permeable to reagents and

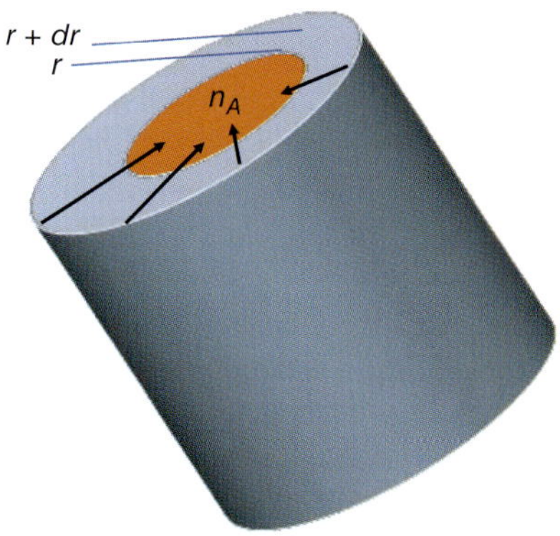

Figure 9.6 Porous cylindrical catalyst pellet with reagent A entry on the exterior surface.

products on one side, cylindrical pellets, and spherical particles. In the case of the permeable single porous pellet on one side, the solution is exactly the same as the problem solved previously but considering "L" at the total thickness.

In the case of cylindrical particles, represented in Figure 9.6, a mass balance of the reactive A would be

$$\text{Input} - \text{Output} + \text{Accumulation} = \text{Generation}$$

$$(n_A \cdot S)]_{r+dr} - (n_A \cdot S)]_r + 0 = r'''_A \cdot dV = -k''' \cdot C_A \cdot dV \tag{9.53}$$

where:

$$S = 2 \cdot \pi \cdot r \cdot h$$

and

$$dV = d(\pi \cdot r^2 \cdot h) = 2 \cdot \pi \cdot r \cdot dr$$

Therefore, substituting the expression of "S" and dividing Eq. (9.53) by $(2 \cdot \pi)$, we get

$$(n_A \cdot r)]_{r+dr} - (n_A \cdot r)]_r = d(N_A \cdot r) = -k''' \cdot C_A \cdot r \cdot dr \tag{9.54}$$

Rearranging:

$$\frac{d(n_A \cdot r)}{dr} = -k''' \cdot C_A \cdot r \tag{9.55}$$

Introducing Fick's law:

$$\frac{d\left(-D_e \cdot \frac{dC_A}{dr} \cdot r\right)}{dr} = -k''' \cdot C_A \cdot r \tag{9.56}$$

and then:

$$\frac{1}{r} \cdot \frac{d\left(\frac{dC_A}{dr} \cdot r\right)}{dr} = \frac{k''' \cdot C_A}{D_e} \tag{9.57}$$

and, operating:

$$\frac{d^2C_A}{dr^2} + \frac{1}{r} \cdot \frac{dC_A}{dr} - \frac{k''' \cdot C_A}{D_e} = 0 \tag{9.58}$$

This second-order differential equation is of the type:

$$y''(x) + \frac{y'(x)}{x} + q \cdot y(x) = 0$$

This is an ordinary differential equation with a complex function for the concentration of reactant, resulting:

$$\frac{C_A(r)}{C_{As}} = \frac{J_0\left(R \cdot \sqrt{\frac{k'''}{D_e}} \cdot r\right)}{J_0\left(R \cdot \sqrt{\frac{k'''}{D_e}}\right)} \tag{9.59}$$

The function J_0 is called the modified Bessel function of the first kind of order zero. "R" represents the total radius of the cylindrical particle. Let us call $L = R/2$, and then:

$$\frac{C_A(r)}{C_{As}} = \frac{J_0(2 \cdot m \cdot L \cdot r)}{J_0(2 \cdot m \cdot L)} \tag{9.60}$$

As we have defined the effectiveness as the quotient between the actual reaction rate and those we would have if no diffusional limitations are present:

$$\eta = \frac{(-\bar{r}_A{'''})}{(-r_A{'''})_{\text{no diffusion problems}}} \tag{9.46}$$

We can then write

$$(-\bar{r}_A{'''}) = \frac{\int_0^V (-r_A{'''}) \cdot dV}{V} = \frac{\int_0^V k''' \cdot C_A(r) \cdot dV}{V} \tag{9.61}$$

Then, for calculating the effectiveness factor we must do

$$\eta = \frac{\frac{\int_0^V k''' \cdot C_A(r) \cdot dV}{V}}{k''' \cdot C_{As}} = \frac{\int_0^V k''' \cdot C_{As} \cdot \frac{J_0(2 \cdot mL \cdot r)}{J_0(2 \cdot mL)} \cdot dV}{V \cdot k''' \cdot C_{As}} = \frac{1}{m \cdot L} \cdot \frac{J_1(2 \cdot mL)}{J_0(2 \cdot mL)} \tag{9.62}$$

The solution of the integral of the Bessel function can be consulted in different books treating mathematics. If we do the balance in a spherical geometry, following the same procedures, the expression for the effectiveness factor is

$$\eta = \frac{(-\bar{r}_A)}{(-r_A)_{\text{no diffusion problems}}} = \frac{\int_0^V k \cdot C_A(r) \cdot dV}{\int_0^V k \cdot C_{As} \cdot dV} \tag{9.63}$$

Evaluation of the integral in the denominator is straightforward, and obviously equals $\left(\frac{4}{3} k C_{As} \pi R^3\right)$. The integral in the numerator is more complicated to evaluate and it is easier to calculate the observed rate from the flow into the pellet. A balance of the reactant over the whole spherical pellet can be written as

$$\int_0^V k''' \cdot C_A(r) \cdot dV = -(-D_e 4\pi R^2)\left(\frac{dC}{dr}\right)_{r=R} \tag{9.64}$$

That is, the observed reaction rate equals the rate of diffusion from the pellet surface into the pellet. The minus sign outside the parenthesis is needed because

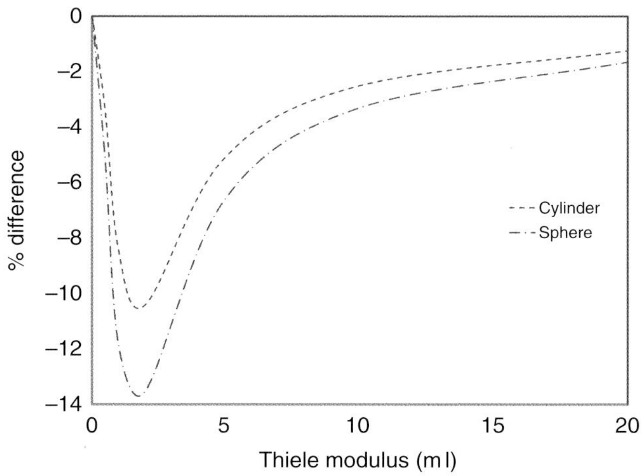

Figure 9.7 Error implied in the use of the approximation Eq. (9.50) instead of Eq. (9.62) (cylinder) and Eq. (9.65) (sphere).

the flow into the pellet has a negative radial direction. Differentiation is generally easier to do than integration, so from Eq. (9.64) we can get

$$\eta = \frac{\frac{\int_0^V k'''\cdot C_A(r)\cdot dV}{V}}{k''' \cdot C_{As}} = \frac{1}{m \cdot L} \cdot \left(\frac{1}{\tanh(mL)} - \frac{1}{3 \cdot mL} \right) \tag{9.65}$$

On this occasion, for the spherical particles we must use $L = R/3$.

As we see, the use of these characteristic lengths greatly simplifies the calculation of the efficiency factors. One more approximation is usually done frequently: since Eqs. (9.50), (9.62), and (9.65) give approximately the same values of η for a given value of $(m{\cdot}L)$, the simplest Eq. (9.50) is taken as valid, making for a reasonably low error. In Figure 9.7, we can check the relative error implied in the use of Eq. (9.50) instead of the corresponding equation for cylindrical and spherical geometries. As we can see, the maximum error is approximately −14% at intermediate values of $(m{\cdot}L)$. Errors are negative, i.e. the corresponding effectiveness factor calculated using (9.50) is higher than that of Eqs. (9.62) and (9.65).

In this way, we see that the effectiveness factor for different geometries can be calculated in a totally equivalent form to the case of plane geometry, only taking care to choose the correct value of L. Actually, for a given catalyst with arbitrary geometry, or for a catalyst of various mixed forms, the use of Eq. (9.50) is extended by particularizing the characteristic length using the relation:

$$L = \frac{\text{Volume of the particles}}{\text{External surface available for entry and difussion of the reagent}} \tag{9.66}$$

As mentioned before, this is valid only for first-order kinetics, which was assumed for deducting the equations. If we pretend to generalize the Thiele modulus in such a way that (Eq. (9.50)) continues being correct, we must use the following definition:

$$\Phi = m \cdot L = \frac{(-r'''_{As}) \cdot L}{\left[2D_e \int_{C_{Ae}}^{C_{As}} (-r'''_A) \cdot dC_A\right]^{1/2}} \tag{9.67}$$

where C_{Ae} is the equilibrium concentration of the reaction, being 0 when treating an irreversible reaction. This generalized modulus is if the reaction is first order and reversible *with rapid diffusion of the products and, therefore, with constant concentration inside the particle,* converted to

$$\Phi = m \cdot L = L \cdot \sqrt{\frac{k'''}{D_e \cdot X_{Ae}}} \tag{9.68}$$

X_{Ae} being the molar conversion of A in equilibrium with the reaction products (in the case where the diffusion of the products is also important, the corresponding equations must be presented and resolved numerically).

For irreversible reactions of nth order, the corresponding expression is

$$m \cdot L = L \cdot \left[\frac{k''' \cdot C_{As}^{n-1}(n+1)}{2 \cdot D_e}\right]^{\frac{1}{2}} \tag{9.69}$$

where C_{As} the concentration of C_A on the outer surface of the particle.

9.5.3 Limits of Thiele Modulus and Weisz Modulus

As commented on before, a low value of the Thiele modulus indicates a system with no diffusion problems, i.e. a system where the effectiveness factor is unity. A value of $\Phi = mL < 0.4$ is generally accepted to represent a system where the resistance to internal diffusion is negligible. Another important module giving an idea of the rate of diffusion vs. chemical reaction is the Weisz modulus, defined by

$$We = \frac{(-r'''_A)_{\text{Observed}} \cdot L^2}{D_e \cdot C_{As}} \tag{9.70}$$

In the case where $mL < 0.4 \rightarrow \eta \approx 0.95\text{–}1 \rightarrow \eta(mL)^2 \ll 1$, we will have $(-r_A)_{\text{Observed}} = k \cdot C_{As}$ and the Weisz modulus is

$$We = \frac{k''' \cdot C_{As} \cdot L^2}{D_e \cdot C_{As}} < 0.15 \tag{9.71}$$

On the other hand, in the case where the diffusion in the pores is important, we have $mL > 4$, and $\eta = 1/(mL)$ so $(-r'''_A)_{\text{Observed}} = k''' \cdot \frac{C_{As}}{m \cdot L}$ and then:

$$We = \frac{k''' \cdot C_{As} \cdot L^2}{m \cdot L \cdot D_e \cdot C_{As}} = k''' \cdot \frac{L}{D_e} > 4 \tag{9.72}$$

Example 9.3 ***Diesel Cracking***

A diesel is cracked at 630 °C and 1 atm by passing vaporized feed through a packed bed of silica-alumina spherical particles with radius 0.088 cm. For a feed of 60 cm^3 liquid diesel/(cm^3 catalytic reactor h), the conversion of the diesel is 50%. The density of the diesel is 0.869 g/cm^3, the average molecular weight is 255 g/mol, the overall density of the filler bed (catalyst mass/filling volume) is 0.7 g/cm^3, the catalyst density is 0.95 g/cm^3, the effective diffusivity of the diesel in the catalyst is 8·10^{-4} cm^2/s. The average concentration of diesel in the volume considered is 0.6·10^{-5} mol/cm^3. Admitting a first-order reaction:

(a) Calculate the average reaction rate (generation/volume catalytic reactor).
(b) Analyze if there is influence of internal diffusion.
(c) Calculate the efficiency factor.
(d) Estimate the value of the kinetic constant.

Solution

(a) In the data, it is said that a feed rate of 60 h^{-1} produces a conversion of $X_A = 0.5$. The average reaction rate would be given by

$$(-r''_A) = 60 \frac{\text{cm}^3 \text{ diesel}}{\text{cm}^3 \text{ reactor h}} \cdot X_A \cdot \frac{\rho_{\text{diesel}} \text{ g/cm}^3}{MW_{\text{diesel}} \text{g/mol}}$$

$$(-r''_A) = 60 \frac{\text{cm}^3_{\text{diesel}}}{\text{cm}^3_{\text{reactor}} \text{ h}} \cdot 0.5 \cdot \frac{0.869 \text{ g/cm}^3}{255 \text{ g/mol}} = 0.1022 \frac{\text{mol}}{\text{cm}^3_{\text{reactor}} \text{ h}}$$

(b) For this purpose, we will calculate the Weisz modulus:

$$We = \frac{(-r'''_A) \cdot L^2}{D_e \cdot C_A}$$

For a spherical particle, we have that $L = R/3$, and then:

$$(-r'''_A) = (-r''_A)(1-\varepsilon) \frac{\text{cm}^3_{\text{reactor}}}{\text{cm}^3_{\text{cat}}} = 0.1022 \frac{\text{mol}}{\text{cm}^3_{\text{reactor}} \text{ h}} \cdot \frac{1}{0.7} \frac{\text{cm}^3_{\text{reactor}}}{\text{g}_{\text{cat}}}$$

$$\times 0.95 \frac{\text{g}_{\text{cat}}}{\text{cm}^3_{\text{cat}}} = 0.1387 \frac{\text{mol}}{\text{cm}^3_{\text{cat}} \text{ h}}$$

$$We = \frac{(-r'''_A) \cdot (R/3)^2}{D_e \cdot C_A} = \frac{0.1387 \frac{\text{mol}}{\text{cm}^3_{\text{cat}} \, h} \cdot \frac{1 \text{ h}}{3600 \, s} \cdot \left(\frac{0.088}{3}\right)^2 \text{cm}^2}{8 \cdot 10^{-4} \frac{\text{cm}^2}{\text{s}} \cdot 0.6 \cdot 10^{-5} \frac{\text{mol}}{\text{cm}^3}} = 6.2$$

As the Weisz modulus is higher than 4, the influence of the internal diffusion is quite important.

(c) In order to calculate the efficiency factor, as the internal diffusion is important, we can do

$$\eta = \frac{\tanh(mL)}{mL} \cong \frac{1}{mL}$$

$$We = \frac{(-r'''_A) \cdot L^2}{D_e \cdot C_A} = \frac{\frac{k''' \cdot C_{As}}{mL} \cdot L^2}{D_e \cdot C_A} = \eta \cdot (mL)^2 \cong mL$$

where $m^2 = k'''/D_e$. And then, with the value of the Thiele's modulus: $\eta \cong \frac{1}{6.2} = 0.161$

(d) We would like to estimate the kinetic constant, so:

$$(-r_A''') = k''' \cdot C_A \cdot \eta \quad \text{i.e. } k''' = \frac{(-r_A''')}{C_A \cdot \eta}$$

$$= \frac{0.122 \frac{\text{mol}}{\text{cm}^3 \text{ reactor h}}}{0.6 \cdot 10^{-5} \frac{\text{mol}}{\text{cm}^3} \cdot 0.161} = 1.05 \cdot 10^5 \text{ h}^{-1}$$

Example 9.4 ***Fixed-Bed Reactor Design***

In a stirred tank experimental reactor with 10 g of 1.2 mm spherical catalyst particles and a feed of 5 l/s of pure A at 1 atm and 333 °C, a conversion of 70% is obtained for the first-order reaction:

$$A \rightarrow R \quad \Delta H_r = 0$$

We want to design a fixed-bed reactor (considered ideal piston flow) of 1.4 cm particles. Calculate the amount of catalyst needed in this bed. Data: Solid density = 2000 kg/m³ $D_e = 10^{-6}$ m³/(m$_{cat}$ s)

Solution

From the data at the experimental tank reactor, we can calculate the value of the kinetic constant, as in the CSTR we have

$$n_{A0} - n_A = (k' \cdot C_A) \cdot W \text{ i.e. } k' = \frac{n_{A0} - n_A}{C_A \cdot W}$$

The inlet molar feed rate can be calculated from the data:

$$n_{A0} = \frac{P \cdot Q}{R \cdot T} = \frac{1 \text{ atm} \cdot 5 \text{ l/s}}{0.082 \frac{\text{atm l}}{\text{mol K}} \cdot 606 \text{ K}} = 0.1 \frac{\text{mol}}{\text{s}}$$

$$n_A = 0.1 \cdot (1 - 0.7) = 0.03 \frac{\text{mol}}{\text{s}}$$

$$C_A = \frac{n_A}{Q} = \frac{0.03}{5} = 0.006 \frac{\text{mol}}{\text{l}}$$

$$k' = \frac{n_{A0} - n_A}{C_A \cdot M} = \frac{(0.1 - 0.03)}{0.006 \cdot 10} = 11.66 \frac{\text{l}}{\text{s g}_{\text{cat}}} = 1.166 \cdot 10^{-3} \frac{\text{m}^3}{\text{s g}_{\text{cat}}}$$

For the bed of big particles, considering a PFR:

$$r_A = \frac{dn_A}{dM}, \text{i.e.} - (k' \cdot C_A \cdot \eta) \cdot dW = -n_{A0} \cdot dX_A$$

$$+ (k' \cdot C_{A0}(1 - X_A) \cdot \eta) \cdot dW = +Q \cdot C_{A0} \cdot dX_A$$

$$W = \int_0^{X_A \text{final}} \frac{Q}{(k' \cdot (1 - X_A) \cdot \eta)} \cdot dX_A$$

The value of the efficiency factor can be estimated: $\eta = \frac{\tanh(mL)}{mL}$, with values:

$$m = \sqrt{\frac{k'''}{D_e}} \quad \text{and} \quad L = R/3$$

But in the calculation of "m" the correct units must be used:

$$m = \sqrt{\frac{k'''}{D_e}} = \sqrt{\frac{1.166 \cdot 10^{-3}\ \frac{m^3}{s\,g} \cdot \frac{10^3 g}{kg} \cdot 2000 \frac{kg_{cat}}{m^3_{cat}}}{10^{-6}\ \frac{m^3}{m_{cat}\ s}}} = 48\,305\ \text{m}^{-1}$$

and $L = 4.6 \cdot 10^{-3}$ m, so $m \cdot L = 225.4$ and $\eta = 4.43 \cdot 10^{-3}$

We can now calculate the weight of the catalyst required:

$$W = \int_0^{X_{A\text{final}}} \frac{Q}{(k' \cdot (1 - X_A) \cdot \eta)} \cdot dX_A = -\frac{Q}{(k' \cdot \eta)} \ln(1 - X_{A\text{final}}) = 116.4\ \text{g}$$

Example 9.5 ***Investigating the Controlling Step***

In an experiment to investigate the controlling step of a chemical reaction of decomposition of a substance "A," with a catalyst, we found that:

- ✓ Spherical particles of $d_p = 2.4$ mm of diameter
- ✓ Effective diffusion coefficient, $D_e = 5 \cdot 10^{-5}\ \text{m}^3/(\text{h m}_{cat})$
- ✓ Mass transfer coefficient, $k_C = 300\ \text{m}^3/(\text{h m}^2{}_{cat})$
- ✓ First-order reaction, with $C_A = 20\ \text{mol/m}^3$ (at 1 atm and 336 °C)
- ✓ Observed reaction rate, $(-r_A)_{obs} = 0.042\ \text{mol}/(\text{h m}^2{}_{cat})$

Do the needed calculations to investigate which factors are affecting the reaction rate.

Solution

First, we can estimate the rate of external diffusion of the reactive:

$$r_A = k_c(C_{Ag} - C_{As}) \cong k_c C_{Ag} = 300 \cdot 20 = 6000 \frac{\text{mol}}{\text{h m}^2_{cat}}$$

This rate represents the maximum transfer through the external film in the catalytic system. This is quite high and, at first sight, probably there will not be external diffusion problems, but this must be compared to the other steps.

The Weisz modulus can be calculated to estimate the internal diffusion problems. For that, the reaction rate should be expressed in the correct units:

$$g_A = r_A \cdot S_p = r_A \cdot \left(4\pi \left(\frac{d_p}{2}\right)^2\right) = 0.042 \cdot \left(4\pi \left(\frac{2.4 \cdot 10^{-3}}{2}\right)^2\right)$$
$$= 7.6 \cdot 10^{-7}\ \text{mol}\,A/\text{h}$$

$$r'''_A = \frac{g_A}{V_p} = \frac{g_A}{\left(\frac{4}{3}\pi\left(\frac{d_p}{2}\right)^3\right)} = \ldots = 105 \frac{\text{mol}\,A}{\text{h m}^3_c}$$

The concentration in the catalyst surface is the same as that in the bulk fluid if no problems of external diffusion are present. Then:

$$We = \frac{-r'''_A \cdot L^2}{D_e \cdot C_{As}} = \frac{105\ \frac{\text{mol A}}{\text{h m}^3_{cat}} \cdot (1.2 \cdot 10^{-4})^2\ m^2_{cat}}{5 \cdot 10^{-5} \frac{\text{m}^3}{\text{h m}_{cat}} \cdot 20 \frac{\text{mol A}}{\text{m}^3}} = 0.0162 \ll 0.15$$

and there are no internal diffusion problems.

As the observed reaction rate is much lower than that of the external diffusion, it is only possible that the chemical reaction is dominating the process. In this way, the lowest rate is that of the chemical reaction and equals $0.042\frac{\text{mol A}}{\text{h·m}_c^2}$.

Example 9.6 ***Catalytic Reactor with Spherical Particles***

A 50 ml volume reactor is completely filled with spherical catalyst particles 3 mm in diameter. The porosity of the bed is 0.6. The reactor can be considered as an ideal piston flow. The decomposition of pure A follows the kinetics of first order:

$$(-r_A) = k \cdot C_A \ldots (\text{mol}/(\text{s m}^2_{\text{cat}}))$$

where C_A = concentration of A (mol/m^3), $k = 3{\cdot}10^{-6}$ m^3/(s m$^{-2}{}_{\text{cat}}$)

For the catalyst, the following data is available: density = 1500 kg/m^3; specific surface = 400 m^2/g; diffusion coefficient of reactant = $2{\cdot}10^{-6}$ m^2/s

(a) Indicate the expression to calculate the efficiency factor and the design equation of an isothermal piston flow reactor with a certain amount of this catalyst.
(b) In the given situation, the conversion of A to the reactor is 0.4. Calculate the efficiency factor of the catalyst and the volumetric flow rate that passes through the reactor.
(c) The reaction tube is subjected to a vibration system, causing the density of the bed to increase, decreasing the porosity to 0.52. Calculate the conversion that will result.
(d) The 3 mm particles are replaced by the same catalyst mass, but now with a diameter of 0.03 mm. The porosity of this bed is also 0.52. What is the new conversion?

Solution

(a) The design equation for a PFR is given by

$$+(k' \cdot C_{A0}(1 - X_A) \cdot \eta) \cdot dW = +Q \cdot C_{A0} \cdot dX_A$$

and integrating:

$$W = \frac{Q}{k' \cdot \eta} \int_0^{X_{Af}} \frac{dX_A}{(1 - X_A)} = -\frac{Q}{k' \cdot \eta} \cdot \ln(1 - X_{Af})$$

Note that the kinetic constant in the mass balance done must be expressed in units of flow rate/(mass of catalyst), i.e. m^3/(s g). The effectiveness factor can be calculated using the following expression:

$$\eta = \frac{\tanh(m \cdot L)}{m \cdot L}$$

$$m \cdot L = \sqrt{\frac{k'''}{D_e}} \cdot \frac{R}{3}$$

In that expression, the kinetic constant should be given in units of (time)$^{-1}$.

(b) We know that the conversion is $X_A = 0.4$ and:

$$k''' = 3\cdot 10^{-6}\frac{\text{m}^3}{\text{s m}^2_{\text{cat}}}\cdot 400\frac{\text{m}^2_{\text{cat}}}{\text{g}_{\text{cat}}}\cdot 1500\cdot 10^3\frac{\text{g}_{\text{cat}}}{\text{m}^3_{\text{cat}}} = 1800\ \text{s}^{-1}$$

$$m\cdot L = \sqrt{\frac{1800\ \text{s}^{-1}}{2\cdot 10^{-6}\ (\text{m}^2/\text{s})}}\cdot\frac{1.5\cdot 10^{-3}\ (\text{m})}{3} = 15$$

$$\eta = \frac{\tanh(15)}{15} = 0.0667$$

We need to know the amount of catalyst in the reactor bed:

$$W = 50\ \text{ml}_{\text{bed}}\cdot 10^{-6}\frac{\text{m}^3_{\text{bed}}}{\text{ml}_{\text{bed}}}\cdot(1-0.6)\frac{\text{m}^3_{\text{cat}}}{\text{m}^3_{\text{bed}}}\cdot 1500\frac{\text{kg}_{\text{cat}}}{\text{m}^3_{\text{cat}}} = 0.03\ \text{kg}_{\text{cat}} = 30\ \text{g}_{\text{cat}}$$

Using the data:

$$k' = 3\cdot 10^{-6}\frac{\text{m}^3}{\text{s m}^2_{\text{cat}}}\cdot 400\frac{\text{m}^2_{\text{cat}}}{\text{g}_{\text{cat}}} = 12\cdot 10^{-4}\frac{\text{m}^3}{\text{s g}_{\text{cat}}}$$

$$W = -\frac{Q}{k'\cdot\eta}\cdot\ln(1-X_{Af}) = 30\,g$$

$$= -\frac{Q\ (\text{m}^3/\text{s})}{12\cdot 10^{-4}\ \text{m}^3\text{s}^{-1}\text{g}^{-1}\cdot 0.0667}\cdot\ln(1-0.4)$$

Resulting $Q = 4.7\cdot 10^{-3}\ \text{m}^3/\text{s}$.

(c) In this case, the lower porosity will produce an increase in the void volume, but the amount of catalyst will be the same, so the conversion does not vary.

(d) If the particles had a diameter of 0.03 mm:

$$L = 0.5\cdot 10^{-6}\ \text{m}\rightarrow m\cdot L = 0.15\rightarrow \eta = 0.992\rightarrow 30$$

$$= -\frac{4.7\cdot 10^{-3}}{12\cdot 10^{-4}\cdot 0.992}\cdot\ln(1-X_A)$$

From that equation, $X_A = 0.9994$

Example 9.7 ***Reaction with Internal Diffusion Resistance***

In the design of the reactor presented in Example 9.1, it has been found that the internal diffusion stage plays an important role in the reaction rate under certain conditions.

The system consists of spherical particles 6 mm in diameter with an internal surface area of 400 m^2/g, in this case presenting reactant A with an effective diffusivity of $2.66\cdot 10^{-8}$ m^2/s.

Calculate the variation of the conversion of A with the length of the reactor for a total length of 0.1 m under isothermal conditions. To carry out the design, it will be necessary to determine the variation of the Thiele module and the efficiency factor of the catalyst along the bed.

Solution

In this case, the reaction rate is modified by the internal diffusion, so:

$$(-r'_A) = k'\cdot\eta\cdot C_A^2$$

$$m \cdot L = \left(\frac{R_p}{3}\right) \cdot \left(k''' \cdot C_A \cdot \frac{3}{2 \cdot D_e}\right)^{\frac{1}{2}}$$

$$\eta = \frac{\tanh(mL)}{mL}$$

The problem is very similar to that presented in Example 9.1, and just the previous equations must be included. A valid program for that calculation is given. For changing the units of the constants, the following relation will be used:

$$k'''(\mathrm{m}_L^3/(\mathrm{m}_{\mathrm{cat}}^3\ \mathrm{s})) = k'(\mathrm{m}_L^3/(\mathrm{kg}_{\mathrm{cat}}\ \mathrm{s})) \cdot \rho_{\mathrm{cat}}\ (\mathrm{kg}_{\mathrm{cat}}/\mathrm{m}_{\mathrm{cat}}^3)$$

Program Listing 9.3

```
clear all
clear all

u0=3;% m/s
porosity=0.4;
dcat=2000;% kg/m3 catalyst
k0=7.707e5;% m6/mol·kg·s

E_R=8656.1; % K

P=500000; % Pa
Ca0=P/(8.314*523); % mol/m3

T=250+273; % K
k=k0*exp(-E_R/T); % m6/mol·kg·s

De=2.66e-8;% diffusivity, m2/s
dp=6e-3;% diameter of particles, m
XA(1)=0; % Initial value of conversion
L(1)=0; % Initial value of length
incx=0.0001; % Increment of length
n=0.1/incx; % Number of points
CA=Ca0;
mL=(dp/6)*(k.*CA*3/(2*De)).^0.5;
effectivity=tanh(mL)/mL;

for i=2:n
    L(i)=L(i-1)+incx;
    XA(i)=XA(i-1)+(dcat*(1-porosity)*k*Ca0*(1-XA(i-
1))^2/(u0*(1+2*XA(i-1))^2))*incx;
    CA=Ca0*(1-XA(i))/(1+2*XA(i));
    mL=(dp/6)*(k*dcat.*CA*3/(2*De)).^0.5;
    effectivity=tanh(mL)./mL
end
```

```
figure(1)
plot(L,XA)
xlabel('Length (m)')
ylabel('Molar conversion')
```

The variation of the conversion with length is given in the following figure. It is quite similar to that presented in Example 9.1, but the conversion increases slowly. This is due to the correction of the constant done by the effectivity, which is approximately 0.5 in this design (it can be checked along the program; Program listing 9.3).

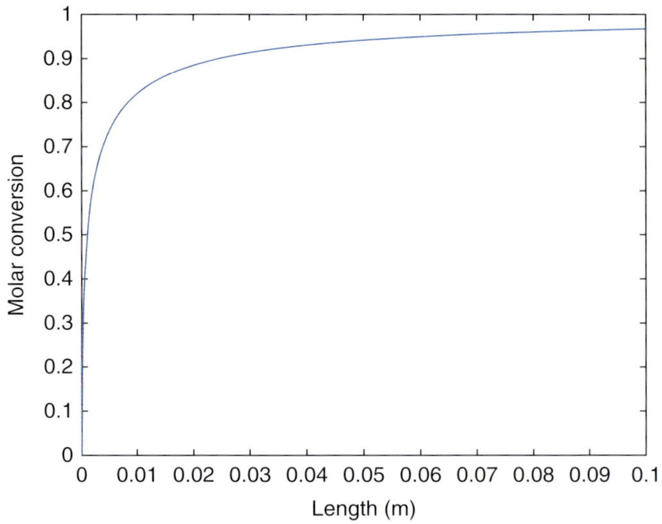

9.6 Combination of Resistances

In the general case, all steps in the chemical reaction process can affect the reaction rate, so let us see what happens if there is no step controlling the rate.

The flow of reactant through the external layer could be expressed using Eq. (9.25):

$$(-r_A) = -\frac{g_A}{S_{\text{ex}}} = k_C \cdot (C_{Ag} - C_{As}) \quad (\text{kmol}/(\text{s m}^2)) \tag{9.25}$$

But in the general case, C_{As} does not equal zero. At the same time, the diffusion thought the pores and chemical reaction rate is given by:

$$(-\bar{r}_A) = kC_{As}\eta \tag{9.73}$$

Obviously, there is no accumulation of reactant species A in the external surface, so these two rates must be equal. Therefore:

$$(-r_A) = k_C \cdot (C_{Ag} - C_{As}) = kC_{As}\eta \tag{9.74}$$

and doing a combination of resistances, we get

$$(-r_A) = \frac{(C_{Ag} - C_{As})}{\frac{1}{k_C}} = \frac{C_{As}}{\frac{1}{k\eta}} = \frac{C_{Ag}}{\frac{1}{k_C} + \frac{1}{k\eta}} \tag{9.75}$$

For the last equality, we have added the numerators and the denominators of the previous quotients. Note that the final equation has two summands in the denominator, corresponding to the resistance to the diffusion through the external layer $\left(\frac{1}{k_C}\right)$ and that of the combined step of pore diffusion and reaction $\left(\frac{1}{k\eta}\right)$. In the case that k_C is infinity (no resistance to external diffusion), $\left(\frac{1}{k_C}\right) = 0$ and the reaction rate is given by (9.74). Contrarily, if $k\eta$ is infinity, Eq. (9.75) equals Eq. (9.25) with $C_{As} = 0$ (logically, as the diffusion and reaction is infinitely fast).

9.7 Monolithic Catalytic Reactors

As stated in Chapter 8, many industrial operations take place with catalysts supported on ceramic materials, or similar. This is a quite different situation compared to those mentioned in previous sections, as the "characteristic length" of these catalytic structures is not easily visualized.

The cordierite that forms the monolith cannot be used to support the catalysts due to its low surface ratio and because it is not able to efficiently disperse the metal particles. In this sense, the first step of the process is to adsorb a thin layer (called "coating" or "washcoat") of a stable material of high surface, on which the metal particles will be dispersed. A diagram of the structure is shown in Figure 9.8. A good approximation is achieved when half the width of the coating is used as the characteristic length of the catalyst washcoat.

Example 9.8 ***Pore Diffusional Resistance For a Monolith Catalyst***

Elimination of SO_2 by reduction is an important process in industrial systems. It has been reported as a first-order reaction kinetics, i.e. $r = k{\cdot}C_{SO_2}$ for a 5% V_2O_5/TiO_2 catalyst. The value of the kinetic constant is 0.566 $cm^3/(g_{cat}\ s)$ at 500 K and the activation energy found was 110 kJ/mol. Calculate the effectiveness factor and reaction rate at the inlet of a commercial unit containing an extruded cellular monolith consisting of 5% V_2O_5/TiO_2 with a channel wall thickness of 1.35 mm and open frontal area of 64% under typical commercial operating inlet conditions of 400 ppm of SO_2, 350 °C and 1 atm. Assume that a catalyst in the square monolith walls can be treated as a flat plate. Additional data: particle density = 1.48 g/cm^3; $D_e = 0.07\ cm^2/s$. Assume that the film mass transfer resistance is negligible.

Solution

For calculating the effectiveness factor, both the diffusion and the chemical reaction rate are needed at the system temperature; then:

$$k'_{623K} = k'_{500K} \exp\left\{\frac{E}{R}\left(\frac{1}{623} - \frac{1}{500}\right)\right\} = 105\ cm^3/(g_{cat}\ s)$$

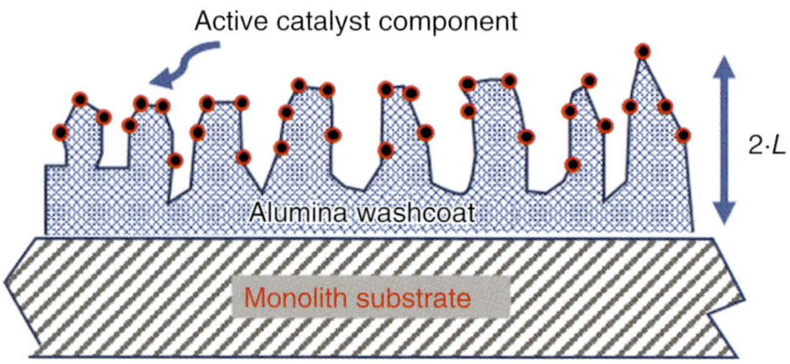

Figure 9.8 Monolithic catalyst structure.

The equivalent length can be calculated:

$$\frac{V_{cat}}{S_{cat}} = \frac{\text{length} \cdot \text{width} \cdot \text{thickness}}{2 \cdot (\text{length} \cdot \text{width})} = \frac{\text{thickness}}{2}$$

where the factor of 2 accounts for each wall having two sides. The Thiele modulus is then:

$$\Phi = m \cdot L = \sqrt{\frac{k''''}{D_e}} \cdot \left(\frac{\text{thickness}}{2}\right)$$

$$= \sqrt{\frac{105\frac{\text{cm}^3}{\text{g}_{\text{cat}}\ \text{s}} \cdot 1.48\frac{\text{g}_{\text{cat}}}{\text{cm}^3}}{0.07\frac{\text{cm}^2}{\text{s}}}} \cdot \left(\frac{1.35 \cdot 10^{-1}\ \text{cm}}{2}\right) = 3.18$$

And the effectiveness factor is $\eta = \frac{\tanh(mL)}{mL} = \frac{\tanh(3.18)}{3.18} = 0.31$

For calculating the reaction rate:

$$C_{SO_2} = \frac{p_{SO_2}}{RT} = 4.82 \cdot 10^{-9} \frac{\text{mol}}{\text{cm}^3}$$

$$(-r'_{SO2}) = k' \cdot C_{SO_2} \cdot \eta = 105 \cdot 4.82 \cdot 10^{-9} \cdot 0.31 = 2.58 \cdot 10^{-7} \frac{\text{mol}}{\text{g}_{\text{cat}}\ \text{s}}$$

Example 9.9 ***De-NOX via SCR***

A monolithic reactor with vanadium oxide catalyst dispersed in a bee-type honeycomb will be designed. The reactor will be used for the elimination (minimum 90%) of the nitrogen oxides of a stream having the following characteristics:

Flow rate = 30 000 Nm^3/h

NOx content = 1000 mg/Nm^3 (95 wt%of these are NO, the rest NO_2)

A current of 24 wt% ammonia will be fed to reduce the NOx by the catalytic reactions:

$$4NO + 4NH_3 + O_2 \rightarrow 4N_2 + 6H_2O$$

$$2NO_2 + 4NH_3 + O_2 \rightarrow 3N_2 + 6H_2O$$

As for the kinetics of the reaction, it has been shown to be of order 1 with respect to the NO component and of zero order with respect to the ammonia component. The rate constant follows the equation:

$$\ln(k''') = (-3266.8/T) + 8.2208 \ldots \text{ with } k''' \text{in s}^{-1} \text{ and } T \text{ in K.}$$

Negligible effects of diffusion to the catalyst surface are observed if the surface rate is greater than 6 m/s. It will be assumed that the reactor operates under an isothermal regime at a temperature of 623 K. Catalysts supported on panels with the following characteristics are available:

Dimensions: 0.15 × 0.15 × 1.35 m
Fraction of holes: 0.7
Ceramic honeycomb thickness: $5.5 \cdot 10^{-4}$ m
Relative dose of catalyst: 5.88%
Diffusivity of NO in the bed: $2.3 \cdot 10^{-6}$ m^2/s

Determine the dimensions of the bed needed to achieve the desired conversion. Simulate the evolution of the NOx concentration with the length of the reactor. Calculate the flow of ammonia needed for that conversion.

Solution

First of all, we will assume there is no influence of the external diffusion on the catalyst. For estimating the effect of internal (pore) diffusion, we will need the value of the kinetic constant at the temperature of the reactor: k''' (623 K) = 19.63 s^{-1}

Bearing in mind that the characteristic length of this catalytic system is $L = \frac{5.5 \cdot 10^{-4}}{2}$ m, we can write

$$\varnothing = m \cdot L = \sqrt{\frac{19.63 \text{ s}^{-1}}{2.3 \cdot 10^{-6}\text{m}^2/\text{s}}}\,\frac{5.5 \cdot 10^{-4}}{2} = 0.803$$

This value of the Thiele modulus can be used to calculate the efficiency factor:

$$\eta = \frac{\tanh(mL)}{mL} = 0.83$$

In these conditions, the reaction rate would be calculated with:

$$(-r'''_{NO}) = k''' \cdot C_{NO} \cdot \eta = 15.773 \cdot C_{NO}$$

In the system, C_{NOin} is the input concentration of NO to the reactor, i.e. $C_{NOin} = 1000$ mg/Nm^3.

We will assume the reactor is plug flow, so in this reactor we can do a mass balance in a differential volume:

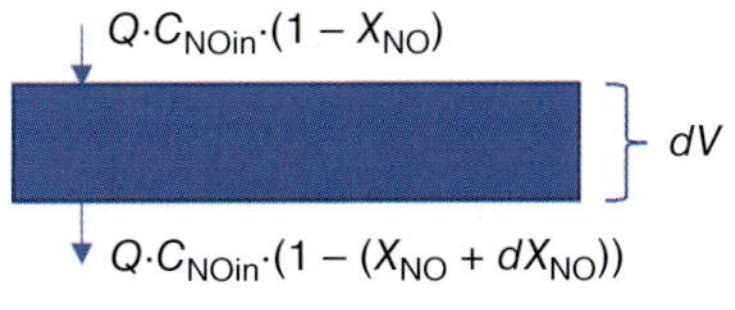

$$n_{NO} + (-r'''_{NO}) \cdot dV = (n_{NO} + dn_{NO})$$
$$Q \cdot C_{NOin} \cdot dX_{NO} = k''' \cdot C_{NOin} \cdot (1 - X_{NO}) \cdot dV$$

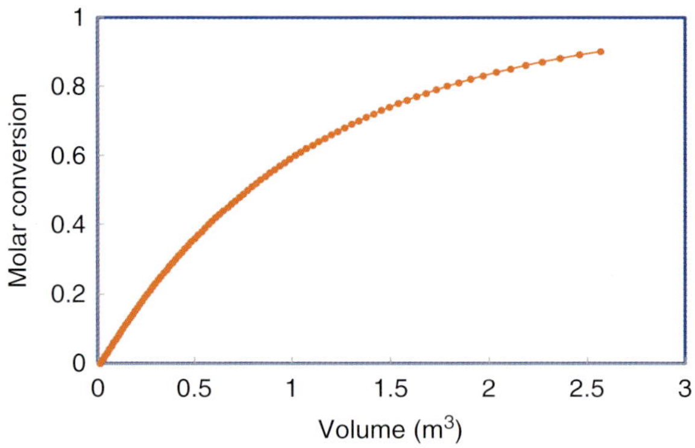

Figure 9.9 Solution to Example 9.7.

Eliminating the concentration and integrating:

$$V = \int_0^{X_{NO}} \frac{Q}{k''' \cdot \eta \cdot (1 - X_{NO})} dX_{NO}$$

The value of Q is known, $Q = 30\,000\,\text{Nm}^3/\text{h} = 62\,720\,\text{m}^3/\text{h}$ (at 623 K) $= 17.42\,\text{m}^3/\text{s}$

Remember that "Nm^3" stands for "normal cubic meter", i.e. measured at 0 °C and 1 atm. The conversion to the corresponding temperature is done by multiplying by T/T_{normal}, in K.

The numerical or analytical integration of the design equation gives the evolution of the conversion with the volume (see Figure 9.9). For a design value of $X_{NO} = 0.9$ the needed volume is 2.61 m^3. This represents the volume of the reactor containing the corresponding catalyst.

The dimension of each cell is $0.15 \times 0.15 \times 1.35\,\text{m} = 0.030\,375\,\text{m}^3$, but not all the volume is active, as one of the data given is the fraction of holes. In this way, the active volume of each cell is 0.030 375*(1 – 0.7) m^3 and the number of cells can be calculated by

$$\text{Number of cells} = \frac{2.61\ \text{active m}^3 \cdot \frac{1\ \text{cell m}^3}{0.7\ \text{active m}^3}}{0.030\,375\ \text{cell m}^3} = 122.51$$

If cells are disposed in a square, the number of cells in each row or column would be $\sqrt{122.51} \approx 11$. A scheme can be found in the following figure, and it contains 121 cells:

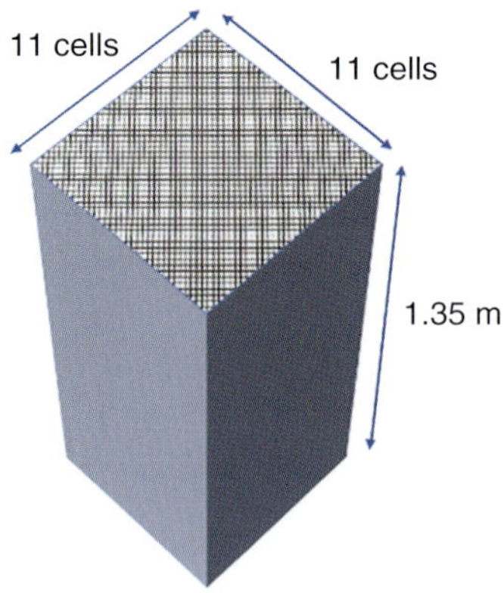

Once we have calculated the needed cell distribution, it is important to check if external diffusion can play a role in the reaction. For this purpose, we can calculate the rate of the gases through the cells. A flow rate of 17.42 m^3/s in a section area of $121 \times (0.15 \times 0.15) = 2.7225\ m^2$. Nevertheless, only 70% of this area is free for flowing (fraction of holes), so must have a lineal velocity of

$$v = \frac{Q}{\text{Section} \cdot 0.7} = \frac{17.42}{1.905} = 9.13\frac{\text{m}}{\text{s}}$$

That is enough according to the requirements. Note that the free area is not calculated subtracting the thickness of the ceramic to the dimension of the cell, and doing the square, as the honeycomb also can contain the monolith substrate as shown in Figure 9.8.

Finally, the calculation of the ammonia feed rate is to be calculated. In the input, we have 31.66 mmol NO/Nm^3 and 1.08 mmol NO_2/Nm^3, in the proportions that the statement indicates. For reacting with these amounts, we will need (31.66 + 2*1.08) = 33.84 mmol NH_3/Nm^3. Having in mind the total flow, this represents 282 mmol NH_3/s, i.e. 4.794 mg/s. Bearing in mind the 24 wt% concentration of this solution, the feed flow needed is approximately 20 g/s.

Example 9.10 ***Oxidation of Ethylene in a Fluidized Bed***
A fluidized-bed reactor will be designed for producing 100 kg/h of ethylene oxide, via the oxidation of ethylene. We will oxidize directly the gas over a silver catalyst. The following conditions will be used:

- ✓ Temperature of the bed = 549 K
- ✓ Pressure of the reactor = 9.65 atm
- ✓ The velocity of the gases will be eight times u_{mf} (minimum velocity of fluidization).
- ✓ The gas entering the reactor has a 6% of oxygen
- ✓ The conversion of the oxygen must be below 40% in order to minimize side reactions.

Two parallel reactions occur in the reactor. On the one hand, the oxidation of ethylene to produce ethylene oxide, and on the other the decomposition of ethylene into final oxidation products:

$$O_2 + 2\,C_2H_4 \rightarrow 2C_2H_4O \quad \Delta H_1 = -2.10 \cdot 10^5 \frac{\text{kJ}}{\text{kmol}}$$

$$O_2 + \frac{1}{3}C_2H_4 \rightarrow \frac{2}{3}CO_2 + \frac{2}{3}H_2O \quad \Delta H_2 = -4.73 \cdot 10^5 \frac{\text{kJ}}{\text{kmol}}$$

Both reactions are first order with respect to the oxygen, and the kinetic constants are

$$k'''_1 = 0.255\,\frac{\text{m}^3_{\text{fluid}}}{\text{m}^3_{\text{catalyst}}\ \text{s}} \qquad k'''_2 = 0.361\,\frac{\text{m}^3_{\text{fluid}}}{\text{m}^3_{\text{catalyst}}\ \text{s}}$$

The catalyst to be used in the reactor will contain a 5% in weight of silver, and an alumina monolith will be the main component. The catalyst will be dispersed into this support in the form of spheres of $75 \cdot 10^{-6}$ m of radius. The effective diffusion coefficient of ethylene in the oxygen is $4.64 \cdot 10^{-6}\ m^2/s$. The density of the catalyst is 1800 kg/m^3.

Calculate the amount of catalyst needed and the molar flow rate profiles of all species through the reactor. *Data:* Molecular weights: $CO_2 = 44$, $C_2H_4O = 44$, $H_2O = 18$, $O_2 = 32$.

Solution

Initially, we will assume the plug flow reactor for the gas and that the catalyst is fixed in a bed. The entering currents will depend on the conversion of the reactor, which must produce

$$n_{B\text{final}} = 100\frac{\text{kg}}{\text{h}} \cdot \frac{1\ \text{kmol}}{44\ \text{kg}}\frac{1}{3600}\frac{\text{h}}{\text{s}} = 0.6313\frac{\text{mol}}{\text{s}}\ \text{ethylene oxide}$$

The high value of velocity of the gases through the bed would indicate that there are no problems of external diffusion, we will have

$$-r_1 = k_1 C_A\, \eta = k_1 C_{A0}(1 - X_A)\, \eta$$

where "A" refers to oxygen. For the effectiveness:

$$\eta = \frac{\tanh(mL)}{mL} \qquad m \cdot L = \left(\frac{R}{3}\right) \cdot \sqrt{\frac{k'''_i}{D_e}}$$

with $i = 1, 2$. The value of L is related to the spherical particles and is calculated using their radius. For the first reaction:

$$m \cdot L = \left(\frac{75 \cdot 10^{-6}}{3}\right) \cdot \sqrt{\frac{0.255}{4.64 \cdot 10^{-7}}} = 0.01\,85$$

and then the effectiveness is approximately unity. The same is true for the second reaction, so internal diffusion does not interfere with the reaction. This is an expected result, as the catalytic particles are so small that it is predictable not to have internal diffusion problems.

Using the plug flow approximation, in a differential volume of catalyst

$$dn_A = (-r_1 - r_2) \cdot dV_{\text{cat}}$$

$$dn_A = (-k'''_1 C_{A0}(1 - X_A) - k'''_2 C_{A0}(1 - X_A)) \cdot dV_{\text{cat}}$$

$$n_{A0} \cdot (-dX_A) = (-k'''_1 C_{A0}(1 - X_A) - k'''_2 C_{A0}(1 - X_A)) \cdot \frac{dW_{\text{cat}}}{\rho_{\text{cat}}}$$

$$\frac{(-dX_A)}{(1 - X_A)} = \frac{C_{A0} \cdot (-k'''_1 - k'''_2)}{n_{A0}} \cdot \frac{dW_{\text{cat}}}{\rho_{\text{cat}}}$$

$$\frac{n_{A0}\rho_{\text{cat}}}{C_{A0}(1 - X_A)(k'''_1 + k'''_2)} dX_A = dW_{\text{cat}}$$

This last equation is the one we must integrate through the reactor (increasing the mass of catalyst) and it will give the change in the molar conversion through the bed.

The molar flow rate of oxygen ("A" in the former equations) can be calculated from the needed production of ethylene oxide. In this way, we can relate the molar flow rates of all species:

Oxygen $\rightarrow n_A = n_{A0}(1 - X_A)$
Ethylene oxide $\rightarrow n_B = 2 \cdot n_{A0} \cdot X_A \cdot S_1$
Carbon dioxide (and water) $\rightarrow n_C = \frac{2}{3} n_{A0}(1 - X_A) \cdot (1 - S_1)$

where S_1 refers to the selectivity of the reaction 1. As the temperature is constant in this design, we can check that the selectivity is also constant:

$$S_1 = \frac{k'''_1 C_A}{k'''_2 C_A} = \frac{k'''_1}{k'''_2} = \frac{0.255}{0.361} = 0.706$$

For $X_{A\text{final}} = 0.4$, $n_{A0} = n_{B\text{final}}/(2 \cdot X_{A\text{final}} \cdot S_1) = 0.6313/(0.8 \cdot 0.706) = 1.117$ mol/s.

Let us use Matlab to solve this example. A valid program to calculate all needed is shown (Program listing 9.4).

Program Listing 9.4

```
%Data
T=549; %K
P=9.65; %atm
Q=100; %kg/h
k1=0.255; %m3fluid/m3cat*s
k2=0.361; %m3fluid/m3cat*s
DH1=-2.1e5; %kJ/kmol
DH2=-4.73e5; %kJ/kmol
R=(75e-6)/2; %m
De= 4.64e-6; %m2/s
denscat=1800; %kg/m3
Cox=0.06;
Rg=0.082/1000; %m3*atm/(K*mol)
MW=44; %g/mol

%Calculation
L=R/3;
m1=sqrt(k1/De);
m2=sqrt(k2/De);
mL1=m1*L;
mL2=m2*L;
Fef1=(tanh(mL1)/mL1);
Fef2=(tanh(mL2)/mL2);

CA0=(P*Cox)/(Rg*T) %mol/m3

nBFinal=Q*(1000/3600)/MW ; %mol/s
S1=k1/k2;

nA0= nBFinal/(2*S1*0.4)
Mcat=0;
n=1000; % Number of points to be used in the integra-
tion
incXA=0.4/(n-1);

% Initial values
XA(1)=0;
```

```
Wcat(1)=0; % weight of catalyst
nA(1)=nA0;
nB(1)=0;
nC(1)=0;


for i=2:n
    XA(i)=XA(i-1)+incXA;
    dWcat=-(((1-XA(i))*nA0*denscat)/(CA0*(-k1*Fef1-
k2*Fef2)));
    Wcat(i)=Wcat(i-1)+dWcat*incXA;
    nA(i)=nA0*(1-XA(i));
    nB(i)=nA0*XA(i)*S1*2;
    nC(i)=(2/3)*nA0*(1-S1)*XA(i);
end

figure(1)
plot(Wcat,XA)
xlabel('Weight of catalyst, kg')
ylabel('Molar conversion of Oxygen')

figure(2)
plot(Wcat,nA,'o',Wcat,nB,'-.',Wcat,nC)
xlabel('Weight of catalyst, kg')
ylabel('Molar flow of the different compounds, mol/s')
legend('Oxygen','Ethylene oxide','Carbon dioxide')
```

This program uses the finite difference method (first order) to do the integration of the ordinary differential equation. In the program, we can choose the number of increments, and the corresponding increment of molar conversion is calculated, and used. The following figures are generated in this program:

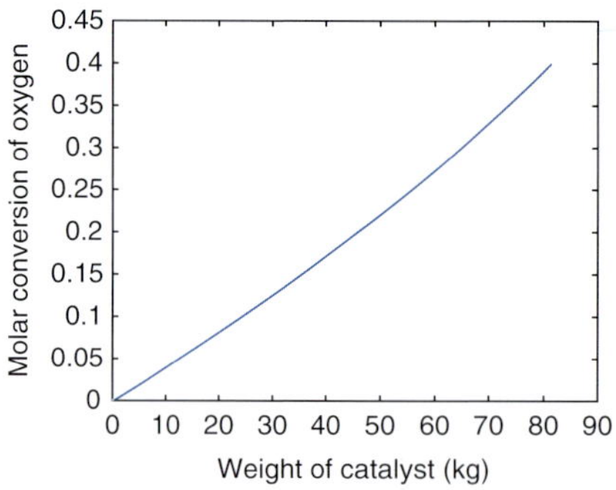

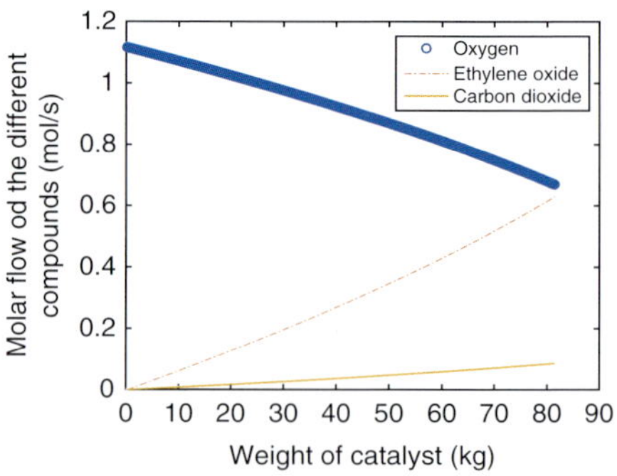

The total amount of catalyst needed in this example is $W = 81.2$ kg. If the reaction were to take place in a fluidized bed, bubbles must be considered, and modification of the previous calculation will be necessary. In this sense, at $u_0 \gg u_{mf}$ generally the model with rapid bubbles would be acceptable. Estimation of the external mass transfer would be necessary too.

9.8 Fluid–Fluid Reactions (Catalyzed and Noncatalyzed)

Industrial application of fluid–fluid reactions is very extended, and almost 90% of them are catalyzed reactions, where three different phases (two fluids and a solid catalyst) are present.

Some of the reasons why fluid–fluid reactions are present in the industry are as follows:

(a) The reaction product of the immiscible reagents is a substance of interest (production reactions).
(b) To facilitate the separation of a component contained in a fluid (purification reactions).
(c) Obtain a more favorable distribution of the product in multiple reactions in a homogeneous phase than could be obtained using a single phase.

There are many examples of applications of this type of reactions in the industry. Among the purification reactions are the elimination of CO_2 in mixtures $CO + H_2$ with potassium carbonate or ethanolamines, and the elimination of H_2S and CO in cracking hydrocarbon gas reactions by ethanolamines or sodium hydroxide. Production reactions include the reaction between CO_2 and an aqueous ammonia stream to give ammonium carbonate, air oxidations of acetaldehydes to give the corresponding acids, chlorination of benzene and other hydrocarbons, absorption of NO_2 in water to give nitric acid, the absorption of SO_3 in H_2SO_4 to give oleum, and many others.

Also important is the reactor design for catalysis involving gaseous reactants. This design requires attention to efficient transfer of reactants from the gas to the liquid phase; mass transfer rates are increased by maximizing interfacial area by sparging (creating small bubbles) and increasing the mass transfer coefficient through higher flow rates and/or stirring, which decreases film thickness.

9.8.1 Transfer Models

For the transfer of component "A" followed by reaction with other species "B," there are three resistances in series: that of the gaseous layer, that of the liquid layer, and that of the bulk of the liquid phase. The reaction, in general, can take place in the liquid layer or in the middle of the liquid phase. Rapid reactions, for example, will occur in a narrow area within the liquid layer, and, on the contrary, slow reactions will take place within the fluid.

When only two fluid phases are present, and there is no need of a catalyst, different steps must be produced for the reaction to take place. Like most heterogeneous reactions, the following steps are present:

1. Diffusion of the reagents towards each other,
2. Chemical reaction itself,
3. Diffusion of the products outside the reaction zone.

Since the substances are in two distinct phases, they must be contacted before reacting, and the overall expression of the reaction rate must consider both the rate of diffusion of the reagents and the rate of the chemical reaction.

The reaction will proceed in one phase or in another depending fundamentally on the solubility of the reactive components in each phase, since solubility determines the displacement of the reactants between the two phases.

Consider that reagent A is initially in phase I, and that reagent B is in phase II. Reagent A is more soluble in phase II, than reagent B in phase I, so the reaction will take place mainly in phase II. In this sense, the reaction we will work with is

$$A\,(\text{phase I} \rightarrow \text{phase II}) + bB\,(\text{phase II}) \rightarrow rR\,(\text{solid, liquid or gas})$$

where "b" is the stoichiometric coefficient of reactant B, and the reaction rate $(-r_A)$ in phase II can have any of the expressions developed and proposed for homogeneous reactions. We will assume that the chemical reaction occurs with first-order kinetics with respect to both reagents, so:

$$(-r''_A)\left(\frac{\text{mol}}{\text{s m}^3_L}\right) = k'' \cdot C_A \cdot C_B \tag{9.76}$$

the units of the constant k'' will be $\text{m}^3{}_L/(\text{mol s})$.

To facilitate the nomenclature, phase I will be considered as a gas and phase II as liquid. If the system were a liquid–liquid one, it would be necessary to replace the nomenclature used by the appropriate one. The solubility of reagent A in phase II is given by the expression (Henry's law if phase I is a gas and phase II is a liquid):

$$p_{Ai} = H_A \cdot C_{Ai} \tag{9.77}$$

The partial pressure of component A in phase I will be designated p_A and C_A is the molar concentration of A (kg/m^3 or kmol/m^3) in phase II. The units of the solubility ratio H_A will depend on the concentration units. The subscript "i" refers to the fact that the equilibrium relation is fulfilled at the interface. For liquid–liquid systems, it would be necessary to substitute p_A for concentrations and the ratio of solubility for the adequate one.

Several models have been proposed to describe the phenomenon that occurs when the two fluid phases come into contact. The most used model is the two-film theory proposed by Lewis–Whitman that proposes the existence of different zones along the total volume of reaction that acts as a set of resistance in series with the movement of the reagents. An outline of this model is observed in Figure 9.10, considering the two fluids moving parallel to the interface.

The different areas in the outline are as follows:

- Resistance to the transfer of reagent A through phase I, equivalent to the resistance that would oppose by molecular diffusion that of a layer adhered to the interface;
- Resistance through the interface (often negligible);
- Resistance to the transfer of reagent A from the interface to the reaction zone, equivalent to that which would oppose a film where the transport regime is molecular diffusion in zone II;
- Resistance to the transfer of reagent B through phase II to the reaction zone.

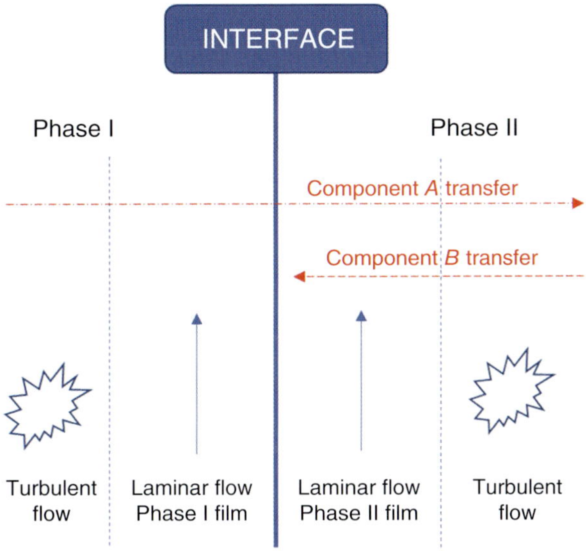

Figure 9.10 Double-layer (two-film) model.

9.8.2 Two-Film Theory

Next, we will study by applying the two-film theory to some cases that are presented according to the relative values of the rate of the chemical reaction and the rate of diffusion. Figure 9.11 shows the most frequent cases, depending on the relative rate of diffusion and reaction. Case A corresponds to a situation where the chemical step is much faster than the diffusion, and assumes that the concentration of reactant A in the liquid phase decreases to zero inside the liquid film. Case B is similar, but the concentration of reactant B is high enough to displace the reaction plane to the interface. Case C represents a situation where the chemical step rate is similar to that of diffusion, and then widening the reaction zone, which is not a plane but a 3D zone. Finally, case D represents the situation where the chemical step is much slower than diffusion, so a constant value of concentration of reactant A inside the liquid phase can be observed. Note that the reaction is displaced to the bulk liquid as the chemical step rate decreases.

9.8.2.1 Case A: Instantaneous Reaction

This case represents more than 90% of the industrial cases in which this type of reactions occurs, since it is usually a diffusion control. The global rate of reaction is given by the rate of diffusion since this is the slowest step. This means that for component A (phase I) to react with component B (phase II), first A must diffuse and find B, so that the reaction plane is located within phase II but very close to the interface, and between the liquid that contains A and the liquid that contains B. Figure 9.11a shows a diagram of this case, applied in a case of gas–liquid reaction.

In this case, when varying the concentration of A or the concentration of B, the plane moves because what is being done is varying the corresponding diffusion rate. Since the limiting step of rate is diffusion, the rate per unit area of interface

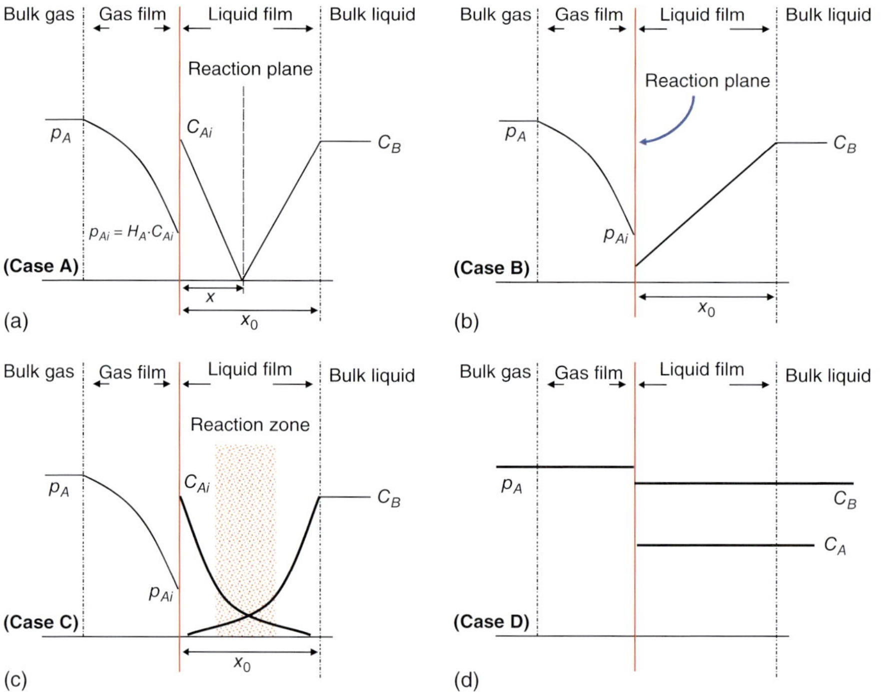

Figure 9.11 Concentration of components being transferred between phases. The interface is indicated by a red line. The boundaries between the well-mixed and the interfacial films are indicated by dotted lines.

($m^2{}_i$) must be used, which is the factor that influences the diffusion, and not the volume of reaction. The expression of the rate equation is

$$-r_A\left(\frac{\text{kmol A}}{\text{s m}_i^2}\right) = -\frac{r_B}{b} = k_{Ag}(p_A - p_{Ai}) = \frac{D_{Al}dC_A}{dx} = \frac{D_{Al}(C_{Ai} - 0)}{x}$$
$$= \frac{D_{Bl}dC_B}{b \cdot dx} = \frac{D_{Bl}(C_B - 0)}{x_0 - x} \cdot \frac{1}{b} \tag{9.78}$$

where k_{Ag} is the individual mass transfer coefficient of reagent A in phase I (for example, in kmol/(s $m^2{}_i$ atm)), and D_{Al} and D_{Bl} are the molecular diffusivities of reagents A and B respectively in phase II (in m^2/s).

The relationship between p_{Ai} and C_{Ai} is given by Henry's law, whether it is between pressures and concentrations for gas–liquid reactions, or Nerst's law if it is a relationship between concentrations for fluid–fluid reactions, as shown in Eq. (9.77).

In the (9.78), the diffusion coefficients have been used according to the double-layer model. The relationship between the diffusion coefficient and the individual coefficient of mass transfer in the same phase, according to the theory of the film, is

$$k_{Al} = \frac{D_{Al}}{x_0} \tag{9.79}$$

Analogously for B, you get:

$$k_{Bl} = \frac{D_{Bl}}{x_0} \tag{9.80}$$

The mass transport coefficients k_{Al} and k_{Bl} have been defined for transport without a chemical reaction (therefore, referred to as the flow along the entire film of thickness x_0), so that in this case the relationship $k_{Ag}(p_A - p_{Ai}) = k_{Al}(C_{Ai} - C_A)$ is fulfilled. When there is a chemical reaction, the expression is modified as previously described (Eq. (9.78)).

From Eqs. (9.79) and (9.80), it follows that:

$$\frac{k_{Al}}{k_{Bl}} = \frac{D_{Al}}{D_{Bl}} \tag{9.81}$$

Since the values of x, x_0, P_{Ai}, and C_{Ai} are not known, an expression of $(-r_A)$ must be found where these variables do not appear. Therefore, it can be written that:

$$(-r_A) = k_{Ag}(p_A - H_A \cdot C_{Ai}) = \frac{k_{Al}C_{Ai}x_0}{x} = \frac{k_{Bl}C_B x_0}{b(x_0 - x)} \tag{9.82}$$

Combining the last two terms:

$$(-r_A) = \frac{\left(k_{Al}C_{Ai} + \left(\frac{k_{Bl}}{b}\right)C_B\right) \cdot x_0}{x + x_0 - x} = k_{Al}C_{Ai} + \left(\frac{k_{Bl}}{b}\right)C_B \tag{9.83}$$

Bearing in mind that the first term can be expressed as

$$(-r_A) = \frac{(p_A - H_A C_{Ai})}{\frac{1}{k_{Ag}}} = \frac{(p_A k_{Al}/H_A - K_{Al}C_{Ai})}{\frac{k_{Al}}{H_A k_{Ag}}} \tag{9.84}$$

and combining Eqs. (9.83) and (9.84):

$$(-r_A) = \frac{\frac{p_A k_{Al}}{H_A} + \frac{k_{Bl}\,C_B}{b}}{1 + \frac{k_{Al}}{H_A k_{Ag}}} \tag{9.85}$$

Finally, the expression of the rate will be

$$(-r_A) = \frac{\left(\frac{p_A}{H_A}\right) + \frac{D_{Bl}}{D_{Al}} \cdot \frac{C_B}{b}}{\frac{1}{H_A \cdot k_{Ag}} + \frac{1}{k_{Al}}} = \frac{\left(\frac{p_A}{H_A}\right) + \frac{k_{Bl}}{k_{Al}} \cdot \frac{C_B}{b}}{\frac{1}{H_A \cdot k_{Ag}} + \frac{1}{k_{Al}}} \tag{9.86}$$

Keep in mind that in the previous expressions, the solution is direct if the value of H_A does not depend on the concentrations. Otherwise, it must be solved by trial and error. A very important aspect that should not be ignored is the reference to the fact that Eq. (9.86) is valid as long as C_B becomes zero within the film, that is, it is in case A of Figure 9.11. If it is not, when in case B of Figure 9.11, the reaction rate is calculated as

$$(-r_A) = k_{Ag} \cdot p_A \tag{9.87}$$

This equation results from imposing the condition $k_{Bl}C_{Bl}b \gg k_{Ag}p_A$ in (9.78), whereby p_{Ai} becomes zero.

A special case that occurs when the resistance of the gas phase is negligible, which means a coefficient $k_{Ag} = \infty$ and $p_A = p_{Ai}$. The equation obtained in this case is

$$(-r_A) = \frac{\left(\frac{p_{Ai}}{H_A}\right) + \frac{D_{Bl}}{D_{Al}} \cdot \frac{C_B}{b}}{\frac{1}{H_A \cdot k_{Ag}} + \frac{1}{k_{Al}}} = k_{Al}C_{Ai}\left(1 + \frac{D_{Bl}}{b \cdot D_{Al}} \cdot \frac{C_B}{C_{Ai}}\right) \tag{9.88}$$

If there were no reaction, i.e. in the case of a pure absorption system, we would have $x_0 = x$ in the scheme of Figure 9.11a, with which, from Eq. (9.82), the rate of the overall process would be given by

$$(-r_A) = k_{Al}(C_{Ai} - 0) \tag{9.89}$$

Therefore, the rate increase factor is defined by the instantaneous reaction E_i, which acts by producing a greater concentration gradient, which is the driving force of diffusion, such as

$$E_i = \frac{\text{reaction rate with instantaneous reaction}}{\text{mass transfer rate}} = \left(1 + \frac{D_{Bl} \cdot C_B}{b \cdot D_{Al} \cdot C_{Ai}}\right) \tag{9.90}$$

So the rate could be expressed as

$$(-r_A) = k_{Al}C_{Ai}E_i \tag{9.91}$$

If we compare this expression with Eq. (9.82), we verify that the equality $E_i = x/x_0$ is met, that is, the factor E_i gives us an idea of how the diffusion layer is enlarged by including the reaction.

In addition to these cases, different situations can be found. Intermediate cases, apart from that shown in Figure 9.11 can be found, depending on the relative rate of diffusion and rate.

9.8.2.2 Analysis of the Controlling Steps: The Hatta Modulus

A dimensionless module has been defined to analyze the effect of the diffusion steps in the reaction. The parameter M_H was developed by Shirôji Hatta in 1932, and serves to deduce when a reaction is slow or fast. For this, only the gas–liquid interface is considered, and it is assumed that the resistance of the gas phase is negligible. The conversion parameter in the film has been defined as

$$M_H^2 = \frac{\text{maximum transfer in the film}}{\text{maximum transfer by difussion in the film}} \tag{9.92}$$

The expression of M_H^2 can be transformed into

$$M_H^2 = \frac{k \cdot D_{Al} \cdot C_B}{k_{Al}^2} \tag{9.93}$$

For liquids, the value of diffusivities are on the order of 10^5 cm^2/s, and x_0 can take values of 10^{-2} cm.

If M_H^2 is much greater than unity, it means that the transfer is very fast, which in turn means that the chemical reaction rate is very high and that the overall rate will depend only on the diffusion, so the reaction is produced entirely in the film and the surface of it is the limiting factor of the rate. From this development, it is deduced that if it wants to favor the reaction, it must be carried out by means of a contact device that develops a high interface area. In these cases, the agitation is an important factor when increasing the contact area by homogenizing the mixture, e.g. the columns of rain or dishes.

On the contrary, if M_H^2 is very small, it means that the chemical reaction is very slow, and that diffusion is not the limiting factor. Therefore, to favor the reaction, all that is needed is a large volume of reaction, e.g. bubble columns.

More precisely, it has been found that:

- If M_H^2 is greater than 4, the reaction takes place in the film and cases A or B considered in Figure 9.11 would be taken.
- If the value of M_H^2 is between 0.0004 and 4, the situation corresponds to the intermediate case C.
- If M_H^2 is much less than 0.0004, the reaction is infinitely slow and corresponds to the case D. The reaction would occur in the liquid core although the main resistance can be anywhere, i.e. in the gas film or in the liquid, but in any case, is negligible.

9.8.2.3 Other Cases in Fluid–Fluid Reactions: The General Rate Equation

As discussed before, there are many possibilities regarding the importance of the different steps in the whole process. The final form of the rate equation will depend on the relative values of the rate constants k_{Ag}, k_{Al}, and k, the concentration of reactants (p_A and C_B), and the Henry (or phase equilibrium) constant. Fortunately, a general expression has been proposed that considers all these aspects:

$$(-r''_A) = \frac{p_A}{\frac{1}{k_{Ag}\cdot a} + \frac{H_A}{k_{Al}\cdot a\cdot E} + \frac{H_A}{(k''\cdot C_B)\cdot f_L}} \tag{9.94}$$

where the rate $(-r''_A)$ is based on the reactor volume (mol/(m_R^3 s)), and the value of "a" is the ratio (surface of the interface)/(reactor volume) (m_i^2/m_R^3). f_L represents the liquid fraction of the reactor, i.e. the amount of liquid in the reactor ($m^3{}_L/m^3{}_R$).

The first summing of the denominator represents the resistance to the transfer in the gaseous layer. In this, k_{Ag} has units of mol/(s m_i^2 atm) (assuming a gaseous phase) and a of m_i^2/m_R^3, so the summing has the units of ((atm s m_R^3)/mol). The second addition takes into account the resistance in the liquid phase and the third, the chemical reaction. This reaction is supposed to be first or pseudo first order, according to the equation: $(-r''_A) = (k''C_B)C_A$. The units of the product $(k''C_B)\cdot f_L$ are $m_R^3/(\text{mol s})\cdot(\text{mol}/m_L^3)\cdot m_L^3/m_R^3 = 1/\text{s}$, or other equivalent units.

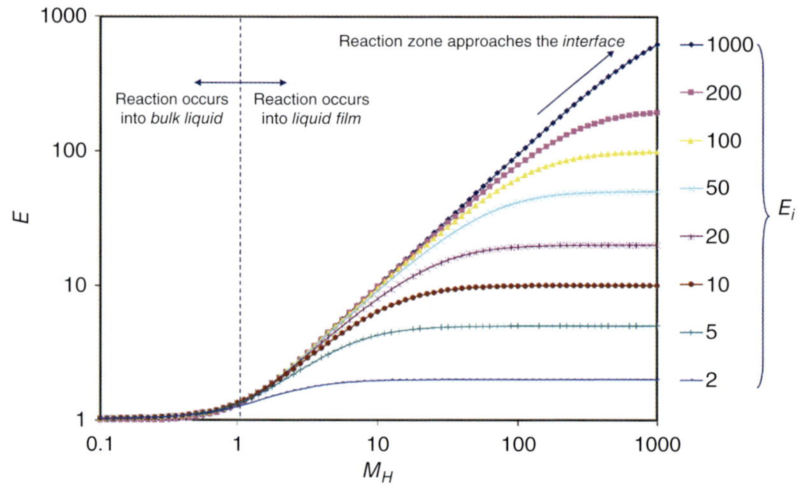

Figure 9.12 General rate equation for different values of M_H and E_i.

Obviously, f_L is not necessary if we know the value of the kinetic constant in $m_L^3/(mol\ s)$ or similar.

The absorption of A is greater when there is a reaction, so for the same concentrations in two different conditions, the E factor of increase is defined as

$$E = \left[\frac{\text{actual adsorption rate of component } A}{\text{adsorption rate of } A \text{ without chemical reaction}}\right]_{\text{at the same } C_{Ai}, C_A, C_{Bi}, C_B} \tag{9.95}$$

Van Krevelens and Hoftijzer deducted numerically a good correlation for the coefficient E as a function of M_H and E_i:

$$E = \frac{\sqrt{M_H^2 \frac{E_i - E}{E_i - 1}}}{\tanh\left(\sqrt{M_H^2 \frac{E_i - E}{E_i - 1}}\right)} \tag{9.96}$$

Note that this equation is nonlinear and must be solved by iterative methods. In Figure 9.12, the equation of a graphic form is shown.

Example 9.11 ***Reaction Rate with No Catalyst***

CO_2 must be separated from air by contact with water in countercurrent at 25 °C. The individual volumetric coefficients for the transfer of CO_2 in the gas and liquid phases are 80 mol/(h l atm) and 25 h^{-1}, respectively. The solubility ratio is given by Henry's law with a value of $H = 30$ (atm l)/mol:

(a) Derive an expression for the calculation of the absorption rate.
(b) If a 2 M solution of NaOH is used, the reaction $CO_2 + OH^-$ is instantaneous and there is a large excess of NaOH relative to CO_2 that can be absorbed, indicate the most appropriate form to express the reaction rate.

Solution

$$(-r_A)\left(\frac{\text{mol CO}_2}{\text{h l}}\right) = k_{Ag} \cdot a \cdot (p_A - p_{Ai}) = k_{Ag} \cdot a \cdot (p_A - H_A \cdot C_{Ai})$$

$$(-r_A)\left(\frac{\text{mol CO}_2}{\text{h l}}\right) = k_{Al} \cdot a \cdot (C_{Ai} - C_A)$$

Both rates must be equal in the stationary regime, so we can derive the equation for the C_{Ai}:

$$C_{Ai} = \frac{k_{Ag} \cdot p_A + k_{Al} \cdot C_A}{H_A \cdot k_{Ag} + k_{Al}}$$

Then it can be used for estimating the reaction rate:

$$(-r_A)\left(\frac{\text{mol CO}_2}{\text{h l}}\right) = k_{Al} \cdot a \cdot \left(\frac{k_{Ag} \cdot p_A + k_{Al} \cdot C_A}{H_A \cdot k_{Ag} + k_{Al}} - C_A\right) = \frac{p_A - H_A C_A}{\frac{1}{k_{Al} \cdot a} + \frac{H_A}{k_{Ag} \cdot a}}$$

(a) So, in the first case: $(-r_A) = \frac{p_A - 30C_A}{\frac{1}{80} + \frac{30}{25}} = 0.8247\,(p_A - 30C_A)$

(b) Now, $C_A = 0$ and then: $(-r_A) = 0.8247 p_A$

Example 9.12 ***Mass Transfer with Instantaneous Reaction***

A fluid–fluid reaction is developed with two reagents *A* (initially in phase I) and *B* (initially in phase II). The reaction will take place in phase II. The concentrations of *A* and *B* in its phases are 1 kg/m^3, and the stoichiometry is 1 mol of *A* reacts with 2 mol of *B* and takes place instantly in phase II. The diffusivities of both reagents in phase II can be considered to be similar. Deduct the controlling stages of the process in the following cases:

Case (A) k_{Ag} (through phase I): 20 m/s
$H_A = 3$ m^3 phase II/m^3 phase I
k_{Al} (through phase II) = 5 m/s

Case (B) k_{Ag} (through phase I): 20 m/s
$H_A = 10$ m^3 phase II/m^3 phase I
k_{Al} (through phase II) = 500 m/s

Solution

Note that in this case, two liquid phases are involved and the units of the coefficients can change. For example, the H_A represents the equilibrium in the interface, so:

$$H_A = \frac{C_{A,\text{phase I}}}{C_{A,\text{phase II}}} \ldots \text{in } \frac{m^3_{\text{phase II}}}{m^3_{\text{phase I}}}$$

From the general rate equation:

$$(-r''_A) = \frac{p_A}{\frac{1}{k_{Ag} \cdot a} + \frac{H_A}{k_{Al} \cdot a \cdot E} + \frac{H_A}{(k'' \cdot C_B) \cdot f_L}}$$

But we have two liquid phases; therefore:

$$(-r''_A) = \frac{C_{A,\text{phase I}}}{\frac{1}{k_{A,\text{phase I}} \cdot a} + \frac{H_A}{k_{A,\text{phase II}} \cdot a \cdot E} + \frac{H_A}{(k'' \cdot C_B) \cdot f_{\text{phase II}}}}$$

In this case, the chemical reaction is instantaneous, so:

$$(-r''_A) = \frac{C_{A,\text{phase I}}}{\frac{1}{k_{A,\text{phase I}} \cdot a} + \frac{H_A}{k_{A,\text{phase II}} \cdot a \cdot E} + 0}$$

and we should compare the two terms in the denominator.

(a) As the reaction is instantaneous, $E \cong E_i$, and then:

$$E = \left(1 + \frac{D_{Bl} \cdot C_{B,\text{phase II}}}{b \cdot D_{Al} \cdot C_{Ai,\ \text{phase II}}}\right)$$

The concentration of "A" at the interface in phase II can be calculated using the value of H_A:

$$('A' \text{ transfered from I to interface}) = ('A' \text{ transfered from interface to II})$$

$$k_{A,\text{phase I}}(C_{A,\text{phase I}} - H_A \cdot C_{Ai,\text{phase II}}) = k_{A,\text{phase II}}(C_{Ai,\text{phase II}} - 0)$$

From this expression, we can calculate $C_{Ai,\text{ phase II}} = 0.308\ \text{kg/m}^3$ and then $E = \dots = 2.62$

The values of the terms in the denominator are then:

$$\frac{1}{k_{Ag} \cdot a} = \frac{1}{20} = 0.05\frac{s}{m}$$

$$\frac{H_A}{k_{Al} \cdot a \cdot E} = \frac{3}{5 \cdot 2.62} = 0.22\frac{s}{m}$$

So, in this case the controlling step is the diffusion through phase II.

(b) In this case also, $E \cong E_i$, but the value results to be 18.48. When comparing $\frac{1}{k_{Ag} \cdot a} = 0.05\frac{\text{s}}{\text{m}}$ and:

$$\frac{H_A}{k_{AL} \cdot a \cdot E} = \frac{3}{500 \cdot 18.48} = 0.00\,108\frac{s}{m}$$

So, in this case the controlling step is the diffusion through phase I.

Example 9.13 ***Membrane Reactor***

In a 1 m^3 membrane reactor, the transport of oxygen from a gaseous (air) to a liquid phase is taking place at 1 atm. To increase the transfer rate, an organometallic molecule (RH) is added to the liquid, with a concentration of 5 mol/m^3. In a simplified way, we have

$$O_2 + RH \rightarrow (RH \cdot O_2) \quad r = k \cdot C_{O_2} \cdot C_{RH}$$

Estimate the rate at which oxygen is consumed in this system, referred to its volume (mol O_2 s^{-1} m$^{-3}{}_{\text{reactor}}$) (Figure 9.13).

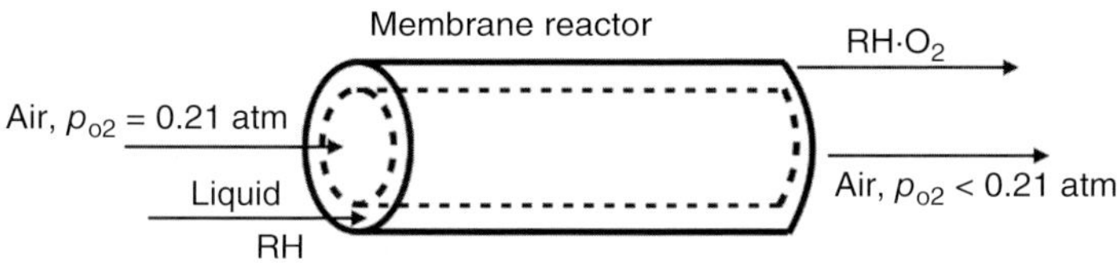

Figure 9.13 Membrane reactor.

Data:

$H_{O_2} = 0.937\ \text{atm m}^3/\text{mol}$	$D_{O_2,L} = 7.1{\cdot}10^{-6}\ \text{m}^2/\text{s}$
$k_{O_2,g} = 0.03\ \text{mol}/(\text{s m}^2\ \text{atm})$	$D_{RH,L} = 8.3{\cdot}10^{-6}\ \text{m}^2/\text{s}$
$k'' = 1.8{\cdot}10^3\ \text{m}^3/(\text{mol s})$	Total membrane thickness = $5{\cdot}10^{-3}$ m
Membrane surface = 7 m^2	

Solution

We will assume that the membrane is a liquid film. The rate of the process will be given by the general rate equation (Eq. (9.94)). In the example, we have

$$p_{O_2} = 0.21\ \text{atm (air)}$$

$$a\,(\text{m}^2_{\text{interface}}/\text{m}^3_{\text{system}}) = 7/1 = 7\ \text{m}^{-1}$$

$$k_{O_2,L} = \frac{D_{O_2,L}}{\text{film thickness}} = \frac{7.1\cdot 10^{-6}}{5\cdot 10^{-3}} = 1.42\cdot 10^{-3}\,\frac{\text{m}}{\text{s}}$$

Using these data, we can estimate the value of the Hatta modulus and then the increment factor "E":

$$M_H = \frac{\sqrt{D_{O_2,L}\cdot k''}}{k_{O_2,L}} = \frac{\sqrt{7.1\cdot 10^{-6}\cdot 1.8\cdot 10^3}}{1.42\cdot 10^{-3}} = 79.6$$

$$E_i = 1 + \frac{D_{RH,L}\cdot C_{RH}\cdot H_{O_2}}{b\cdot D_{O_2,L}\cdot p_{O_2}} = 1 + \frac{8.3\cdot 10^{-6}\cdot 5\cdot 0.937}{1\cdot 7.1\cdot 10^{-6}\cdot 0.21} = 27.08$$

These values of the parameters suggest that the reaction is being produced in a regime where both diffusion and reaction have similar rates. Using the relation:

$$E = \frac{\sqrt{M_H^2\,\frac{E_i-E}{E_i-1}}}{\tanh\left(\sqrt{M_H^2\,\frac{E_i-E}{E_i-1}}\right)} \tag{9.96}$$

We can check that $E = 24.6$ and then calculate the reaction rate (Eq. (9.94)):

$$(-r_A'') = \frac{p_A}{\frac{1}{k_{Ag}\cdot a} + \frac{H_A}{k_{AL}\cdot a\cdot E} + \frac{H_A}{(k''\cdot C_B)\cdot \varepsilon_L}}$$

$$(-r_A'') = \frac{p_A}{\frac{1}{k_{Ag}\cdot a} + \frac{H_A}{k_{AL}\cdot a\cdot E} + \frac{H_A}{(k\cdot C_B)\cdot \varepsilon_L}} = \frac{0.21}{\frac{1}{0.03\cdot 7} + \frac{0.937}{1.42\cdot 10^{-3}\cdot 7\cdot 24.6} + \frac{0.937}{1.8\cdot 10^3\cdot 5}}$$

$$= 0.024\,\frac{\text{mol}}{\text{s m}^3_{\text{reactor}}}$$

Actually, the third adding in the denominator is very small compared to the other two, so the reaction is quite fast compared to diffusion. The result indicates that, bearing in mind the volume of the reactor, 0.024 mol of oxygen is being transferred per second in the whole system if the conditions are maintained along the length.

9.8.3 Gas–Liquid Reactions in Solid Catalysts. General Equation

As mentioned, many fluid–fluid reactions take place over a solid catalyst. These multiphase reactions are of the type:

$$A\,(g \xrightarrow{\text{disolution}} l) + bB\,(l) \xrightarrow{\text{in the catalyst}} \text{products} \tag{9.97}$$

The gaseous reagent must first be dissolved in the liquid, then the reagent must be diffused or moved to the surface of the catalyst for the reaction to take place. Thus, in the expression of rate, the resistance to transfer through the gas–liquid interface and toward the surface of the solid intervenes.

These types of reactions can be carried out in different ways. Packed contact devices use large catalytic solid particles, sludge reactors (or suspended solids, also called slurry type) use very fine suspended catalysts, while fluidized beds use both sizes, depending on the flow rates.

For the chemical reaction step, the following reaction rates will be considered:

$$\left.\begin{aligned} -r_A''' &= k_A''' C_A C_B \\ -r_B''' &= k_B''' C_A C_B \end{aligned}\right\rangle \text{ where } \begin{aligned} -r_A''' &= -\frac{r_B'''}{b} \quad \frac{\text{mol A}}{\text{m}^3_{\text{cat}}\ \text{s}} \\ k_A''' &= \frac{k_B'''}{b} \quad \frac{\text{mol B}}{\text{m}^3_{\text{cat}}\ \text{s}} \end{aligned} \tag{9.98}$$

The development of the rate equation will be based on the two-film theory, using the nomenclature presented in Figure 9.14.

The resistances are shown graphically in Figure 9.15. The expressions of the rate of each of the consecutive steps reflected in this figure can be expressed as (mol/(m$^3{}_R$ s))

- Transfer through the gaseous film: $k_{Ag}a_i(p_A - H_A C_{Ai})$
- Transfer through the liquid film: $k_{Al}a_i(C_{Ai} - C_{Al})$
- Transfer through the solid layer: $k_{Ac}a_c(C_{Al} - C_{As})$
- Surface chemical reaction: $(k_A''' \overline{C_B})\eta_A f_s C_{As}$

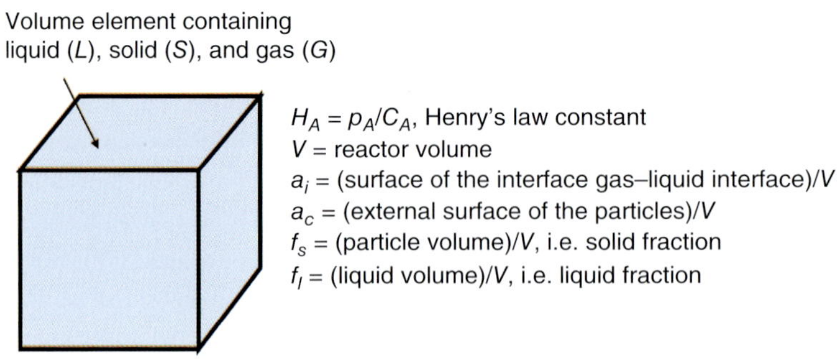

Figure 9.14 Nomenclature used.

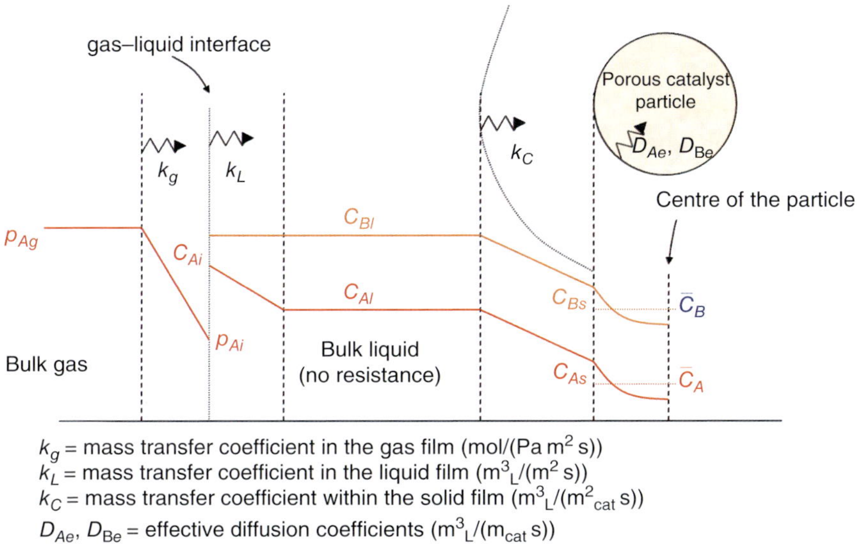

Figure 9.15 Schematic showing the resistances involved in the gas–liquid reaction on the surface of a solid catalyst. Source: Levenspiel (1999). Adapted with permission of John Wiley & Sons.

If they are put in the form of resistances, and equaling, you get

$$(-r''_A)\left(\frac{\text{mol}}{\text{s m}_R^3}\right) = \frac{\frac{p_A}{H_A} - C_{Ai}}{\frac{1}{H_A k_{Ag} a_i}} = \frac{C_{Ai} - C_{Al}}{\frac{1}{k_{Al} a_i}} = \frac{C_{Al} - C_{As}}{\frac{1}{k_{Ac} a_c}} = \frac{C_{As}}{\frac{1}{(k'''_A \overline{C}_B)\eta_A f_s}} \tag{9.99}$$

Combining these quotients, the following general expression of rate can be written, with reference to the reactor volume:

$$(-r''_A) = \frac{p_{Ag}}{\frac{1}{k_{Ag}\cdot a_i} + \frac{H_A}{k_{Al}\cdot a_i} + \frac{H_A}{k_{Ac}\cdot a_c} + \frac{H_A}{(k'''_A\cdot \overline{C}_B)\cdot \eta_A \cdot f_s}} \tag{9.100}$$

For component B, the only resistances are the solid layer and the reaction, so:

$$(-r''_B)\left(\frac{\text{mol}}{\text{s m}_R^3}\right) = \frac{C_{Bl}}{\frac{1}{k_{Bc}\cdot a_c} + \frac{1}{(k'''_B\cdot \overline{C}_A)\cdot \eta_B \cdot f_s}} \tag{9.101}$$

where η_A and η_B are efficiency factors.

Equation (9.100) or Eq. (9.101) would give the reaction rate. Both expressions are related by $(-r''_A) = \frac{(-r''_B)}{b}$ and the terms in parentheses express the rate constant of pseudo first order. Unfortunately, even with all known system parameters (k, a, f, etc.), these expressions cannot be solved without a trial and error procedure, because $\overline{C}_B$ in Eq. (9.100) and $\overline{C}_A$ in Eq. (9.101) are not known. However, generally one works in one of two extreme cases in which simplifications can be made, which are very useful.

In systems with pure liquid B and a slightly soluble gas A, it can be assumed that the concentration of B has the same value at any point, so:

$$C_{Bs} = \overline{C}_{B\text{within the particle}} = C_{Bl}$$

with C_B constant, the reaction becomes of first global order (first order with respect to A) and the previous expressions are reduced to a directly solvable expression, i.e. Eq. (9.100) with known $(k_A''' \cdot \overline{C}_B)$.

On the contrary, in systems with diluted liquid B reagent and a very soluble gas A, and at high pressure, it can be assumed that $C_{Al} = \frac{p_{Ag}}{H_A}$ throughout the reactor. The rate becomes first order with respect to B and is reduced to

$$(-r_B'')\left(\frac{\text{mol}}{\text{s m}_R^3}\right) = \frac{C_{Bl}}{\frac{1}{k_{Bc}\cdot a_c} + \frac{1}{\left(k_B'''\cdot\frac{p_{Ag}}{H_A}\right)\cdot\eta_B\cdot f_s}} \tag{9.101}$$

9.8.3.1 Estimation of the Controlling Resistance in Multiphase Systems

Let us work a little more with the rate equation. From Eq. (9.99), you can easily reach:

$$\frac{p_{Ag}}{-r_A''} = \frac{1}{k_{Ag}a_i} + \frac{H_A}{k_{Al}\cdot a_i} + \frac{H_A}{k_{Ac}\cdot a_c} + \frac{H_A}{(k_A'''\cdot\overline{C}_B)\cdot\eta_A\cdot f_s} \tag{9.102}$$

The first term of this equation is what is called gas resistance R_G and the second is the resistance of the liquid R_L. Considering that for spheres:

$$a_c = d_p/6$$

the resistance of the solid layer is defined as

$$R_s = \frac{H_A}{k_{Ac}\cdot d_p/6} \tag{9.103}$$

and the resistance of the chemical reaction is taken as

$$R_R = \frac{H_A}{(k_A'''\cdot\overline{C}_B)\cdot\eta_A} \tag{9.104}$$

So that Eq. (9.100) remains as

$$(-r_A'') = \frac{p_{Ag}}{R_G + R_L + \frac{R_s+R_R}{f_s}}$$

Reordering:

$$\frac{p_{Ag}}{(-r_A'')} = R_G + R_L + \frac{R_s + R_R}{f_s} \tag{9.105}$$

In this way, a representation of the quotient between the pressure and the reaction rate vs. $1/f_s$ can give an idea of the controlling resistances: on the one hand, the ordinate at the origin is equal to the sum of the resistances of the liquid and gas, and the slope is the sum of R_S and R_R. The representation of this slope in front of the values of d_p will give an idea whether the chemical reaction or the gaseous layer is the controlling one, because of the following:

- A zero slope in this representation $(R_C + R_L)$ vs. d_p, would indicate a process controlled by the superficial chemical reaction.
- A slope 1 indicates limited reaction by internal diffusion.
- A slope of 1.5–2.0 indicates a limitation due to external diffusion.

Example 9.14 ***Catalyst in a Gas–Liquid Reaction***
In a 1 m^3 reactor, a gas–liquid reaction is carried out on a solid catalyst. The reagent of the liquid phase is in excess of that of the gas. The following data is available:

Mass of catalyst (g)	Partial pressure of the gas (atm)	Reaction rate (mol/(h m$^3{}_R$))
1	0.1	1
2	0.15	2
3	0.2	3
4	0.2	4

What can be said about the controlling resistances of the process? Indicate how they could estimate, and, if possible, determine their relative value.

Solution
Using Eq. (9.102):

$$\frac{p_{Ag}}{-r''_A} = \frac{1}{k_{Ag}a_i} + \frac{H_A}{k_{Al}\cdot a_i} + \frac{H_A}{k_{Ac}\cdot a_c} + \frac{H_A}{(k'''_A\cdot \overline{C}_B)\cdot \eta_A \cdot f_s}$$

In this case, the value of f_s (m^3 solid/m^3 reactor) is related to the mass of solid used in each run. We can calculate $-p_{Ag}/r_A$ ((atm h m$^3{}_R$)/mol) and $1/f_s$ (1/g). When plotting the data, we obtain the following:

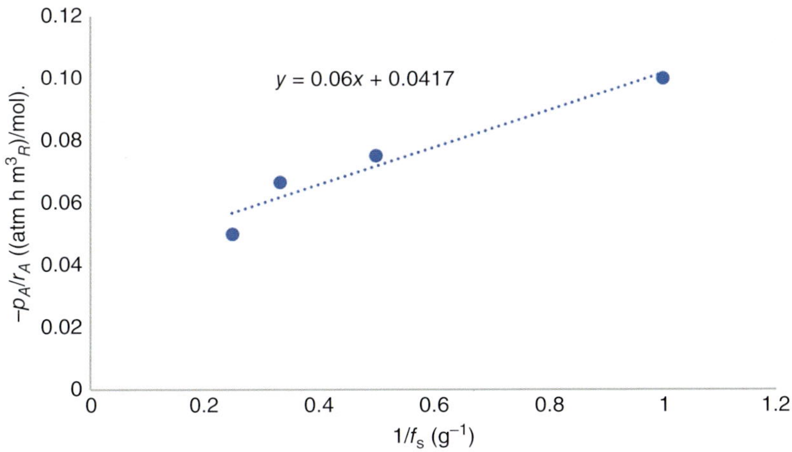

The slope of the plot is 0.06 and represents $(R_S + R_R)$. The value at the origin (0.0417) is related to $(R_G + R_L)$. It is clear that, in this case, $(R_S + R_R)/(R_G + R_L) = 1.44$, so the main resistance is either in the solid layer or in the chemical reaction. If we also had data at different particle diameters, we would distinghish bewteen these two contributions, in the way that representing $(R_S + R_R)$ vs. d_p we will obtain a zero slope if the chemical reaction is the controlling step, and different values for internal and external diffusion control.

9.8.3.2 General *n*-th Order Kinetics

The rate expressions used so far have been developed assuming a first-order reaction with respect to *A* and first order with respect to *B*. But, how are more general forms of rate treated? For example:

$$A(g) + bB(l) \xrightarrow{\text{in the catalyst}} \text{products} \quad (-r''_A) = (-r''_B)/b = k''_A C_A^n C_B^m$$

To be able to combine the chemical step with the steps of transfer of mass in a simple way, we must replace the uncomfortable previous expression of rate with a first-order approximation, in the following way:

$$(-r''_A) = (-r''_B)/b = k''_A C_A^n C_B^m \Rightarrow (-r''_A) = (k''_A \overline{C}_A^{n-1} \overline{C}_B^m) C_A$$

with the average values calculated in the places where the reaction occurs.

This approach is not completely satisfactory, but it is the best thing to do. Thus, instead of the general rate equation, the rate equation to be used in all design expressions will be

$$(-r''_A) = \frac{p_{Ag}}{\frac{1}{k_{Ag}\cdot a_i} + \frac{H_A}{k_{Al}\cdot a_i} + \frac{H_A}{k_{Ac}\cdot a_c} + \frac{H_A}{(k'''_A \overline{C}_A^{n-1} \overline{C}_B^m)\cdot \eta_A \cdot f_s}} \tag{9.100b}$$

Example 9.15 ***n-th Order Reaction Rate***

A solution of acetone ($C_{B0} = 1$ M) with a flow rate of 1 l/s, and pure hydrogen at a total pressure of 1 atm and a flow rate of 0.04 m^3/s are fed by the base of a column of 0.1 m^2 of section, packed with porous nickel catalyst ($d_p = 5$ mm, density = 4500 kg/m^3, load of solid = 0.6 m^3 cat/m^3 reactor, effective diffusivity = $8\cdot 10^{-10}$ $m^3{}_L/(m_{cat}\ s)$), and maintained at 14 °C. Under these conditions, acetone reacts with hydrogen to give propanol:

$$H_2\ (g \rightarrow l) + CH_3COOCH_3\ (l) \xrightarrow{\text{in the catalyst}} CH_3CHOHCH_3$$

If H_2 is designed by "*A*" and acetone by "*B*," the rate is given by

$$-r_A\left(\frac{\text{mol}}{\text{kg}_{\text{cat}}\ \text{s}}\right) = k' C_A^{1/2} C_B^0 \quad \text{and} \quad k' = 2.351 \cdot 10^{-3} \frac{\text{m}_L^3}{\text{kg}_{\text{cat}}\ \text{s}}\left(\frac{\text{mol}}{\text{m}_L^3}\right)^{1/2}$$

What will be rate of that process?

Data:

$$H_A = 36\,845\ \text{Pa m}^3{}_L/\text{mol} \quad k_{Ac}\cdot a_c = 0.05\ \text{m}^3{}_L/(\text{m}^3{}_R\ \text{s})$$

$$\left(\frac{1}{H_A \cdot k_{Ag}\cdot a_i} + \frac{1}{k_{Al}\cdot a_i}\right) = \frac{1}{0.02}\ (s \cdot m_L{}^3)/m_R{}^3$$

Solution

This is a complicated case, because all the equations mentioned are valid only for first-order kinetics, but in this case, it is even difficult to do an approximation. The only thing we can do is

$$(-r'_A) = k' \cdot C_A^{\frac{1}{2}} = \left(k' \cdot C_A^{-\frac{1}{2}}\right) \cdot C_A$$

and consider the term in parenthesis a pseudo first-order kinetic constant.

$$(-r''_A) = \frac{p_{Ag}}{\frac{1}{k_{Ag} \cdot a_i} + \frac{H_A}{k_{Al} \cdot a_i} + \frac{H_A}{k_{Ac} \cdot a_c} + \frac{H_A}{\left(k \cdot C_A^{-\frac{1}{2}}\right) \cdot \eta_A \cdot f_s}}$$

Also, for n-th order kinetics:

$$m \cdot L = L \cdot \left[\frac{k \cdot C_{As}^{n-1}(n+1)}{2 \cdot D_e}\right]^{\frac{1}{2}}$$

$$= \frac{5 \cdot 10^{-3}}{3} \left[\frac{2.35 \cdot 10^{-3} \cdot 2.75^{-\frac{1}{2}} \left(\frac{1}{2} + 1\right)}{2 \cdot D_e}\right]^{\frac{1}{2}} = 64.4$$

And then, $\eta_A = 0.0155$

In this example, $p_A = 1$ atm in all points of the reactor, because the gas is pure hydrogen that does not vary its partial pressure. From all data given in the example, a reaction rate can be calculated:

$$(-r''_A) = 0.0317 \frac{\text{mol}}{\text{m}_R{}^3\text{s}}$$

9.9 Design of Multiphase Reactors

In this section, we describe several types of multiphase reactors and we see the basic equations for their design.

9.9.1 Types of Flow in Multiphase Reactors

The reactions between the components of a gas and a liquid are carried out in a great variety of equipment, which often have a confusing name. This wide variety of systems must fulfill some basic conditions: efficient contact between gas and liquid (and, in many cases, a solid catalyst), limitation in pressure drop, easy heat extraction, and low construction and maintenance costs. Depending on whether the greater resistance to the transfer of mass occurs in the liquid or in the gas, the multiphase reactors operate with the dispersed liquid phase and the continuous gas, or vice versa. The choice of parallel or countercurrent circulation of gas and liquid depends on the availability of driving forces for mass and heat transfers and reaction.

Many textbooks classify the different types of equipment for gas–liquid reactions. The classification is based on geometrical aspects, the presence or absence of solid catalysts, and flow directions. A description of each type can be consulted in classical textbooks such as “Chemical Reaction Engineering,” by O. Levenspiel.

9.9.2 Design Models for Flow in Multiphase Reactors

This section shows some general models that can be used for the design of any of these reactors. The models are based on the plug flow or stirred tank of one or both phases. A distinction is made between the processes taking place without the presence of a reaction (absorption processes) and those where a reactant is present in each phase.

In the reasonings that follow, the following reaction scheme is assumed:

$$A\ (\text{gas}) + bB\ (\text{liquid}) \rightarrow \text{products (R or S)}$$

where component A must diffuse to the phase where B is present, and react. Before continuing, the nomenclature to be followed is presented.

a = Area of interface per unit volume of the tower
f = Fraction in volume of the phase in which the reaction takes place
U = Carrier or component inert in a phase, therefore it is neither reactant, nor product, nor component that diffuses
T = Subscript referred to total moles in the liquid phase
$Y_A = \frac{p_A}{p_U} = \frac{\text{moles of } A}{\text{moles of inerts}}$ in the gas
$X_A = \frac{C_A}{C_U} = \frac{\text{moles of } A}{\text{moles of inerts}}$ in the liquid
G', L' = Molar flow of inert gas and liquid, usually per unit of normal section to the tower (kmol/(s m^2)).
P_T = total pressure
C_T = total concentration

Using this nomenclature (see also Figure 9.16), it is fulfilled that:

$$G' = G\cdot\frac{p_U}{P_T};\quad G'\cdot P_T = G\cdot p_U \tag{9.106}$$

$$L' = L\cdot\frac{C_U}{C_T};\quad L'\cdot C_T = L\cdot C_U \tag{9.107}$$

$$P_T = p_A + p_B + \cdots + p_U \tag{9.108}$$

$$C_T = C_A + C_B + \cdots + C_U \tag{9.109}$$

$$dY_A = d\left(\frac{p_A}{p_U}\right) = \frac{p_U dp_A - p_A dp_U}{p_U^2} \approx \frac{dp_A}{p_U} \tag{9.110}$$

$$dX_A = d\left(\frac{C_A}{C_U}\right) = \frac{C_U dC_A - C_A dC_U}{C_U^2} \approx \frac{dC_A}{C_U} \tag{9.111}$$

The amount of component A accompanying the gas for a given composition, in mol/(s m^2), can be calculated by $G'\cdot Y_A$ or using the relation $G\cdot p_A/P_T$.

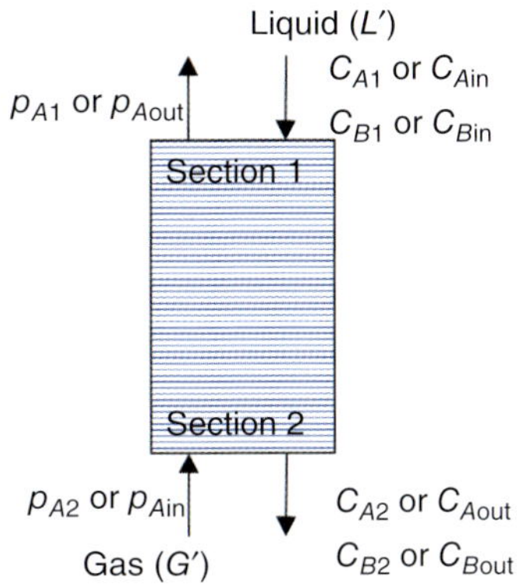

Figure 9.16 Scheme of a multiphase system with the nomenclature used.

The design of the equipment is addressed first when there is no reaction, and thus the reaction is considered. All the equations we present are based on countercurrent flow. In the case of parallel flow, a change of L to $-L$ is required, but the equations are valid.

9.9.3 Case 1: Pure Absorption (No Reaction)

9.9.3.1 Situation 1: Gas and Liquid Phases in Plug Flow

To develop the design equation, the rate equation will be combined with the material balance. Thus, for the steady state in countercurrent operations like the one in Figure 9.16, we have a volume differential element:

$$\text{(A lost by gas)} = \text{(A gained by the liquid)} = \text{(A transferred)}$$
$$= (-r''_A)\,dV \quad (r_A'') \text{ in kmol/(m}^3{}_{\text{equipment}}\text{ s)} \tag{9.112}$$

$$\text{(kmol/s)} \quad G'dY_A = L'dX_A = (-r''_A)dV \tag{9.113}$$

The first two terms can be expressed in different ways:

$$G'dY_A = L'dX_A$$

$$G'\frac{dp_A}{p_U} = L'\frac{dC_A}{C_U} \tag{9.114}$$

and the rate of "reaction," if there is only mass transfer, is

$$(-r''_A) = (-r_A)\cdot a = k_{Ag}\cdot a\cdot(p_A - p_{Ai}) = k_{Al}\cdot a\cdot(C_{Ai} - C_A) \tag{9.115}$$

$(-r_A)$ being in kmol/($m^2{}_i$ s). In this way, the balance in a differential volume is

$$G'dY_A = L'dX_A = (-r_A)\cdot a\cdot dV$$

Integrating for the whole tower gives

$$V = \frac{G'}{a}\int_{Y_{A1}}^{Y_{A2}}\frac{dY_A}{(-r_A)} = \frac{L'}{a}\int_{X_{A1}}^{X_{A2}}\frac{dX_A}{(-r_A)} \tag{9.116}$$

The design of the tower can be done numerically or graphically. For diluted systems, $C_A \ll C_T$ and $p_A \ll \pi$, so $L' \cong L$ and $G' \cong G$, and then the differential balance can be written as

$$\frac{G'}{p_T} \cdot dp_A = \frac{L'}{C_T} \cdot dC_A = (-r_A) \cdot a \cdot dV$$

The mass transfer rate can also be calculated using global mass transfer coefficients, with driving force $(p_A - p_A^*)$ and $(C_A^* - C_A)$, using the equations:

$$\frac{1}{K_{Ag}} = \frac{1}{k_{Ag}} + \frac{H_A}{k_{Al}}$$
$$\frac{1}{K_{Al}} = \frac{1}{k_{Al}} + \frac{1}{H_A \cdot k_{Ag}} \tag{9.117}$$

Then, the mode balance will give

$$V = \frac{G'}{p_T \cdot k_{Ag} \cdot a} \int_{p_{A1}}^{p_{A2}} \frac{dp_A}{(p_A - p_{Ai})} = \frac{L'}{C_T \cdot k_{Al} \cdot a} \int_{C_{A1}}^{C_{A2}} \frac{dC_A}{(C_{Ai} - C_A)}$$

and also:

$$V = \frac{G'}{p_T \cdot K_{Ag} \cdot a} \int_{p_{A1}}^{p_{A2}} \frac{dp_A}{(p_A - p_A^*)} = \frac{L'}{C_T \cdot K_{Al} \cdot a} \int_{C_{A1}}^{C_{A2}} \frac{dC_A}{(C_A^* - C_A)} \tag{9.118}$$

where pressures and concentrations refer to equilibrium, that is

$$p_A^* = H_A \cdot C_A$$
$$C_A^* = p_A/H_A$$

9.9.3.2 Situation 2: Gas and Liquid Phases Completely Mixed

This is the simplest situation from the point of view of the calculation, since the concentrations of A and B are uniform and differential equations are not involved, as there is only one value of the reaction rate throughout the reactor. The continuity equation for A (molar balance of A), for the whole volume can be written as

$$(\text{A lost by gas}) = (\text{A gained by the liquid}) = (\text{A transferred}) = (-r''_A)\, V \tag{9.112}$$

Note that the previous equation is the same as Eq. (9.112), but in this case we can obviate the differential volume as there is only one value of the reaction rate in the whole equipment.

Considering diluted systems, it can be written that:

$$\frac{G'}{p_T}(p_{A2} - p_{A1}) = \frac{L'}{C_T}(C_{A2} - C_{A1}) = (-r''_{A2}) \cdot V \tag{9.119}$$

$(-r''_{A2})$ being the reaction rate in the final conditions of the reaction, which are the conditions present at the reactor. In this equation, $(-r''_{A2}) = (-r_{A2}) \cdot a$.

9.9.3.3 Situation 3: Gas Phase in Plug Flow. Liquid Phase Completely Mixed

Here the gas contacts the liquid in the final conditions, but different rates exist in the reactor, because for the gas, differential elements are necessary. For a differential element:

$$G' dY_A = -r''_{A2}]_{\text{liquid at the exit conditions}} \cdot dV \tag{9.120}$$

Integrating (9.120):

$$V = \frac{G'}{a} \int_{Y_{Ain}}^{Y_{Aout}} \frac{dY_A}{(-r_A)} \tag{9.121}$$

and for diluted gases, these equations can be written as

$$V = \frac{G'}{p_T \cdot a} \int_{p_{Ain}}^{p_{Aout}} \frac{dp_A}{(-r_A)} \tag{9.122}$$

A mass balance throughout the reactor provides:

$$G'(Y_{Ain} - Y_{Aout}) = L'(X_{Aout} - X_{Ain}) \tag{9.123}$$

and then:

$$\frac{G'}{p_T}(p_{Ain} - p_{Aout}) = \frac{L'}{C_T}(C_{Aout} - C_{Ain}) \tag{9.124}$$

In the integration, we must consider that p_A varies along the column or reactor, but C_A does not.

9.9.4 Case 2: Systems with Reaction

In this section, it is assumed that the reaction that is being produced is

$$A(G \rightarrow L) + bB(L) \rightarrow \text{products}\,(L)$$

We will assume that this reaction is fast enough so that there is no unreacted A within the liquid phase or the quantity is negligible for the mass balance.

9.9.4.1 Situation 1: Gas and Liquid Phases in Plug Flow

The design equation is obtained by combining the material balance with the rate equation. Thus, for a differential volume element:

$$\begin{aligned}(A \text{ lost by the gas}) &= (1/b)\,(B \text{ lost by the liquid})\\ &= (A \text{ disappeared by reaction})\end{aligned}$$

which is the same as

$$G'dY_A = -\frac{L'dX_B}{b} = (-r''_A) \cdot dV \tag{9.125}$$

$$G'd\left(\frac{p_A}{p_U}\right) = -\frac{L'}{b}d\left(\frac{C_B}{C_U}\right) = (-r_A) \cdot a \cdot dV \tag{9.126}$$

The value of the reaction rate can be calculated from the equations studied in the previous sections, i.e. the general equation would be used for fluid–fluid reactions (refer Section 9.8.2.3), and in particular systems it could be simplified. In the case of a fluid–fluid reaction taking place in a solid catalyst, the rate reaction equation would consider the stages mentioned in Section 9.8.3.

If the first and third terms of Eq. (9.126) are ordered and integrated:

$$V = \frac{G'}{a} \int_{Y_{A1}}^{Y_{A2}} \frac{dY_A}{(-r_A)} \tag{9.127}$$

and also we can work with the second and third terms of Eq. (9.125):

$$V = \frac{L'}{a \cdot b} \int_{X_{B2}}^{X_{B1}} \frac{dX_A}{(-r_B)} \tag{9.128}$$

The subscripts 1 and 2 refer to the extreme sections of the column (entry and exit). A global mass balance is also fulfilled:

$$G'(Y_{A2} - Y_{A1}) = \frac{L'}{b}(X_{B1} - X_{B2}) \tag{9.129}$$

These equations are simplified in the case of diluted systems.

Note that these equations can be also applied from a fixed section of the tower (point "1" in the equations) to an intermediate point, in such a way that:

$$G'(Y_A - Y_{A1}) = \frac{L'}{b}(X_{B1} - X_B)$$

Example 9.16 ***Liquid–Liquid Transfer with Reaction***
A liquid–liquid reaction is carried out in a packed tower. The heaviest liquid is fed at the top with a flow rate of 10 m^3/min and has a concentration of reagent A of 1 kmol/m^3, while the lighter fluid is fed from the bottom with a flow of 20 m^3/s and with a B concentration of 2 kmol/m^3. According to the reaction scheme, 1 mol of A reacts with 3 mol of B. Calculate the height of the fill tower to get that react 90% of A, admitting an average value for the reaction rate is $5{\cdot}10^{-5}$ kmol B/(s m$^2_{\text{interface}}$), and knowing that the diameter of the tower is 30 cm and the area ratio of Interface/reactor volume is 1200 m$^2_{\text{interface}}$/m^3.

Solution
In this example, $A + 3B \rightarrow P$ and we want $X_A = 0.9$, the reaction rate is

$$(-r_A) = \frac{(-r_B)}{3} = \frac{5 \cdot 10^{-5}}{3} = 1.67 \cdot 10^{-5} \frac{\text{kmol}\,A}{\text{s}\,\text{m}^2_{\text{interface}}}$$

A change in the units is necessary:

$$(-r''_A) = (-r_A) \cdot a = 1.67 \cdot 10^{-5} \frac{\text{kmol}\,A}{\text{s}\,\text{m}^2_{\text{interface}}} \cdot 1200 \frac{\text{m}^2_{\text{interface}}}{\text{m}^3} = 0.02 \frac{\text{kmol}\,A}{\text{s}\,\text{m}^3}$$

In the system, both phases behave as piston flow. If we do a mole balance in a differential volume:

$$(\text{moles of } A \text{ reacted in a differential volume}) = (-r''_A) \cdot dV$$

$$n_{A0} \cdot dX_A = (-r''_A) \cdot dV \quad \text{i.e.} \frac{n_{A0} \cdot dX_A}{(-r''_A)} = dV$$

where n_{A0} is the mole feed rate of component A:

$$n_{A0} = 1 \frac{\text{kmol}\,A}{\text{m}^3} \cdot 10 \frac{\text{m}^3}{\text{min}} \cdot \frac{1}{60} \frac{\text{min}}{\text{s}} = 0.164 \frac{\text{kmol}}{\text{s}}$$

Considering that $(-r''_A)$ is approximately constant and equal to 0.02 $\frac{\text{Kmol A}}{\text{s m}^3}$, we can easily calculate the required volume:

$$V = \int_0^{X_{A\text{final}}} \frac{n_{A0} \cdot dX_A}{(-r''_A)} = 7.5\ \text{m}^3$$

The required variable is the height of the tower, so: $h = 7.5/(\pi \cdot (0.3/2)^2) = 106.1$ m

Example 9.17 ***CO_2 Elimination***

CO_2 will be removed from an air stream at 1 atm by absorption in a 0.5 M solution of NaOH at 20 °C in a 1 m diameter column packed with 25 mm rings. The air enters at a feed rate of 0.02 kmol/s and contains 0.1 mol% CO_2 that must be reduced to 0.005 mol% at the exit. NaOH is supplied to the column at such speed that its concentration does not vary appreciably when going through the reactor. Calculate the height of the packed required. Is a packed bed the best system for this operation? Data:

- ✓ Second-order speed constant for the reaction $CO_2 + OH \rightarrow HCO^-_3$ $k_2 = 9.5 \cdot 10^3$ m³/(kmol s). In a NaOH solution, this reaction is followed by the instantaneous reaction $HCO_3^- + OH^- \rightarrow CO_3^{2-} + H_2O$, leading to a global reaction: $CO_2 + 2NaOH \rightarrow Na_2CO_3 + H_2O$
- ✓ For 0.5 M NaOH at 20 °C: CO_2 diffusivity: $1.8 \cdot 10^{-9}$ m²/s. NaOH diffusivity = 1.7 times that of CO_2.
- ✓ CO_2 solubility: Henry's law constant = 25 bar m³/kmol
- ✓ Effective interfacial area of the packing: 280 m^2_i/m^3_R
- ✓ Coefficients of transfer of mass in the film: Liquid: $k_{Al} = 1.2 \cdot 10^{-4}$ m/s; Gas: $k_{AG} = 0.056$ kmol/(m³ s bar)

Solution

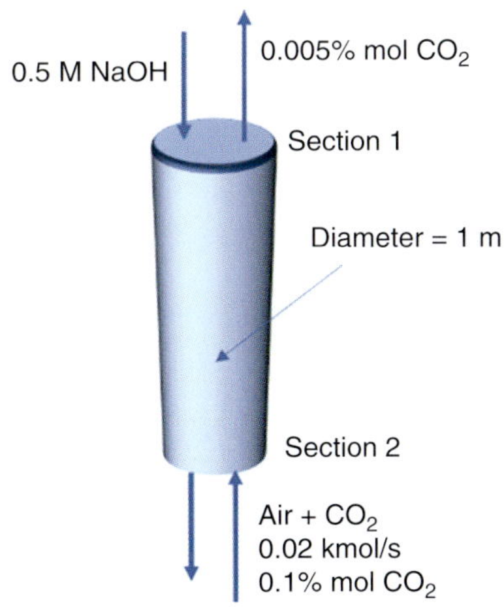

First of all, let us calculate the values of M_H and E_i, just to have an idea of the controlling steps of the reaction.

$$M_H = \frac{\sqrt{(k \cdot C_B) \cdot D_{Al}}}{k_{Al}} = \ldots = 24.4$$

For the calculation of E_i, let us assume $C_{Ai} = C_A{}^*$ (concentration in equilibrium with the maximum value of p_A); in any case, the actual value of C_{Ai} will be lower.

$$E_i = \left(1 + \frac{D_{Bl} \cdot C_B}{b \cdot D_{Al} \cdot C_{Ai}}\right)$$

$$C_{Ai} = \frac{p_{Ai}}{H_A} = \frac{0.001 \cdot 1\ \text{atm}}{25 \frac{\text{atm m}^3}{\text{kmol}}} = 4 \cdot 10^{-5} \frac{\text{kmol}}{\text{m}^3}$$

With these values: $E_i = \left(1 + \frac{D_{Bl} \cdot C_B}{b \cdot D_{Al} \cdot C_{Ai}}\right) = \ldots = 10\,626$
(in any case, as C_{Ai} may be lower, the value of E_i will be higher). This indicates that the reaction is done with both chemical reaction and diffusion interfering in the reaction rate. Let us calculate the "E" value:

$$E = \frac{\sqrt{M_H^2 \frac{E_i - E}{E_i - 1}}}{\tanh\left(\sqrt{M_H^2 \frac{E_i - E}{E_i - 1}}\right)}$$

Solving for $E = 24.37$.

The value of the reaction rate can now be calculated, using Eq. (9.94) and results:

$$(-r_A'') = \frac{p_A}{\frac{1}{k_{Ag} \cdot a} + \frac{H_A}{k_{Al} \cdot a \cdot E} + \frac{H_A}{(k \cdot C_B) \cdot \varepsilon_L}} = \ldots = 0.0206 \cdot p_A \frac{\text{kmol}}{\text{m}^3\ \text{s}}$$

where p_A must be in atm. As we are in a case where both liquid and gas are flowing in a piston flow pattern, we can write:

$$G' dY_A = \frac{L' dX_B}{b} = (-r_A'') \cdot dV$$

where Y_i refers to the mole fraction of the species "i" in its phase. As the concentrations are very low, $Y_A \approx \frac{p_A}{P_T}$ and $Y_B \approx \frac{C_B}{C_T}$. Operating we can write:

$$V = \frac{G'}{0.0206} \int_{Y_{A1}}^{Y_{A2}} \frac{dY_A}{Y_A \cdot p_T}$$

We have $G' = 0.02$ kmol/s, and operating we find $V = 2.89\ \text{m}^3$. For a column of 1 m in diameter (Section $= \pi(1/2)^2 = 0.785\ \text{m}^2$), this means a height of 3.69 m.

9.9.4.2 Situation 2. Gas and Liquid as Mixed Flow

Since the composition is the same at any point of the tower/reactor, only one value of the reaction rate is found, and the balance can be made in the whole system:

(kmol A lost by gas) = $(1/b)$ (kmol B lost by the liquid) = (kmol A missing)

Or in another way:

$$G'(Y_{Ain} - Y_{Aout}) = \frac{L'}{b}(X_{Bin} - X_{Bout}) = -r''_A]_{\text{at the exit conditions}} \cdot V \qquad (9.130)$$

The calculation of V is direct: evaluate the reaction rate from the known compositions and solve (Eq. (9.130)).

9.9.4.3 Situation 3. Gas in Plug Flow. Liquid Mixed Flow

As in the previous case with no reaction, there exist in the system several values of reaction rate, as one of the phases is in plug flow. This indicates that differential balances are needed, although the conditions of the liquid phase will be the same in all points.

In this way, in a differential volume of gas:

(kmol A lost by gas) = (kmol A missing by reaction)

$$G'dY_A = -r''_A]_{\text{exit conditions}} \cdot dV \qquad (9.131)$$

From a balance around the reactor, you get:

$$G'(Y_{Ain} - Y_{Aout}) = \frac{L'}{b}(X_{Bin} - X_{Bout}) \qquad (9.132)$$

Integrating Eq. (9.131) and using Eq. (9.132) is

$$V = \frac{G'}{a}\int_{Y_{Aout}}^{Y_{Ain}} \frac{dY_A}{(-r_A)} \qquad (9.133)$$

For diluted systems, these equations can be simplified in the usual way. The value of $(-r_A)$ is relatable to p_A.

9.9.5 Case 3. Multiphase Reactors

Until this point, we have discussed the form of the kinetic treatment of reactors in which a reaction occurs between a liquid and a gas with no presence of a solid catalyst. When treating with multiphase systems, where a component initially in the gas phase is transferred to the liquid, and then reacts over the surface of a catalyst, the mass and rate balance equations do not vary, not depending on the type of reaction being designed. In this sense, Eqs (9.125–9.133) are valid for the design of fluid–fluid catalyst reactors.

Nevertheless, the reaction rate expressions must consider the presence of the catalysts, and then the general equation studied in Section 9.8.3 is used. On many occasions, these equations are only valid in the cases of excess of A or excess of B, so this should be approximately met in the system to be designed.

Figure 9.17 shows a scheme of the different cases and situations, which can be useful for the design of multiphase reactors.

Example 9.18 ***Ammonia Absorption***

In a packed column for the absorption of ammonia with sulfuric acid, the ammonia partial pressure at the inlet is 0.05 atm in a gas that enters at atmospheric

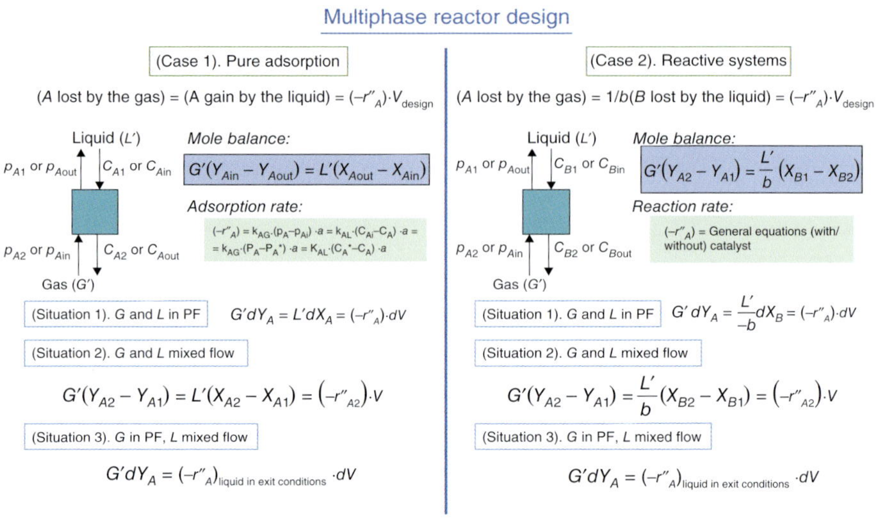

Figure 9.17 Summary of cases and situations studied for the design of multiphase reactors.

pressure. The partial pressure at the exit must be 0.01 atm. The total gas flow is 45 kmol/h. The liquid enters the column at its top and flows in countercurrent with the gas at 9 m^3/h. The input concentration of sulfuric acid is 0.6 kmol/m^3. Consider the isothermal operation at 25 °C. Determine the sulfuric acid outlet concentration and the required interfacial area. The reaction is second order, instantaneous, and irreversible:

$$2NH_3 + H_2SO_4 \rightarrow (NH_4)_2SO_4$$

$$\text{Data}: k_G = 0.35\frac{\text{kmol}}{\text{h atm m}^2} \quad H = 0.0033\ \frac{\text{atm}\cdot\text{m}^3}{\text{kmol}}$$

$$k_L = 0.05\frac{\text{m}}{\text{h}} \quad D_{Al} = D_{Bl}.$$

Solution

Let us first draw a scheme of the system variables.

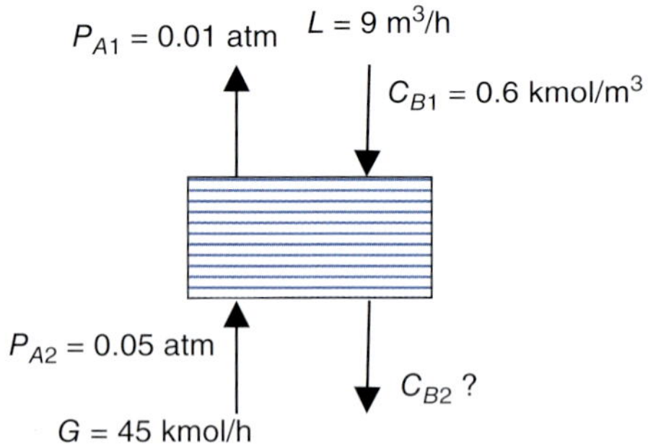

The reaction is of the type $2A + B = C$, and the mole balance around all the reaction system is

$$G'(Y_{A2} - Y_{A1}) = \frac{L'}{b}(X_{B1} - X_{B2})$$

As we are working in a diluted system, the previous equation can be approximated by this:

$$G'(Y_{A2} - Y_{A1}) = (G_2 y_{A2} - G_1 y_{A1}) \approx G(y_{A2} - y_{A1})$$

and, similarly, for the liquid:

$$\frac{L'}{b}(X_{B1} - X_{B2}) \approx \frac{L}{b}(C_{B1} - C_{B2})$$

In the system, we have $b = 0.5$ and then we can calculate the concentration at the exit:

$$G(y_{A2} - y_{A1}) = \frac{L}{b}(C_{B1} - C_{B2})$$

$$45\frac{\text{kmol}}{\text{h}}(0.05 - 0.01) = \frac{9\frac{\text{m}^3}{\text{h}}}{0.5}(0.6 - C_{B2})\frac{\text{kmol}}{\text{m}^3}$$

Then: $C_{B2} = 0.5001\ \text{kmol/m}^3$. Note that this equation can be applied to any point in the column, in such a way that a value of y_A is related to the corresponding value of C_B with this equation.

Now, let us calculate the necessary reactor interface. The rate of the process will be given by

$$(-r''_A) = \frac{p_A}{\frac{1}{k_{Ag}\cdot a} + \frac{H_A}{k_{Al}\cdot a\cdot E} + \frac{H_A}{(k\cdot C_B)\cdot f_L}}$$

As the chemical step is instantaneous ($k \to$ infinity), the reaction will be governed by the diffusion of the reactants. When the reaction is instantaneous, $E = E_i = \left(1 + \frac{D_{Bl}\cdot C_B}{b\cdot D_{Al}\cdot C_A}\right)$, and then previous equation can be transformed into:

$$(-r''_A) = \frac{p_A}{\frac{1}{k_{Ag}\cdot a} + \frac{H_A}{k_{Al}\cdot a\cdot\left(1+\frac{D_{Bl}\cdot C_B}{b\cdot D_{Al}\cdot C_A}\right)}}$$

This equation is expressed in terms of the volume of the reactor. Instead of this, the one needed in the present case is that based on the interface surface, i.e. $(-r_A)\frac{\text{kmol}}{\text{s m}^2_{\text{cat}}} = (-r''_A)\frac{\text{kmol}}{\text{s m}^3_R}\cdot a\left(\frac{\text{m}^2_{\text{cat}}}{\text{m}^3_R}\right)$, and then

$$(-r_A) = \frac{p_A}{\frac{1}{k_{Ag}} + \frac{H_A}{k_{Al}\cdot\left(1+\frac{D_{Bl}\cdot C_B}{b\cdot D_{Al}\cdot C_A}\right)}}$$

Using the data given:

$$(-r_A) = \frac{p_A}{\frac{1}{0.35} + \frac{0.0033}{0.05\cdot\left(1+\frac{C_B\cdot 0.0033}{0.5\cdot p_A}\right)}}$$

As we see, the reaction rate is a function of the pair C_B–p_A that are related with the mass balance. To solve the design, we need to solve the rate equation that, in the case of L and G in plug flow, is

$$G' dY_A = (-r_A) \cdot dS_i$$

$$S_i = \int_{YAin}^{YAout} \frac{G' \cdot dY_A}{(-r_A)}$$

S_i being the interface surface. The design can be finally calculated by means of the following table, for calculating numerically the integral.

y_A	C_B (kmol/m^3)	$(-r_A)$ (kmol/(s $m^2{}_i$))	$G'/(-r_A)$
0.01	0.6	0.003	13069.9
0.015	0.5875	0.005	8728.8
0.02	0.575	0.007	6553.4
0.025	0.5625	0.009	5246.3
0.03	0.55	0.010	4374.0
0.035	0.5375	0.012	3750.5
0.04	0.525	0.014	3282.6
0.045	0.5125	0.015	2918.5
0.05	0.5	0.017	2627.2

Finally, $S_i = 215.5\,\text{m}^2$.

Example 9.19 ***Trickle-Bed Reactor***

A trickle-bed-type reactor will be designed to eliminate 75% of the thiophene present in a hydrocarbon mixture containing 0.012 kmol/m^3 of thiophene. The bed of catalyst particles will operate at a temperature of 200 °C and at a pressure of 40 bar. Under these conditions, it can be assumed that the reaction rate is first order with respect to dissolved hydrogen and independent of the concentration of thiophene. The superficial velocity of the liquid in the reactor will be 0.05 m/s.

Determine the height of the bed needed to provide a thiophene conversion of 0.75. The necessary data are given here:

* Reaction rate constant ... $k = 1.1\cdot10^{-4}\ \text{m}^3{}_{\text{cat}}/(\text{kg}_{\text{cat}}\ \text{s})$
* Density of the catalyst ... $\rho_{\text{cat}} = 960\ \text{kg/m}^3$
* Effectiveness factor = 0.98
* Volumetric transfer coefficients for hydrogen:
  in the liquid film: $k_{Al} \cdot a_i = 0.030\ \text{s}^{-1}$
  in the solid film: $k_C \cdot a_C = 0.5\ \text{s}^{-1}$
* Value of Henry's coefficient: $H_A = 1940\ \text{bar m}^3/\text{kmol}$

Solution

Let us name hydrogen by "A" and thiophene by "B." In this case, we have

$$A + B \rightarrow \text{products} \quad (-r''_A)$$
$$= k \cdot C_A \ldots \text{ With the reaction rate given in Kmol/(s m}^3{}_R)$$

From the general equation:

$$(-r''_A) = \frac{p_{Ag}}{\frac{1}{k_{Ag} \cdot a_i} + \frac{H_A}{k_{Al} \cdot a_i} + \frac{H_A}{k_{Ac} \cdot a_c} + \frac{H_A}{(k'''_A \cdot \overline{C}_B) \cdot \eta_A \cdot f_s}}$$

We can calculate all terms. The first sum in the denominator is zero, as the transfer in the gas phase is quick. The second and third can be easily calculated by

$$\frac{H_A}{k_{AL} \cdot a_i} = \frac{1940 \frac{\text{bar} \cdot \text{m}^3}{\text{kmol}}}{0.03\ \text{s}^{-1}} = 64667 \frac{\text{bar m}^3\ \text{s}}{\text{kmol}}$$

$$\frac{H_A}{k_{Ac} \cdot a_c} = \frac{1940 \frac{\text{bar m}^3}{\text{kmol}}}{0.5\ \text{s}^{-1}} = 3880 \frac{\text{bar m}^3\ \text{s}}{\text{kmol}}$$

and the last term is

$$\frac{H_A}{(k'''_A \cdot \overline{C}_B) \cdot \eta_A \cdot f_s} = \frac{1940 \frac{\text{bar m}^3}{\text{kmol}}}{1.1 \cdot 10^{-4} \frac{\text{m}^3_{\text{cat}}}{\text{kg}_{\text{cat}} \cdot s} \cdot 960 \frac{\text{kg}_{\text{cat}}}{\text{m}^3_{\text{cat}}} \cdot 0.98} = 18746 \frac{\text{bar m}^3\ \text{s}}{\text{kmol}}$$

The reaction rate is then:

$$(-r''_A) = \frac{p_{Ag}}{87\,293} = 1.14 \cdot 10^{-5} \cdot p_{Ag}$$

with the rate in kmol/(s $\text{m}^3{}_R$) if the pressure is in bar.

In this example, the gas is pure hydrogen, so its partial pressure equals the total pressure, and the reaction rate is constant in this particular case:

$$(-r''_A) = 1.14 \cdot 10^{-5} \cdot 40 = 4.58 \cdot 10^{-4} \frac{\text{kmol}}{\text{s m}^3_R}$$

For a conversion of 0.75, we have

$$C_{B\text{out}} = 0.012\ \text{kmol/m}^3 \cdot (1 - 0.75) = 0.003\ \text{kmol/m}^3$$

And we can write

$$L' \cdot C_{B\text{in}} - L' \cdot C_{B\text{out}} = (-r''_A) \cdot V$$
$$L' \cdot 0.012 - L' \cdot 0.003 = 4.58 \cdot 10^{-4} \cdot V \quad \text{i.e.} V/L' = 19.56\ \text{s}$$

As the lineal velocity is 0.05 m/s, we can write

$$\text{Height} = 0.05\ \text{m/s}\ 19.56\ \text{s} = 0.978\ \text{m}$$

Example 9.20 ***Sulfur Oxide Removal***

A column with a packing of active carbon that constitutes a bed is placed in the laboratory. The column is used for the removal of a stream of SO_2 with a content of 5% in air. To do this, the gas is introduced through the upper part of the column next to a stream of water, creating a percolating bed in which plug flow can be considered for both phases. The elimination of SO_2 is produced by oxidation to SO_3 catalyzed by the active carbon surface. SO_3 combines with water to form H_2SO_4 that dissolves in water. In previous tests, it was found that the intrinsic reaction rate is first order with respect to oxygen and independent of the concentration of SO_2.

(a) Deduce the expression for the calculation of the rate of the process from fundamental equations of mass transfer or chemical reaction, making an outline of the reaction stages and explaining all the terms involved in the equation.
(b) Represent qualitatively how each one of the variables of the equation evolves within the height of the column.
(c) Write the mass balance of oxygen in a volume differential of the column.

Solution

The reaction to be considered is

$$SO_2(g) + O_2(g) \rightarrow SO_3(g)$$

That is followed by the combination with water:

$$SO_2(g) + H_2O(l) \rightarrow H_2SO_4\ (g)$$

The reaction is first order with respect to oxygen, so:

$$r_{SO_2} = -k \cdot C_{O_2}$$

Let us call A to the SO_2, B to the water, and C to the SO_3.

(a) Let us first make a drawing of the possible concentration profiles. For the first reaction to occur, available catalyst is needed. The concentration profiles would be like those presented in the next figure, although the relative position of the absolute values of the different species could be different. For convenience, one part of the particle represents the gas–solid phenomena and the other part the liquid–solid reaction, but obviously all particles are reacting at the same time to produce SO_3 and H_2SO_4.

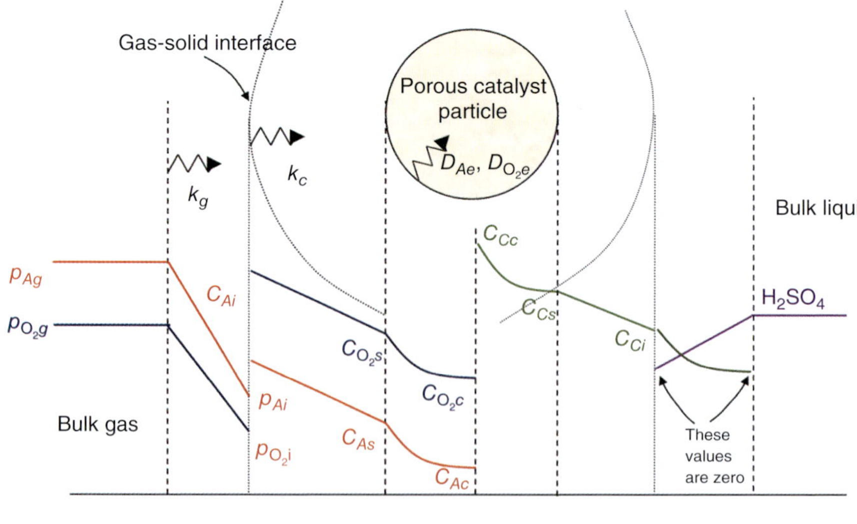

For the $SO_2(A)$, the rate of the process would be

$$(-r_A) = -k_{AG}(p_{Ag} - p_{Ai}) = k \cdot C_{O_2s} \cdot \eta$$

where the first equality is giving the rate of the diffusion through the gas–solid interface, and the second refers to the reaction + pore diffusion, and so the effectiveness factor appears. This effectiveness factor is given by the ability of the oxygen (species appearing in the chemical reaction rate) to diffuse through the catalyst particles.

For the $SO_3(C)$, we can write

$$(-r_C) = k_{Cl}(C_{Cs} - C_{Cc}) = k \cdot C_{O_2s} \cdot \eta$$

So, globally:

$$(-r_A) = k_{AG}(p_{Ag} - p_{Ai}) = k \cdot C_{O_2s} \cdot \eta = -k_{Cl}(C_{Cs} - C_{Cc})$$

(b) Qualitatively, the evolution of the concentrations in the reactor would be those shown in the next scheme.

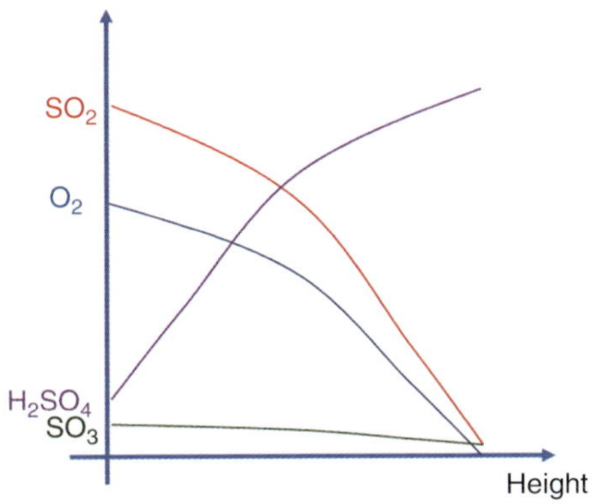

(c) A mass balance in a volume differential can be written as

$$(O_2 \text{ lost by the gas}) = (-r''_A) \cdot dV$$
$$G \cdot C_{O_2\text{in}} \cdot dX_{O_2} = (-r_A) \cdot dV$$

Example 9.21 ***Reaction in a Reactor of Known Volume***

In a reactor of known volume, a gas–liquid reaction is carried out on a solid catalyst. In the reactor, the gas and liquid phases behave according to the complete mixing pattern. If you know the quantity and composition of the liquid foods ($C_{A\text{in}}$, $C_{B\text{in}}$, L') and gas ($p_{A\text{in}}$, G', $p_{B\text{in}} = 0$), propose a method to determine the output compositions. What happens if the two fluids follow the piston flow pattern?

Solution

Assuming a mixed flow pattern, the mass balance can be expressed as

$$\underset{\text{(I)}}{G'(Y_{A\text{in}} - Y_{A\text{out}})} = \underset{\text{(II)}}{\frac{L'}{b}(X_{B\text{in}} - X_{B\text{out}})} = \underset{\text{(III)}}{-r''_A]_{\text{final conditions}} \cdot V}$$

In this case, we know the values of G', L', Y_{Ain} (that is proportional to C_{Ain}), and V. We can calculate the values of the exit conditions following the next calculation:

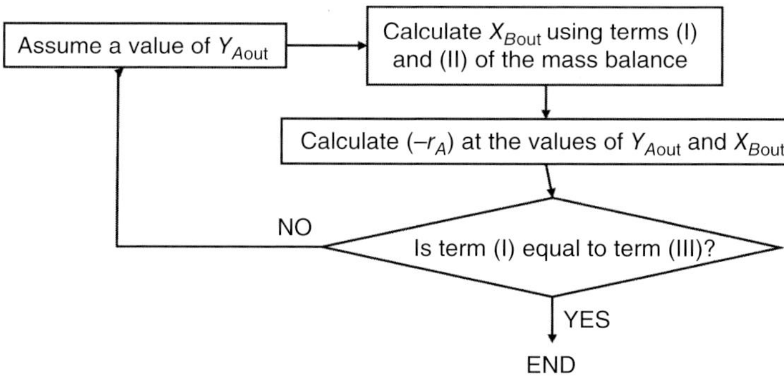

In the case the flow of the fluids follows a plug flow pattern, the mass balance can be only done in a differential volume; therefore:

$$G'dY_A = (-r''_A) \cdot dV$$

From this equation:

$$G' \int_{Y_{Ain}}^{Y_{Aout}} \frac{dY_A}{(-r''_A)} = V \qquad (*)$$

And, on the other hand, we can write the balance over the complete system:

$$G'(Y_{Ain} - Y_{Aout}) = \frac{L'}{b}(X_{Bin} - X_{Bout}) \qquad (**)$$

The procedure to calculate the exit conditions would be as follows:

1. Take values of Y_A from Y_{Ain} to Y_{Aout}.
2. Calculate X_{Bout} using Eq. (**).
3. Integrate Eq. (*) using the tabulated values.
4. If the calculated volume equals the given one, then stop the calculation.
5. If the calculated volume is not equal to V, repeat the calculation with a new value of Y_{Aout}.

Example 9.22 ***Fluid–Fluid Reaction Over Catalyst***

In a certain chemical plant, a reversible isomerization in fluid phase is carried out

$$A \overset{\text{reversible}}{\longleftrightarrow} B$$

on a solid catalyst in a packed-bed tubular reactor. If the reaction is so rapid that the mass transfer between the surface of the catalyst and the bulk fluid limits the rate, show that the kinetics is described in terms of the volumetric concentrations C_A and C_B by

$$(-r''_A) = \frac{k_B\left(C_A - \frac{C_B}{K}\right)}{\frac{1}{K} + \frac{k_B}{k_A}}$$

where $-r_A''$ = moles of A that react per unit area of catalyst per unit time, k_A, k_B = mass transfer coefficients of A and B, and K = equilibrium constant of the reaction

Solution

Assuming the following reaction rate constants:

$$A \underset{k_2}{\overset{k_1}{\rightleftarrows}} B$$

In this reversible system, we can write

$$(-r''_A) = k_1 C_{As} - k_2 C_{Bs}$$

where the concentrations are evaluated at the surface of the catalyst.

Using the relationship $K = k_1/k_2$, we have

$$(-r''_A) = k_1\left(C_{As} - \frac{C_{Bs}}{K}\right)$$

On the other hand, the mass transfer will have the following expressions:

$$\text{For the component } A \rightarrow \frac{\text{Moles of } A \text{ transferred}}{\text{time} \cdot \text{surface}} = k_A(C_A - C_{As})$$

$$\text{For the component } B \rightarrow \frac{\text{Moles of } B \text{ transferred}}{\text{time} \cdot \text{surface}} = k_B(C_{Bs} - C_B)$$

From all three equations, which must be equal, we can eliminate the unknowns C_{As} and C_{Bs}, and then:

$$(-r''_A)\left[\left(\frac{1}{k_A} + \frac{1}{k_1}\right) + \frac{1}{K \cdot k_B}\right] = C_A - \frac{C_B}{K}$$

As the system is limited by mass transfer, we can do $k_1 \gg k_A$, and then:

$$(-r''_A)\left[\frac{1}{k_A} + \frac{1}{K \cdot k_B}\right] = C_A - \frac{C_B}{K}$$

So, finally:

$$(-r''_A) = \frac{k_B\left(C_A - \frac{C_B}{K}\right)}{\frac{1}{K} + \frac{k_B}{k_A}}$$

Example 9.23 ***Catalytic Reaction with Expansion of the Gas***
In a catalytic system, a cracking reaction is taking place:

$$A \rightarrow 4R$$

With the conditions indicated in the figure, calculate the weight of the catalyst needed for the conversion of the reactant.

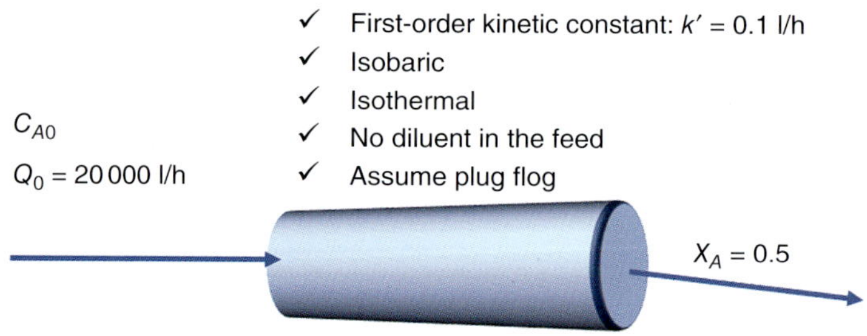

Solution
The kinetics in this system are

$$(-r'_A) = k'C_A = k'C_{A0}\frac{(1-X_A)}{(1+\varepsilon_A X_A)}$$

As we can see, the denominator accounts for the expansion of the gas through the reactor assuming constant pressure. The expansion coefficient, ε_A, can be calculated from the stoichiometric coefficients of the reaction, and is $\varepsilon_A = 4 - 1 = 3$.

The performance equation for a plug flow reactor is

$$(-r'_A) \cdot dW = n_{A0} \cdot dX_A$$

n_{A0} being the moles of "A" at the input of the system per unit time, and dW the weight of the catalyst in the volume differential. And then:

$$W = \int_0^{X_A} \frac{n_{A0} \cdot dX_A}{(-r'_A)} = n_{A0}\int_0^{X_A} \frac{dX_A}{k'C_{A0}\frac{(1-X_A)}{(1+\varepsilon_A X_A)}}$$

$$= \frac{n_{A0}}{k' \cdot C_{A0}}\int_0^{X_A} \frac{(1+\varepsilon_A X_A) \cdot dX_A}{(1-X_A)}$$

$$= \frac{Q_0}{k'}\int_0^{X_A} \frac{(1+\varepsilon_A X_A) \cdot dX_A}{(1-X_A)} = \frac{20\,000}{0.1}\int_0^{0.5} \frac{(1+3X_A) \cdot dX_A}{(1-X_A)}$$

This integral can be done numerically, following the procedure shown in previous chapters, and finally we get $W = 255$ kg of catalyst.

Example 9.24 ***Second-Order Isomerization***
An isomerization and second-order reaction proceeds at atmospheric pressure in a packed bed with spherical alumina pellets containing metal oxides deposited by impregnation. The process is carried out in the gas phase, at atmospheric pressure, in the absence of inert (only the two isomers) at 600 K. The effective diffusivity of a compound into the other isomer is $2.98 \cdot 10^{-7}$ m^2/s, and the

molecular weight is 60. The pellets have the following properties: radius 9 mm; density 1.2 g/cm^3; internal surface 100 m^2/g; and internal porosity 0.60. The rate constant for this process is 3.86 m$^6{}_{\text{fluid}}$/(m$^3{}_{\text{particle}}$ kmol s). Calculate the efficiency factor and the particle size for which the influence of internal diffusion is negligible.

Solution

The Thiele modulus can be estimated from

$$m \cdot L = L \cdot \left[\frac{k''' \cdot C_{As}^{n-1}(n+1)}{2 \cdot D_e}\right]^{\frac{1}{2}} \tag{9.69}$$

In the present case we have

$$C_{As} = P_A/RT = 1/(0.082\,600) = 2.02\ \text{mol/l} = 0.0202\ \text{mol/m}^3$$

In this case the order is 2, then:

$$m \cdot L = L \cdot \left[\frac{3 \cdot k''' \cdot C_{As}}{2 \cdot D_e}\right]^{\frac{1}{2}} = \left(\frac{9 \cdot 10^{-3}}{3}\right)\left[\frac{3 \cdot 3.86 \cdot 0.0202}{2 \cdot 2.98 \cdot 10^{-7}}\right]^{\frac{1}{2}} = 266$$

The efficiency factor is then (Eq. (9.50)):

$$\eta = \frac{\tanh(m \cdot L)}{m \cdot L} = \frac{\tanh(2.66)}{2.66} = 0.37$$

On the other hand, if we want to avoid internal diffusion effects, we must have a Thiele modulus higher than 0.4, i.e.:

$$0.4 < L \cdot \left[\frac{3 \cdot k''' \cdot C_{As}}{2 \cdot D_e}\right]^{\frac{1}{2}} = \left(\frac{R}{3}\right)\left[\frac{3 \cdot 3.86 \cdot 0.0202}{2 \cdot 2.98 \cdot 10^{-7}}\right]^{\frac{1}{2}}$$

The minimum value of radius accomplishing the inequality is 2.08 mm.

Example 9.25 ***Isomerization in a Fluidized Bed Reactor***

An isomerization reaction is going to take place in a fluidized-bed-type reactor. In the absence of diffusional effects, it has been deduced that the rate constant of the process is 0.15 (kmol A/(kg$_{\text{cat}}$ h))/((kmol A)/m$^3{}_{\text{gas}}$). The catalyst density is 12 g/cm^3. The flow rate of the fluid is 10 m^3/h. Calculate the following:

(a) In the absence of diffusional effects, both external and internal, the mass of catalyst needed to obtain a reagent conversion of 90%, for different cases of initial composition of A in the feed mixture: 10%, 20%, and 30% molar.
(b) If the fluidization was not very effective, and the individual external transfer coefficient is estimated to be 5·10^{-4} (kmol A/(h m$^2{}_{\text{cat}}$)/(kmol A/m$^3{}_{\text{fluid}}$), calculate the mass of catalyst of diameter equal to 0.1 mm if you want to reach the 90% conversion, for different cases of initial composition of A in the feed mixture: 10%, 20%, and 30% molar.
(c) With a catalyst of radius 10 mm and being the effective diffusivity of A of 2·10^{-6} m^2/s, calculate the mass of catalyst necessary to achieve the degree of conversion of 90%, for different cases of initial composition of A in the feed mixture: 10%, 20%, and 30% molar.

Solution

(a) In the fluidized bed, a good approximation is considering the solid as a completely mixed flow, but the gas flow is particularly due to the formation of bubbles. Nevertheless, a good consideration can be to assume a plug flow in the gas phase, so:

$$n_{A0} \cdot dX_A = (-r'_A) \cdot dW$$

$$n_{A0} \cdot dX_A = \left(k' \cdot \frac{n_{A0} \cdot (1 - X_A)}{Q}\right) \cdot dW$$

$$\frac{dX_A}{(1 - X_A)} = \left(\frac{k'}{Q}\right) \cdot dW$$

$$\int_0^{X_A} \frac{dX_A}{(1 - X_A)} = \left(\frac{k'}{Q}\right) \cdot W$$

The value of k' represents the kinetic constant, coinciding with the value calculated under chemical control. On the other hand, at the end of the reactor we have $X_A = 0.9$ and the value of W can be calculated, being: $W = 153.31$ kg The mass of catalyst needed in this system is not dependent on the concentration of reagent, as a first-order reaction is considered.

(b) In the case of a system with a clear effect of external diffusion, we can consider

$$(-r''_A) = k_c(C_{Ag} - C_{AS}) \cdot a_c \approx k_c \cdot C_{Ag} \cdot a_c$$

The value of k_C given in the data should be corrected with the surface/mass ratio of the catalyst. Assuming spherical particles:

$$a_c = \frac{S}{W} = \frac{S}{V \cdot \rho} = \frac{4\pi R^2}{\frac{4}{3}\pi R^3 \cdot \rho} = \frac{3}{R} = 50\ \text{m}^2/\text{kg}$$

The mass balance is similar to the former question, so:

$$\int_0^{X_A} \frac{dX_A}{(1 - X_A)} = \left(\frac{k_c \cdot a_c}{Q}\right) \cdot W$$

and doing the corresponding calculations, we have $W = 921$ kg regardless of the concentration of reagent.

(c) In that case, we probably will have problems in the internal diffusion. In this situation, the calculation of the effective factor is needed:

$$\eta = \frac{\tanh(mL)}{mL} \quad \text{with } m \cdot L = L \cdot \left[\frac{k'''}{D_e}\right]^{\frac{1}{2}} = \left(\frac{1 \cdot 10^{-2}}{3}\right)\left[\frac{0.15}{2 \cdot 10^{-6}}\right]^{\frac{1}{2}} = 0.9128$$

The efficiency factor is then $\eta = 0.7914$
And the mass of catalyst needed can be calculated as

$$\int_0^{X_A} \frac{dX_A}{(1 - X_A)} = \left(\frac{k \cdot \eta}{Q}\right) \cdot W$$

$W = 153.31/0.7914 = 193.7$ kg.

Example 9.26 ***Sucrose Acid Treatment***
The following first-order reaction has been studied:

$$C_{12}H_{22}O_{11}\ \text{(sucrose)} + H_2O\ \text{(acid medium)}$$
$$\Rightarrow C_6H_{12}O_6\ \text{(glucose)} + C_6H_{12}O_6\ \text{(fructose)}$$

From the experimental results obtained with spherical particles of resin at 50 °C, it was deduced that the effective diffusivity was on the order of $1.8{\cdot}10^{-11}$ m^2/s and the first-order kinetic constant is 0.0193 (kmol sucrose/(s $m^3{}_{\text{particle}}$))/((kmol sucrose/($m^3{}_{\text{disol}}$)). Calculate: (a) the critical size from which the internal diffusion begins to have an effect and (b) the critical size from which internal diffusion plays a dominant role.

Solution
The critical sizes will be calculated using the limits:

- Negligible internal diffusion effect → $m{\cdot}L < 0.4$
- Important internal diffusion effects → $m{\cdot}L > 4$

As stated before, this is equivalent to

$$\text{Negligible internal diffusion effect} \rightarrow We = \frac{-r'''_A L^2}{D_e C_{As}} < 0.15 \tag{9.134}$$

$$\text{Important internal diffusion effects} \rightarrow We = \frac{-r'''_A L^2}{D_e C_{As}} > 4 \tag{9.135}$$

(a) In the case where the diffusion effects are negligible, we have

$$We = \frac{-r'''_A L^2}{D_e C_{As}} \cong \frac{(k''' \cdot C_{AS})L^2}{D_e C_{As}} = \frac{k''' \cdot L^2}{D_e} < 0.15$$

From this relation:

$$L < \sqrt{\frac{0.15 \cdot D_e}{k'''}} = \sqrt{\frac{0.15 \cdot 1.6 \cdot 10^{-11}}{0.0193}} = 1.18 \cdot 10^{-5}\ \text{m}$$

For spherical particles, $L = R/3$, so $R < 3.54{\cdot}10^{-5}$ m = 0.0354 mm

(b) In diffusion effects are much important, we will have

$$(-r'''_A) = k''' \cdot C_{As} \cdot \eta \cong \frac{k \cdot C_{As}}{m \cdot L}$$

In this way:

$$We = \frac{-r'''_A L^2}{D_e C_{As}} = \frac{\frac{k''' \cdot C_{As}}{m \cdot L} \cdot L^2}{D_e C_{As}} > 4$$

and then:

$$L > \sqrt{\frac{4 \cdot D_e \cdot m \cdot L}{k'''}} = \sqrt{\frac{4 \cdot 1.8 \cdot 10^{-11} \cdot 4}{0.0193}} = 1.22 \cdot 10^{-4}\ \text{m}$$

So, $R > 3.66{\cdot}10^{-4}$ m = 0.366 mm

Example 9.27 *Chlorine Removal*

A chemical plant is removing traces of chlorine from a waste gas stream by passing it through a solid granular absorbent in a packed tubular bed. A conversion of 63.2% is currently being achieved, but it is believed that a higher removal could be achieved if the flow were increased by a factor of 4, the diameter of the particles was reduced by a factor of 3, and the length of the tube packed increased by 50%. What percentage of chlorine would be eliminated with the proposed scheme? Consider that the chlorine transferred to the absorbent is completely eliminated by a practically instantaneous chemical reaction. Consider that the filling are perfect spheres. In the bed, you can apply the Thoenes and Kramers equation:

$$\left(\frac{k_c \cdot d_p}{D_{AB}(1-\varepsilon)}\right) = \left(\frac{\rho \cdot u \cdot d_p}{\mu\,(1-\varepsilon)}\right)^{1/2} \cdot \left(\frac{\mu}{\rho\, D_{AB}}\right)^{1/3}$$

Solution

Solving for k_C:

$$k_c = \left[\left(\frac{\rho}{\mu \cdot (1-\varepsilon)}\right)^{0.5} \cdot \left(\frac{\mu}{\rho \cdot D_{AB}}\right)^{0.33} \cdot D_{AB}(1-\varepsilon)\right] \cdot \left(\frac{u}{d_p}\right)^{0.5}$$

The term in the square bracket is a constant, so k_C is proportional to $\left(\frac{u}{d_p}\right)^{0.5}$, in such a way that:

$$k_C = (\text{constant}) \cdot \left(\frac{u}{d_p}\right)^{0.5}$$

In this way, if we call k_{C1} and k_{C2} the values of k_C corresponding to the first and second situation, respectively, we can write

$$k_{C1} = (\text{constant}) \cdot \left(\frac{u_1}{d_{p1}}\right)^{0.5}$$

$$k_{C2} = (\text{constant}) \cdot \left(\frac{4u_1}{d_{p1}/3}\right)^{0.5} = \sqrt{12} \cdot k_{C1}$$

Let us now do a molar balance in the reactor bed, considered as plug flow:

$$dn_A = (-r'_A) \cdot dW$$

W being the weight of catalyst needed and $(-r'_A)$ the reaction rate based on the solid catalyst weight. We will relate the amount of catalyst with the length of the reactor, as one of the changes suggested is to increase reactor length. It is easy to see that:

$$W = \text{catalyst weight} = \rho_{\text{cat}} \cdot S_{\text{bed}} \cdot L \cdot (1-\varepsilon)$$

S_{bed} being the cross-sectional area of the bed, and $(1-\varepsilon)$ is the porosity of the catalyst. Considering a first-order reaction, we can write

$$\frac{dn_A}{dM} = \frac{dn_A}{\rho_{\text{cat}} \cdot S_{\text{bed}} \cdot (1-\varepsilon) \cdot dL} = \frac{n_{A0} \cdot dX_A}{\rho_{\text{cat}} \cdot S_{\text{bed}} \cdot (1-\varepsilon) \cdot dL} = (-r'_A)$$
$$= k_C \cdot (C_{Ag} - C_{As}) = k_C \cdot (C_{Ag})$$

In the last equality, $C_{AS} = 0$ as the reaction is quick and it is not possible to have a reactant in the particle. Reordering and integrating through the whole bed:

$$n_{A0} \cdot dX_A = k_C \cdot (C_{A0}(1 - X_A)) \cdot \rho_{\text{cat}} \cdot S_{\text{bed}} \cdot (1 - \phi) \cdot dL$$

$$\int_0^{X_A} \frac{n_{A0} \cdot dX_A}{(C_{A0}(1 - X_A))} = \int_0^{X_A} \frac{u_0 \cdot S_{\text{bed}} \cdot dX_A}{(1 - X_A)} = k_C \cdot \rho_{\text{cat}} \cdot S_{\text{bed}} \cdot (1 - \varepsilon) \cdot L$$

Finally, we have

$$\ln\left(\frac{1}{1 - X_A}\right) = \frac{k_C \cdot \rho_{\text{cat}} \cdot (1 - \phi) \cdot L}{u_0}$$

Comparing both situations:

$$\frac{\ln\left(\frac{1}{1-X_{A1}}\right)}{\ln\left(\frac{1}{1-X_{A2}}\right)} = \frac{\frac{k_{C1} \cdot \rho_{\text{cat}} \cdot (1-\phi) \cdot L_1}{u_{01}}}{\frac{k_{C2} \cdot \rho_{\text{cat}} \cdot (1-\phi) \cdot L_2}{u_{02}}} = \frac{\frac{k_{C1} \cdot L_1}{u_{01}}}{\frac{\sqrt{12} \cdot k_{C1} \cdot L_2}{u_{02}}} = \frac{\frac{L_1}{u_{01}}}{\frac{\sqrt{12} \cdot 1.5 \cdot L_1}{4 \cdot u_{01}}} = 0.7698$$

Taking the given value $X_{A1} = 0.632$ we can calculate: $X_{A2} = 0.7129$.

Example 9.28 ***Fluid–Fluid Reaction in a Tower***

Absorption of a gaseous compound A is occurring in a liquid containing a compound B which reacts according to

$$A\ (\text{g} \rightarrow \text{L}) + B\ (\text{L}) \rightarrow R\ (\text{L})$$

The rate equation of the chemical stage is: $(-r_A) = kC_A C_B$ and the reaction occurs in a packed bed in which:

$k_{Ag} = 0.001\ \text{mol/(h m}^2)$
$k_{Al} = 100\ \text{m}^3{}_L/(\text{m}^2{}_i\ \text{h})$
$a = 100\ \text{m}^2{}_i/\text{m}^3{}_R$
$f_L = 0.01\ \text{m}^3{}_L/\text{m}^3{}_R$
$D_{Al} = D_{AB} = 10^{-6}\ \text{m}^2/\text{h}$
$k = 10\ \text{m}^3{}_L/(\text{mol h})$
$H_A = 10^5\ (\text{Pa m}^3{}_L)/\text{mol}$

Check how the reaction rate varies, as well as the influence of each resistance throughout the reactor, for a reactor in which the liquid is fed at a concentration of $100\ \text{mol/m}^3$ and is countercurrent to a gas that passes through it as a piston flow, making the design so that the partial pressure of component A decreases from 100 to 1 Pa. Previous data obtained in a pilot plant indicates that the optimum ratio between G' and L' fulfills:

$$\frac{G'/P_T}{L'/C_T} = 1\ \text{mol/(m}^3\ \text{Pa)}$$

Solution

The objective is to obtain the reaction rate throughout the reactor when the gas pressure varies from 100 to 1 Pa. For this, the equation based on the sum of the resistances corresponding to the gaseous layer, to the liquid layer and to the chemical reaction is used:

$$(-r_A'') = \frac{p_A}{\frac{1}{k_{Ag}\cdot a} + \frac{H_A}{k_{Al}\cdot a\cdot E} + \frac{H_A}{(k''\cdot C_B)\cdot f_L}} \tag{9.94}$$

where the increment factor is given by

$$E = \frac{\sqrt{M_H^2 \frac{E_i - E}{E_i - 1}}}{\tanh\left(\sqrt{M_H^2 \frac{E_i - E}{E_i - 1}}\right)} \tag{9.96}$$

$$E_i = \left(1 + \frac{D_{Bl}\cdot C_B}{b\cdot D_{Al}\cdot C_{Ai}}\right) \tag{9.90}$$

The concentration of the gas in the interface can be calculated according to Henry's law:

$$p_{Ai} = H_A C_{Ai} \tag{9.77}$$

and the Hatta modulus is

$$M_H^2 = \frac{k\cdot D_{Al}\cdot C_B}{k_{Al}^2} \tag{9.93}$$

A molar balance of the reactant in the reactor gives

$$G'(Y_{\text{Ain}} - Y_{\text{Aout}}) = \frac{L'}{b}(X_{\text{Bin}} - X_{\text{Bout}}) \tag{9.132}$$

Bearing in mind that $Y_A \approx p_A/P_T$ and $X_A \approx C_B/C_T$:

$$C_{\text{Bout}} = C_{\text{Bin}} - \frac{\frac{G'}{P_T}}{\frac{L'}{b\cdot C_T}}(p_{\text{Ain}} - p_{\text{Aout}})$$

From the data, we know that $\frac{G'/P_T}{L'/C_T} = 1$, and $b = 1$, so, in the present example:

$$C_{\text{Bout}} = C_{\text{Bin}} - (p_{\text{Ain}} - p_{\text{Aout}})$$

Note that in this problem we cannot calculate the volume of the system, because we do not know the value of G' to use Eq. (9.133).

We will assume constant values of $(-r_A'')$ at small sections of the reactor, corresponding each one to a value of p_A. For the calculation of $(-r_A'')$, a nonlinear equation given in Eq. (9.96) must be solved at each value of p_A. A Matlab program will be used in the calculation, as shown in Program listing 9.5.

Program Listing 9.5

```
clear all
clear all
close all
pA0=100; % Pa
CB0=100; % mol/m3
```

```
kAg = 0.001;% mol/(h · m2)
kAl = 100;% m3L/(m2sup · h)
a = 100;% m2sup/m3reactor
fL = 0.01;% m3L/m3reactor
DAB = 10e-6;% m2/ h
k = 100000; % m3liq/(mol·h)
HA = 10e5; %Pa·m3liq/mol

CBv=[];
rAv=[];
pAv=100:-1:1;
R1v=[];
R2v=[];
R3v=[];

for pA=pAv
    CB=CB0-(pA0-pA);
    MH=(k*DAB*CB)^0.5/kAl;
    Ei=1+CB/(pA/HA);
    % Calculation of the 'E'
    dif=10; % Value needed for starting calculation
    E=1; % Value needed for starting calculation
    while dif>0.0001
        value= sqrt(MH^2*(Ei-E)/(Ei-1));
        Enew = value/tanh(value);
        dif=(E-Enew)^2;
        E=Enew;
    end
    % Calculation of the reaction rate
    R1=1/(kAg*a);
    R2=HA/(kAl*a*E);
    R3=HA/(k*CB*fL);
    rA=pA/(R1+R2+R3);
    CBv=[CBv CB];
    rAv=[rAv rA];
    R1v=[R1v R1/(R1+R2+R3)]; % Relative value of R1
compared to the total
    R2v=[R2v R2/(R1+R2+R3)]; % R2
    R3v=[R3v R3/(R1+R2+R3)]; % R3
end

figure(1)
plot(pAv,R1v,'--',pAv,R2v,'*',pAv,R3v)
legend('R_g_a_s','R_l_i_q_u_i_d','R_r_e_a_c_t_i_o_n')
xlabel('P_A (Pa)')
ylabel('Relative resistance')
```

```
figure(2)
plot(pAv,rAv)
xlabel('P_A')
ylabel('Reaction rate (mol/(h·m^3))')
```

The following plots are generated:

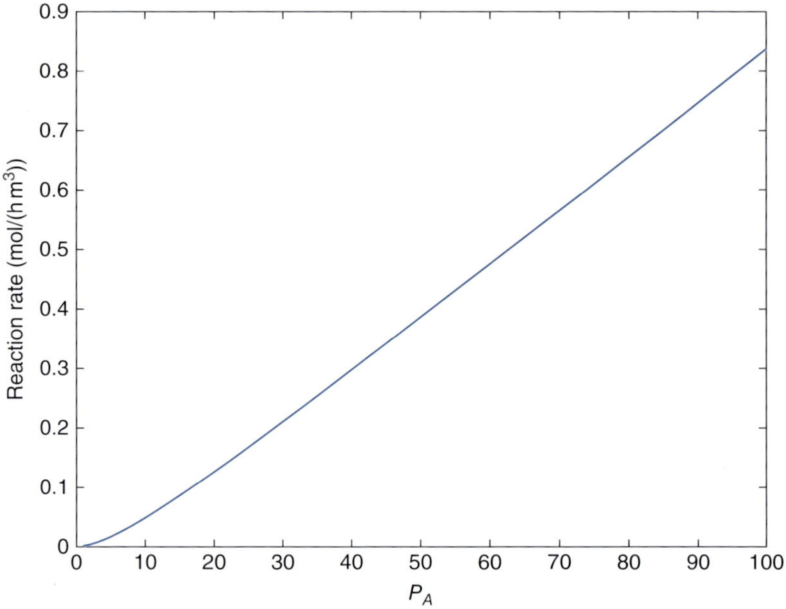

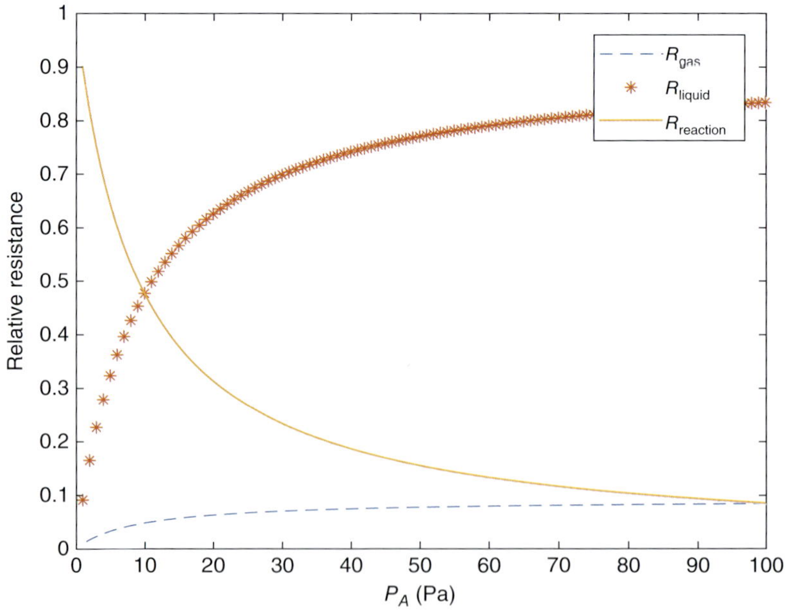

Bibliography

Carberry, J.J. (1976). *Chemical and Catalytic Reaction Engineering*. McGraw-Hill.

Conesa Ferrer, J.A. and Font Montesinos, R. (2002). *Reactores Heterogeneos*. Universidad de Alicante.

Coulson, J.M. and Richardson, J.F. (1979). *Chemical Engineering III. Chemical Reactor Design*, 2e. Oxford: Pergamon Press.

Gates, B.C. (1992). *Catalytic Chemistry*, 1992. Wiley.

Doraiswamy, L.K. and Sharma, M.M. (1984). *Heterogeneous Reactions: Analysis, Examples, and Reactor Design*. Wiley.

Fogler, H.S. (1998). *Elements of Chemical Reaction Engineering*, 3e. Upper Saddle River, New Jersey: Prentice Hall.

Jackson, R. (1977). *Transport in Porous Catalyst*. Elsevier.

Levenspiel, O. (1999). *Chemical Reaction Engineering*, 3rde. New York, NJ: Wiley.

10

Biochemical Reactors

10.1 Introduction

The term "fermentation", in its strict sense, refers exclusively to the production of alcohol from sugar. Etymologically, it means bubbling or boiling action and was effectively used for the first time in the production of wine. At present, the meaning of the word is broader: fermentation is a molecular transformation in which an organic matter is converted into a product by the direct action of microorganisms or by the action of enzymes.

The aim of the fermentation can be the production of microorganisms (for example, the production of bread yeast), the disappearance of the substrate (water purification), the production of a product, or a combination of the previous objectives. In the cases in which the microbial mass produced is a residual product, it is used as a feed for animals, due to its high protein content.

Biotechnology is important in the production of several substances, and in the pharmaceutical, chemical, and food industries, as well as in the application of microorganisms in the treatment of contaminated wastewater. In these processes, the microorganisms transform certain unwanted products into harmless matter.

The biological reactions, under the point of view of kinetics and thermodynamics, present a series of own characteristics that differentiate them substantially from other processes of chemical transformation. We can remark on some of them:

1. They are generally slow processes, compared to chemical reactions. For this reason, its time constants are usually measured in hours, and even days.
2. The fermentation processes have an autocatalytic character, since the microorganisms (product) act as catalysts of the process.
3. The specific activity of biocatalysts is usually low compared to that of chemical catalysts. They are extraordinarily dependent on environmental conditions (T, pH, oxygen, etc.).
4. They may present problems of inhibition by product and/or substrate depending on the operating conditions and work concentrations.
5. The nature of the biocatalyst can be significantly modified over a relatively long process. Thus, both a loss of activity (due to problems of inhibition, deactivation of extracellular enzymes, etc.) and an activation thereof

Chemical Reactor Design: Mathematical Modeling and Applications,
First Edition. Juan A. Conesa.

(by adaptation of the microorganisms to the new environmental conditions) can take place.

6. The distribution of products can be strongly conditioned by the environmental operating conditions since, in many cases, a microorganism can derive its metabolism in different directions, depending on the pH, dissolved oxygen, etc.
7. Reaction enthalpies are low, so the design of heat exchange equipment should not be a major problem.
8. The activation energy of the reaction is important so, especially in some temperature ranges, the reaction rate is strongly dependent on the reaction temperature. On the other hand, there are areas where the deactivation is also very dependent on the temperature.
9. The operating temperature and pressure are moderate.
10. Good solid/liquid/gas contact is necessary to achieve effective action.
11. Both the gas supply and the disposal of the gas produced can be problematic.
12. The rheological properties of the fluid may vary during the fermentation process. Furthermore, sometimes the fluids present a non-Newtonian behavior.
13. Fermentation operations often require separation or, at least, management of solid mass.

However, the efficiency of the fermentation processes is usually limited by different causes, among which they have a greater importance:

1. Low productivity due to a discontinuous operation,
2. Substrate inhibition problems,
3. Inhibition problems by product,
4. Low cellular or enzymatic concentration in the bioreactor,
5. Limitations for mass transfer.

10.2 Enzymatic Catalysis

From the structural point of view, enzymes are proteins, and from the kinetic point of view, they are catalysts. The detailed knowledge of the catalytic function of an enzyme requires the maximum knowledge of the structure of the enzymatic molecule. This is not only to know which amino acids and in what number are present in a given molecule but also the sequence of the amino acids in the polypeptide chain. And, finally, the spatial conformation, which has a great importance on the catalytic action.

Amino acids are not the only constituents of proteins (enzymes), as many enzymes contain other organic or inorganic groups. These parts of the conjugated proteins are called prosthetic groups.

10.2.1 Characteristics of Enzymatic Catalysis. The Active Center

Enzymatic catalysis involves the formation of a complex between the reagent (substrate) and the enzyme, in a process of equilibrium. This complex (ES) is usually called the Michaelis complex. The formation of the product is according to the following reaction scheme:

$$E + S \Leftrightarrow E \cdot S \rightarrow E \cdot P \Leftrightarrow E + P$$

The existence of the $E{\cdot}S$ complex has been verified by different techniques such as X-ray crystallography, spectroscopy and electronic spin resonance.

The substrate binds to a specific region of the enzyme, called the active site, in each catalytic cycle, and the catalysis occurs only at that site. The formation of the complex is due, in many cases, to the action of weak forces of attraction (ionic effects, hydrogen bridges, and hydrophobic attractions between nonpolar groups), although there are cases in which a covalent bond is involved.

A fundamental characteristic of enzymes as catalysts is their specificity, which differentiates them from synthetic catalysts. Most catalysts used in the chemical industry are nonspecific, that is, they catalyze similar reactions involving different types of reagents. Although some enzymes are nonspecific, most catalyze a single reaction for certain substrates. A simple model has been proposed to explain this specificity depending on the existence of the active site, and the formation of the enzyme–substrate complex, which is illustrated in the Figure 10.1.

This model states that the active site is the geometric complement of the substrate and only the substrates that have the appropriate complementary form can form the Michaelis complex. This hypothesis has been confirmed from the knowledge of the tertiary structure of the enzymes.

The degree of specificity in the catalytic action of enzymes is a consequence of their biological action in the cell. The specificity can be absolute: there are enzymes that only catalyze a reaction from a single substrate. It is evident that the enzymes that catalyze a certain reaction in the cellular metabolism have to discriminate the substrates on which they act. Other enzymes (for example, proteases), have as a function the hydrolysis of proteins breaking the peptide bonds and their specificity is much lower, acting on a relatively wide range of proteins, although with preferences in the recognition of certain amino acids on both sides of the protein peptide bond. Many enzymes are stereospecific (recognize only one molecular configuration) and regiospecific (they recognize a group among others of the same molecule), which is one of its main advantages over chemical catalysts.

Another characteristic of the enzymes that differentiate them from the usual chemical catalysts is the frequent need for cofactors. A cofactor is a non-protein

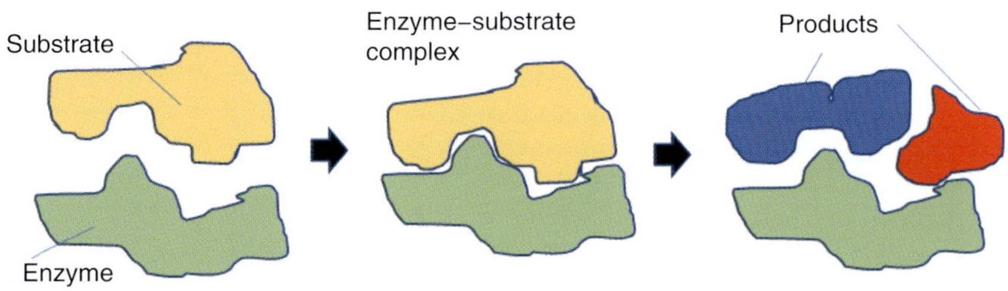

Figure 10.1 Biochemical route in catalyzed enzymatic reactions.

substance that is combined with a protein that does not act as a catalyst alone (apoenzyme) to form a catalytically active complex. In biochemistry, when the apoenzyme is linked to the cofactor it is called a holoenzyme, although in this topic it is simply called an enzyme. There are two large groups of cofactors:

(a) Metal ions, such as Fe^{2+}, Fe^{3+}, Mn^{2+}, Zn^{2+}, etc.,
(b) Organic molecules called coenzymes (two of the most common coenzymes are NAD or FAD).

Finally, when the cofactor is irreversibly bound to the enzyme, the name prosthetic group is used, as discussed earlier.

10.2.2 Kinetics of Enzymatic Reactions

In this section, we study the expressions of the reaction rate for enzymatic reactions. The simplest reactions, consisting of the conversion of a single substrate into a single product, for which the formation of an enzyme–substrate complex is postulated, as discussed, is taken up first. Next, other reaction mechanisms that will lead to more complex kinetic expressions are presented.

10.2.2.1 Kinetics of Reactions with a Single Substrate. Michaelis–Menten Equation

Michaelis and Menten proposed a model to relate the reaction rate to the substrate concentration. Assuming the existence of a single central complex, the reaction scheme would be

$$E + S \underset{k_{-1}}{\overset{k_1}{\rightleftarrows}} ES \xrightarrow{k_2} E + P$$

In fact, this scheme of reactions can be written more adequately in the following way:

$$E^f + S^f \underset{k_{-1}}{\overset{k_1}{\rightleftarrows}} ES \xrightarrow{k_2} E^f + \beta P$$

According to this mechanism, the free enzyme E^f and the free substrate S^f are reversibly bound to form the enzyme–substrate complex ES, which leads, irreversibly, to the obtaining of β moles of the product P. In a batch reactor of constant volume, the equations of balance of the different species are the following:

$$\frac{dE^f}{dt} = k_{-1}(ES) + k_2(ES) - k_1 E^f S^f \tag{10.1}$$

$$\frac{dS^f}{dt} = k_{-1}(ES) - k_1 E^f S^f \tag{10.2}$$

$$\frac{d(ES)}{dt} = -k_{-1}(ES) - k_2(ES) + k_1 E^f S^f \tag{10.3}$$

$$\frac{dP}{dt} = \beta k_2(ES) \tag{10.4}$$

where E, S, (ES), and P represent the molar concentrations of the different species. From a mass balance, the initial concentrations of enzyme and substrate can be expressed as follows:

$$E_0 = E^f + (ES) \tag{10.5}$$

$$S = S^f + (ES) + \frac{P}{\beta} \tag{10.6}$$

Applying the hypothesis that the complex (ES) is in a pseudo stationary state ($d(ES)/dt = 0$), from Eq. (10.1):

$$k_1 E^f S^f = k_{-1}(ES) + k_2(ES)$$

$$\frac{E^f S^f}{(ES)} = \frac{k_{-1} + k_2}{k_1} = K_M \tag{10.7}$$

K_M is known as the Michaelis constant. Substituting E^f for its value of the initial concentration of E, we can obtain an expression for the concentration of the enzyme complex-substrate:

$$K_M = \frac{(E_0 - (ES)) \cdot S^f}{(ES)} \quad \text{i.e. } (ES) = \frac{E_0}{1 + \frac{K_M}{S^f}} \tag{10.8}$$

Substituting in the product balance equation:

$$\frac{dP}{dt} = \beta \frac{k_2 E_0 S^f}{K_M + S^f} \tag{10.9}$$

The second hypothesis that is introduced is to suppose that the molar concentration of the enzyme is much smaller than that of the substrate ($E_0 \ll S_0$), so it can be considered that the concentration of free substrate is approximately equal to that of the substrate not converted S^f or S:

$$\frac{dP}{dt} = \beta \frac{k_2 E_0 S}{K_M + S} \tag{10.10}$$

The reaction rate r is based on

$$r = -\frac{dS}{dt} = \frac{1}{\beta}\frac{dP}{dt} \tag{10.11}$$

Finally, the Michaelis–Menten equation is obtained:

$$r = \frac{k_2 E_0 S}{K_M + S} \tag{10.12}$$

This expression is usually written by grouping the term $k_2{\cdot}E_0$ as the maximum reaction rate r_m. The Michaelis–Menten equation predicts the first-order kinetic behavior for low substrate concentrations and zero order for high concentrations.

10.2.2.2 Meaning of the Parameters of the Michaelis Equation

The Michaelis constant (K_M) corresponds to the substrate concentration at which the reaction velocity is half the maximum velocity ($r_m = k_2{\cdot}E_0$). Furthermore, for small values of k_2, K_M represents the constant of the equilibrium of dissociation of the $E{\cdot}S$ complex ($K_M = k_{-1}/k_1$), in such a way that a high value of K_M indicates a low affinity of the $E{\cdot}S$ complex and then it is easily dissociated to E^f and S^f. The contrary is true for low values of K_M.

In the Michaelis–Menten model, the maximum reaction rate per molecule of enzyme is r_m/E_0. On the other hand, k_{cat} is defined as the turnover number, the maximum number of substrate molecules that can be converted into product per unit of time and per active site of the enzyme. If the enzyme has a single active center, k_{cat} matches with k_2; for an enzyme containing "n" active centers, it can be calculated as $r_m/(n{\cdot}E_0)$.

The quotient k_{cat}/K_M can be interpreted from the Michaelis equation, written for low substrate concentrations such as

$$r = \frac{k_{cat}}{K_M} \cdot E_0 \cdot S \tag{10.13}$$

This quotient represents a pseudo second-order rate constant. Although it is not an actual rate constant, the value of this quotient is used to determine the affinity of the enzyme for different substrates, in the case of two substrates in competition.

Example 10.1 ***Metabolism of Ethanol***

A patient in a hospital presents an alcohol content in the blood of 0.2 mg/dl, and affirms that he had just taken a cup of white wine three days ago. It is reasonable that this cup represents 250 ml of wine, what would produce an initial content of alcohol in blood of 25 mg/dl. The patient's liver only metabolizes ethanol at a maximum rate of 1 mg/(dl h), and it is known that the enzyme involved in this metabolism presents a value of K_M of 10 mg/dl. Assuming a Michaelis–Menten kinetics, try to find out if the patient is lying or, on the contrary, telling the truth.

Solution

In this situation, the metabolism follows the equation:

$$r = \frac{dS}{dt} = r_{max}\frac{S}{K_M + S} = 1\frac{\text{mg}}{\text{dl h}}\frac{S}{\left(S + 10\frac{\text{mg}}{\text{dl}}\right)}$$

where the substrate S is the alcohol. The initial amount is 25 mg/dl and we must check if the level goes down to 0.2 mg/dl in three days (72 h).

We can use a simple finite differences method, in such a way that:

$$\frac{S^t - S^{t+1}}{\Delta t} = 1\frac{\text{mg}}{\text{dl h}}\frac{S^t}{\left(S^t + 10\frac{\text{mg}}{\text{dl}}\right)}$$

$$S^{t+1} = S^t - \frac{\Delta t \cdot 1 \cdot S^t}{(S^t + 10)}$$

Choosing an adequate value of time increment (for example, 0.5 h), we will have enough precision to give a response. This can be done both in Matlab® or using a spreadsheet. Using the last method, we get the following figure, where we can check the patient is right and his level of alcohol in the blood corresponds with that expected at $t = 72$ h.

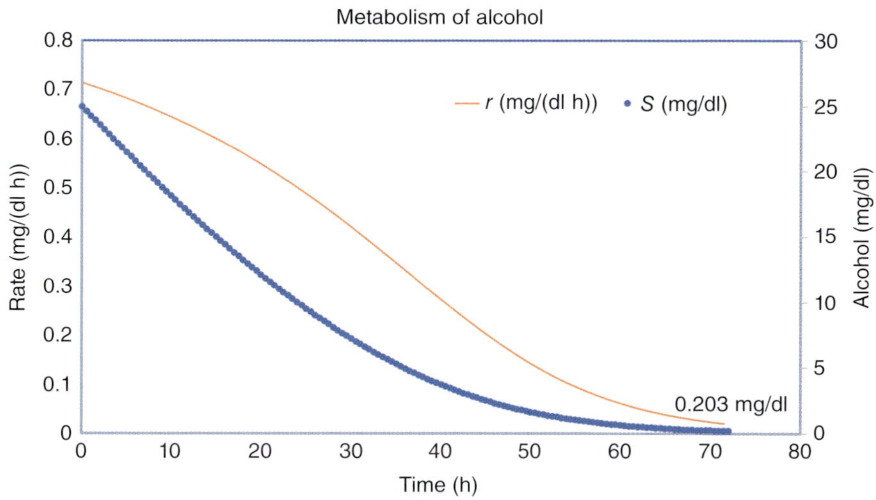

10.2.3 Enzymatic Reactions with Inhibition

Inhibitors are those substances that cause an enzymatic reaction to proceed more slowly. This effect can occur in several ways:

(a) By means of a substrate analog that binds (reversibly or irreversibly) to the active center, reducing the enzymatic activity.
(b) By binding an inhibitor to another part of the enzyme molecule, causing conformational changes that reduce the ability of the enzyme to bind to the substrate.

The kinetics of the reactions with inhibition depends on the mechanism of the process, distinguishing between reversible or irreversible inhibition depending on whether the original activity can be recovered or not:

(a) Irreversible inhibition. The inhibitor combines with the enzyme to give a stable but inactive complex.

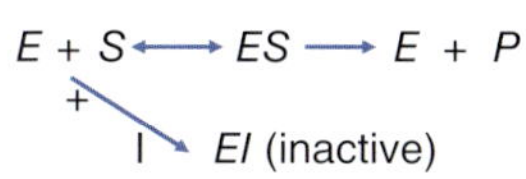

(b) Reversible inhibition.

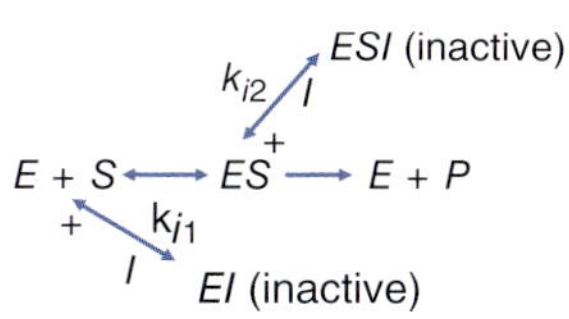

The so-called pure types of inhibition have been considered, with K_i as the dissociation constant of the *El* or *EIS* complex, with "I" being the inhibitor concentration. In fact, there are also situations of intermediate inhibition, which can be explained by the general inhibition equation:

$$r = \frac{r_m \cdot S}{K_M\left(1 + \frac{1}{k_{i1}}\right) + S\left(1 + \frac{1}{k_{i2}}\right)} \tag{10.14}$$

in which k_{i1} and k_{i2} are the equilibrium constants of dissociation of *El* and *EIS*, respectively.

(c) Inhibition by the substrate. In some cases of enzymatic reactions with a single substrate, a behavior different from the hyperbolic law proposed by the Michaelis equation is observed, so that the reaction velocity presents a maximum with the substrate concentration.

Several mechanisms have been proposed to explain the behavior of inhibition by the substrate: (i) the formation of a second enzyme–substrate complex with two substrate molecules bound to the enzyme, (ii) the binding of a fraction of substrate molecules to the enzyme in a place other than the active center.

From any of the proposed mechanisms, the same kinetic equation is obtained:

$$r = \frac{r_m \cdot S}{K_M + S + K'S^2} \tag{10.15}$$

10.2.4 Enzymatic Reactions with More than One Substrate

Many enzymes catalyze reactions between two or more substrates to give two or more products. A particular case of this type of reactions are those catalyzed by enzymes that use cofactors (for example, NAD+ in oxidoreductases), where the cofactor plays the role of a second substrate.

Various mechanisms have been proposed to explain this type of reactions, most of which can be classified into two types:

(a) All substrates bind to the enzyme before catalysis (in the case of two substrates, it is called the mechanism of formation of a ternary complex).
(b) A product is released from a first binary complex before binding to the second substrate (for example, in group transfer reactions).

Restricting the discussion to enzymatic reactions with two substrates, in any of the indicated cases, the treatment by means of the hypothesis of the pseudo-stationary state of the intermediates leads to kinetic equations that can be summarized in the following general form:

$$r = \frac{r_m \cdot A \cdot B}{K'_A \cdot K_B + K_B A + K_A B + A \cdot B} \tag{10.16}$$

where A and B are the molar concentrations of the two substrates involved, A and B. The particular expression corresponding to the values of the constants varies according to the mechanism that takes place.

10.2.4.1 Case 1. Enzymatic Reactions with Two Substrates by Formation of a Ternary Complex

In this case, the ternary complex can be formed independently of the binding order of the substrates (random) or only in a certain order (ordered mechanism).

(a) *Random mechanism*: The substrates A and B form a complex with the enzyme. Either of these can be combined with the other substrate to form the EAB ternary complex, which leads to the X and Y products and regenerates the enzyme:

$$\left.\begin{array}{l} E + A \underset{k_{-1}}{\overset{k_1}{\rightleftharpoons}} EA + B \underset{k_{-2}}{\overset{k_2}{\rightleftharpoons}} EAB \\ E + B \underset{k'_{-1}}{\overset{k'_1}{\rightleftharpoons}} EB + A \underset{k'_{-2}}{\overset{k'_2}{\rightleftharpoons}} EAB \end{array}\right] \xrightarrow{k_3} E + X + Y$$

Working with a multisubstrate, reactions can be quite complicated when treating to represent the schemes of the reactions. Throughout history, several systems have been proposed, but the most successful has been the one proposed by Cleland. The input of substrates and the output of products are represented by vertical arrows indicating the entry direction or output of the reaction. Using the Cleland's symbols, the random mechanism can be written as

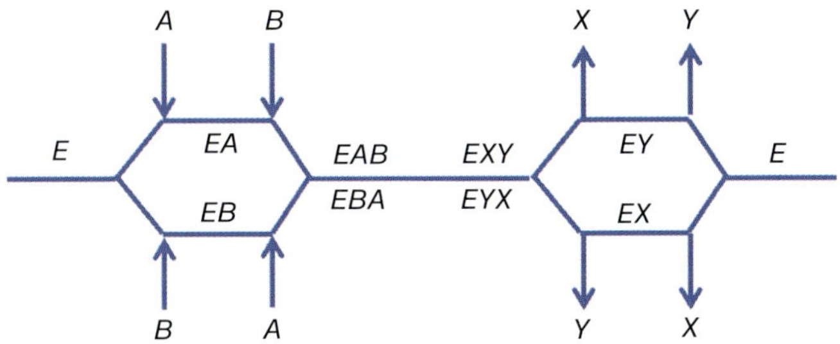

The starting conditions for deriving the rate equation with two or more substrates are the same as we used for the derivation of the rate equation in a simple system: (i) we will assume the formation of a rapid equilibrium between the E and the substrates (A and B); (ii) the equation of conservation tells us that (E_0) equals the sum of all the chemical species in which the enzyme; and (iii) we will assume the existence of steady-state conditions. Using Michaelis–Menten kinetics for the formation of the complexes, we can then write

$$K_A = \frac{[E]\cdot[A]}{[EA]} \tag{10.17}$$

$$K_B = \frac{[E]\cdot[B]}{[EB]} \tag{10.18}$$

$$K_M^{AB} = \frac{[EA]\cdot[B]}{[EAB]} \tag{10.19}$$

$$K_M^{BA} = \frac{[EB]\cdot[A]}{[EAB]} \tag{10.20}$$

$$E_0 = [E] + [EA] + [EB] + [EAB] \tag{10.21}$$

In that equations, K_M^{AB} is the concentration of substrate "B" that produces half of the maximum rate when the substrate "A" is in excess, and K_M^{BA} is the concentration of substrate "A" that produces half of the maximum rate when substrate "B" is in excess. Note that the following is true:

$$\frac{K_A}{K_B} = \frac{K_M^{BA}}{K_M^{AB}}$$

From the previous equations, we can deduce that:

$$E_0 = [E] + \frac{[E][A]}{K_A} + \frac{[E][B]}{K_B} + \frac{[E][A][B]}{K_A \cdot K_M^{AB}}$$

And, bearing in mind that the observed reaction rate is

$$r = k_3 \cdot [EAB] \tag{10.22}$$

we can finally write

$$\begin{aligned} r = k_3 \cdot [EAB] &= k_3 \cdot E_0 \cdot \frac{[A][B]}{K_A K_M^{AB} + [A]K_M^{AB} + [B]K_M^{BA} + [A][B]} \\ &= r_{\max} \cdot \frac{[A][B]}{K_A K_M^{AB} + [A]K_M^{AB} + [B]K_M^{BA} + [A][B]} \end{aligned} \tag{10.23}$$

In the present case, using the previous definitions, $k_{cat} = k_3$, if the runs are performed with a constant concentration of "A," but not saturating the environment, we can observe that the following is fulfilled:

$$r = r'_{\max} \cdot \frac{[B]}{K' + [B]} \tag{10.24}$$

where

$$r'_{\max} = k_3 \cdot E_0 \cdot \frac{[A]}{K_M^{BA} + [A]} \tag{10.25}$$

$$K' = K_M^{AB} \cdot \frac{(K_A + [A])}{K_M^{BA} + [A]} \tag{10.26}$$

If we represent the reaction rates vs. concentration of the substrate, we will obtain graphs similar to those shown in Figure 10.2. Plotting the value of $r'_{\max}$ obtained at different rates, $[A]$we can obtain the value of K_M^{BA} as the value of concentration giving half of the maximum reaction rate. On the other hand, at

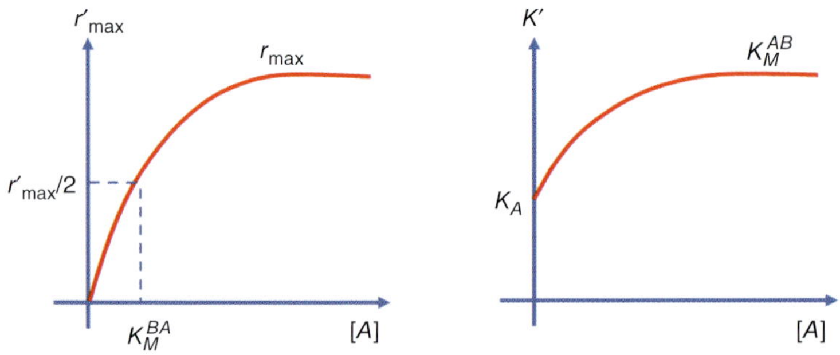

Figure 10.2 Reaction rate and value of K' obtained vs. concentration of one of the substrates.

$[A] = 0$, the graph K' vs. $[A]$ gives us the K_A, and the value of K' at $[A]$ very high is K_M^{AB}.

The same is true if the substrate with constant concentration is "B" (obviously changing "A" by "B" in the expressions). Furthermore, if the concentration of substrate "A" is saturating (i.e. the reaction rate does not increase with an ulterior increase of $[A]$), we can see that:

$$r = r_{max} \cdot \frac{[B]}{K_M^{BA} + [B]} \tag{10.27}$$

(b) *Ordered mechanism*: A substrate (A) initially binds with the enzyme forming the EA complex. The second substrate only joins the EA complex to lead to the EAB ternary complex:

$$E + A \underset{k_{-1}}{\overset{k_1}{\rightleftarrows}} EA + B \underset{k_{-2}}{\overset{k_2}{\rightleftarrows}} EAB \xrightarrow{k_3} E + X + Y$$

Using the Cleland' symbols, we can say:

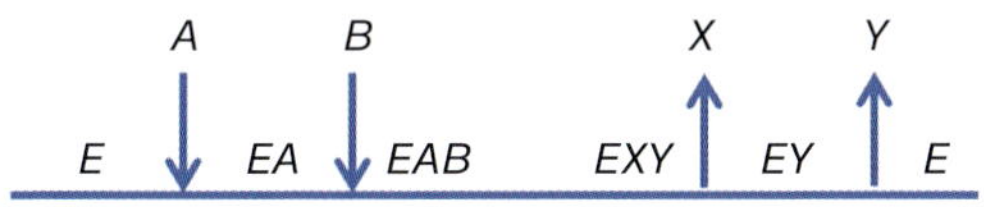

Using Michaelis–Menten kinetics for the formation of the complexes, we can then write

$$K_A = \frac{[E] \cdot [A]}{[EA]} \tag{10.17}$$

$$K_M^{AB} = \frac{[EA] \cdot [B]}{[EAB]} \tag{10.19}$$

$$E_0 = [E] + [EA] + [EAB] \tag{10.28}$$

From the previous equations, we can deduce that:

$$r = k_3 \cdot [EAB] = k_3 \cdot E_0 \cdot \frac{[A][B]}{K_A K_M^{AB} + [A]K_M^{AB} + [A][B]}$$
$$= r_{max} \cdot \frac{[A][B]}{K_A K_M^{AB} + [A]K_M^{AB} + [A][B]} \tag{10.29}$$

In the present case, using the previous definitions, $k_{cat} = k_3$, if the runs are performed with a constant concentration of "A," but not saturating the environment, we can observe that the following is fulfilled:

$$r = r_{max} \cdot \frac{[B]}{K' + [B]} \tag{10.30}$$

where

$$K' = \left(\frac{K_A}{[A]} + 1\right) \cdot K_M^{AB} \tag{10.31}$$

Now we can construct the graphs shown in Figure 10.3 to calculate the values of the constants.

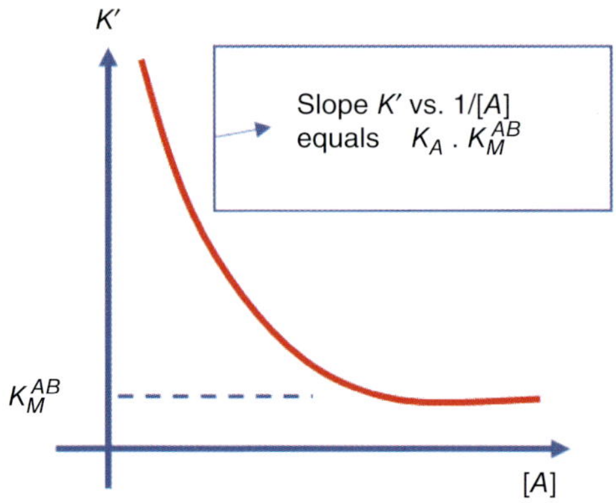

Figure 10.3 K' vs. concentration of substrate in a bisusbtrate ordered mechanism.

Nevertheless, in that case, the equations are not true for the other substrate concentration constant. If the substrate with constant concentration is "B":

$$r = r'_{\max} \cdot \frac{[A]}{K' + [A]} \tag{10.32}$$

where

$$r'_{\max} = k_3 \cdot E_0 \cdot \frac{[B]}{K_M^{AB} + [B]} \tag{10.33}$$

$$K' = K_M^{AB} \cdot \frac{K_A}{K_M^{BA} + [B]} \tag{10.34}$$

Graphs similar to the other cases can be constructed to calculate the corresponding values of the constants.

10.2.4.2 Case 2. Enzymatic Reactions with Two Substrates Without Formation of a Complex

The mechanism is called ping-pong, forming a complex of the enzyme with a substrate (EA), which releases the product X, remaining as EA (for example, an acyl-enzyme) that is attacked by the second substrate (for example, transferring the acyl group).

$$E + A \underset{k_{-1}}{\overset{k_1}{\rightleftharpoons}} EA \xrightarrow{k_2} EA' + X$$

$$EA' + B \xrightarrow{k_3} EA'B \xrightarrow{k_4} E + Y$$

This is

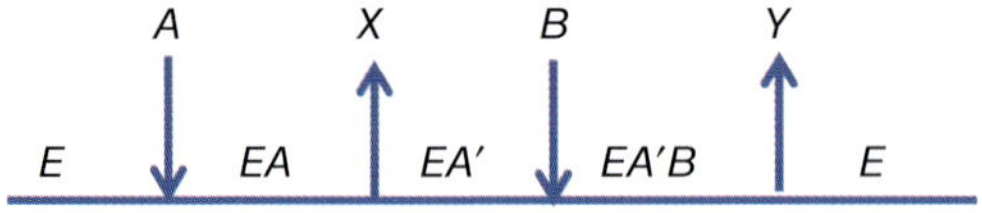

In that case:

$$r = k_4 \cdot [EB] \tag{10.35}$$

and $k_{cat} = k_4$.

Using the Michaelis–Menten constants:

$$K_M^A = \frac{k_{-1} + k_2}{k_1} = \frac{[E][A]}{[EA]} \tag{10.36}$$

$$K_M^A = \frac{k_{-3} + k_4}{k_3} = \frac{[EA'][B]}{[EA'B]} \tag{10.37}$$

$$E_0 = [E] + [EA] + [EA'] + [EA'B] \tag{10.38}$$

And solving for the rate equation, we find

$$r = r_{\text{max}} \cdot \frac{[A][B]}{\left(\frac{k_4}{k_2}\right) K_M^A[B] + [A]K_M^B + [A][B]\left(1 + \frac{k_4}{k_2}\right)} \tag{10.39}$$

where $r_{\text{max}} = k_4 \cdot E_0$. For $[A]$ the constant we can see that:

$$r = r'_{\text{max}} \cdot \frac{[A]}{K' + [A]} \tag{10.40}$$

with:

$$r'_{\text{max}} = r_{\text{max}} \cdot \frac{[A]}{\left(\frac{k_4}{k_2}\right) K_M^A + [A]} \tag{10.41}$$

and

$$K' = K_M^B \cdot \frac{[A]}{\left(\frac{k_4}{k_2}\right) K_M^A + [A]} \tag{10.42}$$

10.2.4.3 Strategies to Distinguish the Previous Cases

Representing the value of the observed constant K' vs. the concentration of both substrates, we can easily distinguish the previous cases. Figure 10.4 presents a scheme of the expected graphs for the different mechanisms.

Example 10.2 ***Calculating Kinetic Law***

In different experiments trying to calculate the kinetic law of the reduction of pyruvate (A) and hydroxybutyrate (B) with lactate dehydrogenase (LDH, EC 1.1.1.27), we find the following reaction rates at the different concentrations:

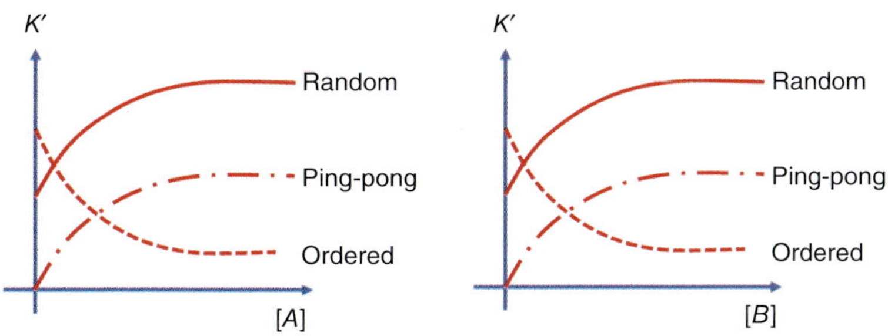

Figure 10.4 Exemplification of K' vs. concentration for the different mechanisms.

[A] (g/l)	[B] (g/l)	r (g/(l s))	[A] (g/l)	[B] (g/l)	r (g/(l s))
0.05	0.1	0.34	0.5	0.1	1.35
0.05	1	1.93	0.5	1	3.34
0.05	1.5	2.34	0.5	1.5	3.57
0.05	2	2.61	0.5	2	3.66
0.05	2.5	2.82	0.5	2.5	3.76
0.05	3	2.94	0.5	3	3.81
0.1	0.1	0.58	2	0.1	1.78
0.1	1	2.51	2	1	3.60
0.1	1.5	2.90	2	1.5	3.76
0.1	2	3.11	2	2	3.77
0.1	2.5	3.28	2	2.5	3.87
0.1	3	3.37	2	3	3.85

Considering that all runs were performed at the same initial concentration of enzyme, determine the mechanism of the transformation and the corresponding kinetic constants.

Solution

First of all, let us show in a graph the measured values of "r" at the different runs. As the runs are grouped in different concentrations of pyruvate (A), let us use the following plot:

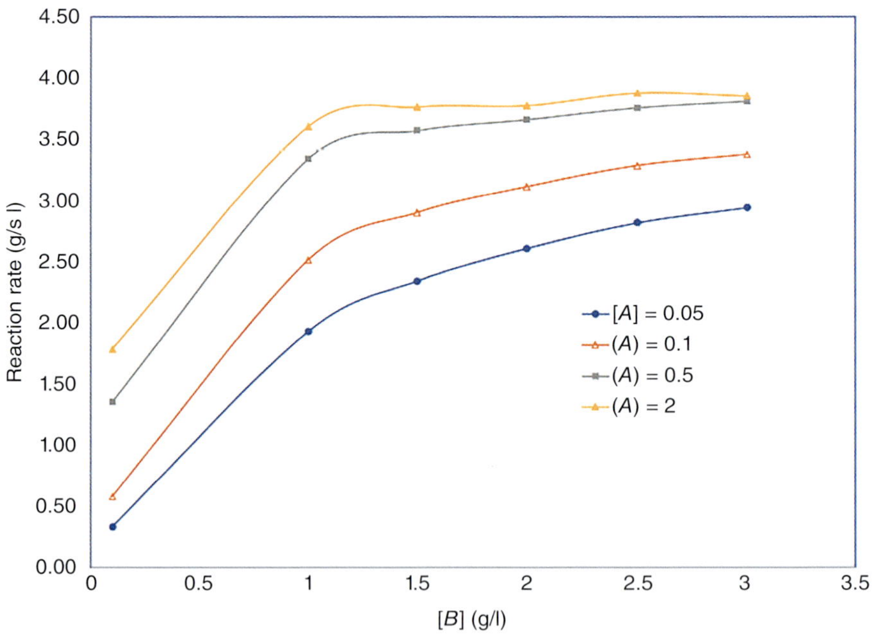

Considering Eqs. (10.24) or (10.30) or (10.40), we see that, in any case, we can do:

$$\frac{1}{r} = \frac{K'}{r_{\max}} \cdot \frac{1}{[B]} + \frac{1}{r_{\max}}$$

Using the data in the statement, the following is obtained:

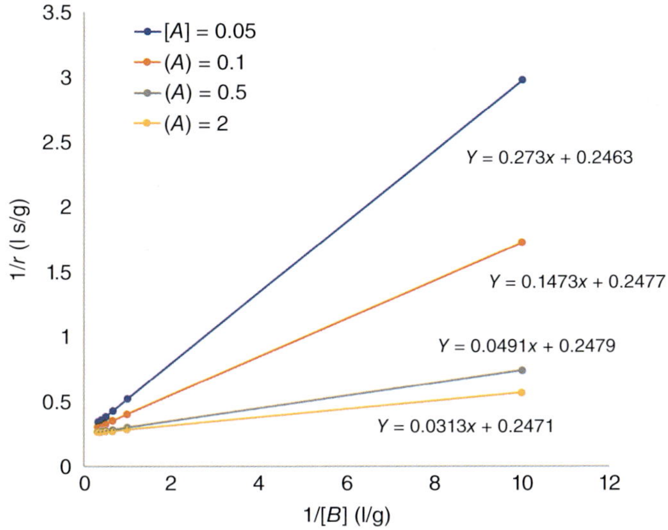

From these lines, we can obtain the best values of K' and r_{max}, for each of the concentrations of "A" used, and we obtain

[A] (g/L)	k′ (g/l)	$1/r_{max}$ ((l s)/g)
0.05	1.10	0.246
0.1	0.60	0.248
0.5	0.20	0.248
2	0.13	0.247

The value of r_{max} is then 4 g/(l s) approximately. Plotting the values of K', we obtain

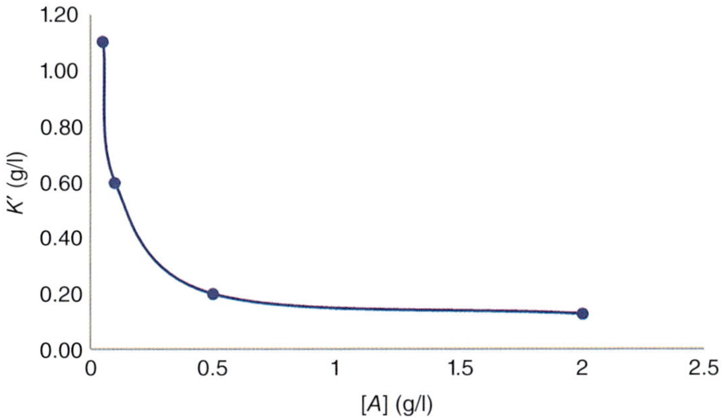

This is typical of an ordered mechanism, as mentioned before, with

$$K' = \left(\frac{K_A}{[A]} + 1\right) \cdot K_M^{AB}$$

At the highest concentration $[A]$, we see that $K_M^{AB} = 0.13$ g/l and the slope of the plot K' vs. $1/[A]$ at the initial stages ($[A]$ close to zero) is $K_M^{AB} \cdot K_A = 0.051$ so $K_A = \frac{0.051}{0.13} = 0.39$ g/l.

10.3 Microbial Kinetics

A cell growth process involves the consumption of the energy supplying substrates and the raw mass needed for the synthesis of cellular mass and other metabolic products. From a macroscopic viewpoint, this requires that the microenvironment contains all the elements necessary for the formation of additional cellular mass and that, in addition to this, the energy consumed by substrates must be greater than the energy of the cells or products. It is to be fulfilled, in addition to this, that all elements supplied as nutrients are present in a manner consistent with enzymatic mechanism of the cell.

Cell growth obeys the laws of conservation of matter. The atoms of carbon, nitrogen, oxygen, and other elements are reordered in the metabolic processes so that the total amount incorporated matches the amount eliminated from the environment. Furthermore, the amount of some of the metabolic products formed or heat generated by cell growth is often proportional to the consumed amount of any of the substrates or the formed amount of a particular product. All these properties make possible the approach of energy and mass balances in processes of cell growth, expressed generally as

$$\text{Source of C} + \text{N} + \text{source } O_2 + \text{minerals} + \text{specific nutrients}$$
$$\rightarrow \text{cell mass} + \text{products} + CO_2 + H_2O$$

The stoichiometric treatment of the cell growth processes, consumption of substrates, and production of substances and heat is done in Section 10.3.1. In the analysis in Section 10.3.2, we study the kinetics to which these processes take place.

10.3.1 Stoichiometry of the Microbian Growth

From a quantitative point of view, the required amounts of nutrients can be determined from the stoichiometry of the growth and product formation. For example, the reaction aerobically consuming glucose to synthetize cellular matter is

$$C_6H_{12}O_6 + aO_2 + bNH_3 \rightarrow cCH_xO_yN_z + dCO_2 + eH_2O$$

In order to calculate the stoichiometric coefficients, it is necessary to know the elemental composition of the microorganism ($CH_xO_yN_z$). This composition can be determined experimentally by an elemental analysis, obtaining the proportions of the various elements.

10.3.2 Stoichiometry of Product Formation

In a growth medium, a large variety of products that accumulate intracellularly or are released into the medium can be produced.

If the main product is a consequence of primary metabolism, a chemical reaction similar to the former one can be written, introducing the product indicated as CH_vO_w (if the product contains nitrogen, it should be also considered):

$$a\mathrm{CH}_m\mathrm{O}_n + b\mathrm{O}_2 + c\mathrm{NH}_3 \rightarrow d\mathrm{CH}_x\mathrm{O}_y\mathrm{N}_z + e\mathrm{CO}_2 + f\mathrm{H}_2\mathrm{O} + g\mathrm{CH}_v\mathrm{O}_w$$

10.3.2.1 Yields

The limiting substrate concept gives the possibility to define process yields. From experimental data of the amount of substrate consumed and the biomass obtained, it can be seen that there is a relationship between them that, in many cases, is a direct proportionality. The biomass–substrate yield is defined as the quotient between the increase in cell mass obtained and the substrate consumption (usually, a carbon source):

$$Y_{S/X} = \frac{\Delta X}{\Delta S} \tag{10.43}$$

The yield has units derived from those used for the measurement of the biomass and the substrate (g dry weight of biomass/g of substrate, g/mol, etc.). Overall yields, calculated from experimental data growth, are dependent on the carbon source used and the operation conditions, and may vary throughout the process.

10.3.2.2 Theoretical Yield Obtained from Stoichiometric Coefficients

As the biomass–substrate yield is defined, yields can be defined for other pairs of parameters of the process, as indicated in Table 10.1.

Using the definitions of the yields, we can relate the rate of the formation or consumption of different species, so that, once having characterized the behavior of one, the other can be calculated:

$$r_S = -\left(\frac{1}{Y_{\frac{X}{S}}}\right) \cdot r_X = -\left(\frac{1}{Y_{\frac{P}{S}}}\right) \cdot r_P = -\left(\frac{1}{Y_{C/S}}\right) \cdot r_C \tag{10.44}$$

where r represents the rate of disappearance of the substrate and r_x, r_p, and r_c are the rate of formation of biomass, product, and CO_2, respectively.

10.3.3 Cell Growth, Substrate Consumption, and Product Formation

As mentioned in the introduction of this chapter, the process of cell growth is very complex, since it is the result of the interaction between a population of cells and their environment.

In a batch reactor, the cell growth occurs within the reactor. But in a definite moment, as a consequence of the absence of regeneration of the culture medium,

Table 10.1 Yield definitions.

Symbol	Definition
$Y_{X/S}$	Molar growth rate: gram of dry biomass per mole of substrate consumed
$Y_{X/O}$	gram of dry biomass per gram of oxygen consumed or per mole of oxygen consumed
$Y_{P/S}$	gram or mole of product per gram or mole of substrate consumed
$Y_{C/S}$	mole of CO_2 produced per mole of substrate consumed

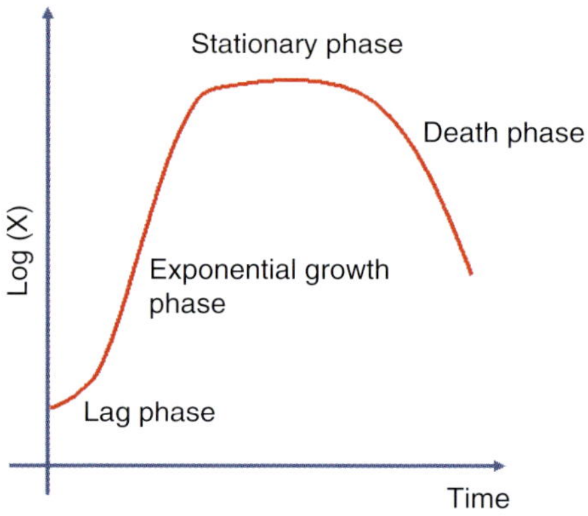

Figure 10.5 Phases of the cell growth.

the growth stops when it finds a limitation (consumption of an essential nutrient, accumulation of a toxic product for metabolism, and limitation by oxygen are the more frequent). Figure 10.5 shows the typical phases of batch cell growth.

Firstly, there is a lag phase, in which a substantial increase in the number of cells does not occur, and that is required for adaptation to the medium and culture conditions. Then the exponential growth phase begins, in which growth is produced at a specific rate constant. In this phase, it can be considered that the growth is balanced, in the sense that there are no limitations and the metabolic and physiological behavior of the cells therein is basically constant. The equation describing the increased cell concentration is in this case:

$$\frac{dX}{dt} = \mu \cdot X \tag{10.45}$$

with a certain concentration of cells at the end of the lag phase (t_{lag}), which depends on the inoculum used (X_0).

$$X = X_0 \text{ at } t = t_{\text{lag}}$$

The integrated form of these equations is

$$\ln\left(\frac{X}{X_0}\right) = \mu(t - t_{\text{lag}}) \tag{10.46}$$

$$X = X_0 e^{\mu(t-t_{\text{lag}})}$$

The usual way to determine the value of μ from the experimental data of batch growth is a semilog representation of X vs. time.

10.3.3.1 Kinetics of Growth

As mentioned, although the cell growth is a complex phenomenon, often we can obtain a reasonably good overall description using relatively simple equations. Among them, the most usual is the Monod equation, which describes the cell growth as a function of the availability of a limiting substrate, and that can be expressed as

$$\text{Substrate } (S) + \text{Cells } (X) \to \text{More cells } (X) + \text{Product } (P)$$

$$r_X = \frac{dX}{dt} = \mu_m \frac{S \cdot X}{K_S + S} \tag{10.47}$$

where r_X is the growth rate of the cells, μ_m is the maximum specific rate of growth, and K_S is the Monod constant. It is quite common to express the equation in the function of the specific growth rate:

$$\mu = \frac{1}{X} \cdot \frac{dX}{dt} = \mu_m \frac{S}{K_S + S} \tag{10.48}$$

in which μ_m is the maximum value that the growth rate can reach, when $S \gg K_S$ and the concentrations of the rest of nutrients have not changed significantly. K_S is the value of the concentration of the limiting nutrient, the specific growth rate of which is half of the maximum.

The Monod equation is very simple, although it is not always possible to obtain a good representation of the data of the growth of a microorganism. Therefore, other models have been developed, but we focus on the Monod that is more general.

The generalized Monod equation proposed by Han and Levenspiel tries to cover most of the situations, in particular the inhibition by the substrate, the product, or the cells. Its form is

$$\mu = \mu_m \left(1 - \frac{C_i}{C_i^*}\right)^n \frac{S}{K_S\left(1 - \frac{C_i}{C_i^*}\right)^m + S} \tag{10.49a}$$

$$\mu = k_{\text{obs}} \frac{S}{k_{\text{Sobs}} + S} \tag{10.49b}$$

C_i is the concentration of the inhibitor, C_i^* is the critical concentration of the inhibitor to completely stop the process (in this case, cell growth). n and m are constants, usually related to the toxic power of the inhibitor. Substituting in equation or generalized C_i by S, P, or X, it could model the inhibition by the substrate, product, or the cells themselves. As can be seen, when $C_i \ll C_i^*$, the equations are reduced to that of Monod.

10.3.3.2 Kinetics of Maintenance

Although the Monod kinetics is one of the most widely used and provides a reasonable description of cell growth in the function of the substrate available for most cases, these equations do not include the death phase. This can be changed with the addition or an extra term that reflects the consumption of cellular mass to produce energy for maintenance:

$$\mu = \mu_m \frac{S}{K_S + S} - k_d \tag{10.50}$$

The constant k_d is known as the coefficient of endogenous breathing. At high growth rates, the breathing virtually does not affect the formation of biomass; however, in an environment with low levels of nutrients, it becomes significant and must be included in the kinetics expressions. In any case, the rate of cell death can be calculated as

$$r_d = k_d \cdot X \tag{10.51}$$

Example 10.3 ***Bacterial Growth***
A strain of Papallacta bacteria is going to be used as precursor of an enzymatic broth, which studies show can be used for the degradation of cellulosic compounds and hydrocarbons. The chosen microorganism is a thermophilic bacterium isolated from the thermal sources of Papallacta that, according to elemental analysis, fits the formula $C_{4,17}H_{7,21}O_{1,79}N$. The microbiological reports indicate that the source of nitrogen for growth is nitrate, but it is also possible to grow on ammonium, and it also indicates that the bacteria can grow in an aerobic or anaerobic medium (facultative bacteria, which indicates that the electron acceptor may or may not be oxygen) and use as a source of carbon (electron donor) monosaccharides of 5 or 6 carbon atoms, which could well come from cane juice or molasses. In a 1-l laboratory batch reactor, the kinetics of bacterial growth is determined using glucose as substrate. The following data is obtained:

Time (h)	Concentration biomass (g/l)	Concentration glucose (g/l)
0	0.64	30
5	1.95	27.4
10	4.21	23.6
15	5.54	21
20	6.98	18.4
25	9.5	14.8
30	10.3	13.3
35	12	9.7
40	12.7	8
45	13.1	6.8
50	13.5	5.7
55	13.7	5.1

Determine the values of K_S and μ_m of the Monod equation.

Solution
From the data presented, we can calculate the rate of biomass production and substrate (glucose) consumption using the finite differences approximation, i.e.:

$$\frac{dX}{dt} = \frac{X^{t+1} - X^t}{\Delta t} \quad \text{and} \quad \frac{dS}{dt} = \frac{S^{t+1} - S^t}{\Delta t}$$

Although this is usually valid for small values of Δt, in the data we have only time increments of five or more hours, so we have no choice. Using the approximation, we can calculate the first two columns of the following table. The last two columns are calculated as indicated in the table:

Time (h)	$r_X = (1/V) \cdot dX/dt$ (g/(h l))	$r_S = -(1/V) \cdot dS/dt$ (g/(h l))	$\mu = r_X/X$ (h^{-l})	r_X/r_S (g cells produced/g substrate)
0	0.262	0.52	0.4094	1.98
5	0.452	0.76	0.2318	1.68
10	0.266	0.52	0.0632	1.95
15	0.288	0.52	0.0520	1.81
20	0.504	0.72	0.0722	1.43
25	0.16	0.3	0.0168	1.88
30	0.34	0.72	0.0330	2.12
35	0.14	0.34	0.0117	2.43
40	0.08	0.24	0.0063	3.00
45	0.08	0.22	0.0061	2.75
50	0.04	0.12	0.0030	3.00

As we can see, the rate of production of cells is continuously increasing, in such a way that a phenomenon similar to autocatalysis is found. The amount of cells produced by a gram of substrate is also continuously increasing. Following Monod kinetics, we have that (from Eq. (10.48)):

$$\frac{1}{\mu} = \frac{1}{\mu_m} + \frac{K_S}{\mu_m} \cdot \frac{1}{S}$$

So, a graph $1/\mu$ vs. $1/S$ will give the desired parameters. In the following plot, we can see the corresponding data graphically:

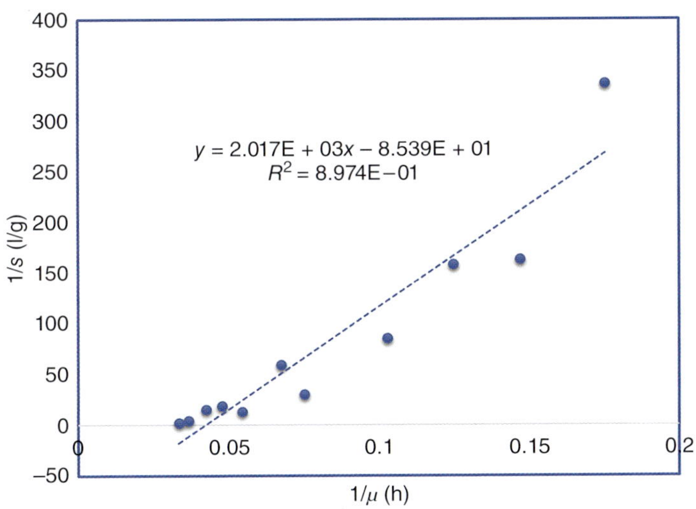

We can see that the data seems to fit the Monod kinetics, as a straight line is more or less obtained, but from the slope and ordinate (fitted using least-squares approximation), a negative value of the maximum growth rate is found. This is obviously not possible, and a better approximation is needed. From the data in the former table, the maximum value of μ is approximately $0.41\ \text{h}^{-1}$, which we will use as the value of μ_m. Making the line obtained to pass through the point (0, 0.41), the slope changes slightly and becomes 1225: therefore:

$$\frac{K_S}{\mu_m} = 1225\ (\text{g h})/\text{l}$$

So we can take a value of $K_S = 1225 \times 0.41 = 501.5$ g/l. With these values of the constants, let us finally see the fit by representing the experimental and calculated values of the growth rate vs. concentration of glucose.

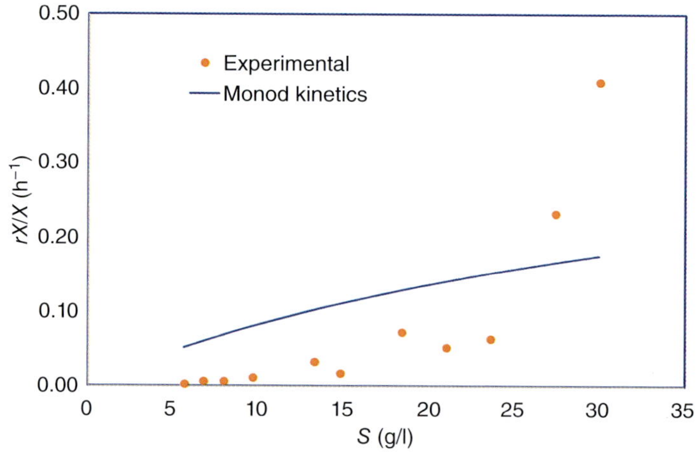

We see that the approximation gives values of the growth rate in the order of those experimental, but the model is not good enough to represent this bioreaction.

10.4 Immobilization of Enzymes and Cells: Mass Transfer Effects

The immobilization of a biocatalyst, whether it is an enzyme or a cell, consists of its location in a defined region of space, while maintaining a desired catalytic activity and, in the specific case of the cells, very possibly also its viability. A general characteristic of any system with immobilized biocatalysts is that the transport of substrates and products is produced by diffusional mechanisms.

The main methods of immobilization of biocatalysts can be divided into five large groups, depending on the mechanism on which they are based:

1. *Adsorption*: The immobilization is produced by interaction of the ionic type or weak attraction forces on the surface of the support.

2. *Covalent link*: The immobilization is also performed on the surface of a support, but the interaction between the biocatalyst and the support is by covalent bonding.
3. *Cross-links and self-immobilization*: In this group, there is no support itself. The particles of the immobilized biocatalyst are achieved by direct interaction, either by covalent binding or, in the case of some cells, by flocculation processes.
4. *Entrapment*: Formation of a three-dimensional structure, in the presence of the biocatalyst, which is trapped uniformly inside it.
5. *Systems with membranes (microencapsulation, preformed membranes)*: In both cases, the biocatalyst is immobilized inside a space limited by a membrane. In the first case, the microcapsules are produced in the presence of the biocatalyst, which is incorporated in its interior. In the second case, the biocatalyst is introduced into a system with preformed membranes.

The immobilization of biocatalysts presents a series of advantages:

- It allows the continuous use of the biocatalyst, since it is easily retained inside the reactor, while maintaining a continuous flow of liquid in and out. This means that in the case of cells, the continuous reactors can operate with flow rates higher than those corresponding to the wash limit observed when operating with free cells. In the case of batch reactions, the immobilization favors the reuse of the biocatalyst.
- Another important aspect is that with the immobilization the biocatalyst concentration can be significantly increased, with respect to a process in which it is in suspension. With this it is deduced that in systems operating continuously, the immobilized biocatalysts allow to increase sensibly the productivity of the corresponding bioreactors, since they allow to operate at a higher concentration of catalyst (and, therefore, greater conversion of substrate), and with higher flow rates.

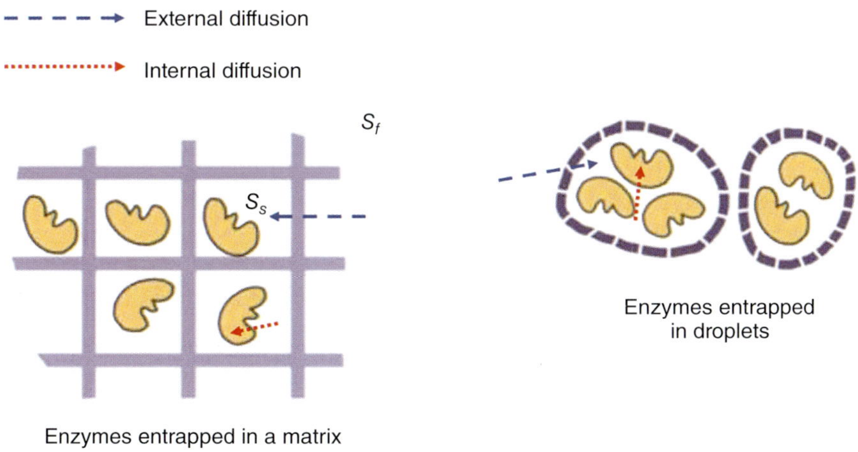

Figure 10.6 Scheme of immobilization of enzymes in different structures.

In the case of immobilized enzymes, it is common to observe an increase in their stability and resistance to denaturation and proteolysis when they are immobilized. In the case of cells, the effects of accidental contamination of the process are usually reduced, given the population of contaminating cells that are in suspension. And at a much lower concentration than the population of immobilized cells, it can be eliminated much more easily from the reactor.

On the other hand, the use of immobilized biocatalysts also presents some drawbacks:

- In the first place, it is necessary to consider that the activity of a cell or an enzyme can be directly affected by the conditions under which the immobilization process is carried out, losing part of its activity or its viability.
- The rate of diffusion of substrates and products within the system of immobilized biocatalysts can limit their catalytic activity and their efficiency, when this rate is slower than the rate of the transformation that is being carried out.
- Finally, from the point of view of the overall process, it must be borne in mind that the immobilization means the incorporation of a new stage to it, and therefore an increase in its complexity, which has to be compensated for by clear productivity increases, improvements in the separation between the product and the biocatalyst, and longer operation times.

The decision to use enzymes or cells as the main groups of immobilized biocatalysts depends fundamentally on the type of conversion that is to be carried out. When it is a conversion involving a single stage, it is normally more appropriate to use purified enzymes, since their cost can be reasonable and they have sufficiently good activities and stabilities. In addition, these preparations are simple and well-defined systems. In those more complex transformations, in which multiple stages are required, involving a large number of enzymes and their cofactors, the use of whole cells is indicated. In this case, it is not necessary to carry out a preliminary extraction and purification of the enzymes, which is often expensive. On the other hand, the cells have other enzymes and active metabolic routes in addition to those that intervene in the transformation of interest, which can cause the appearance of unwanted by-products, or even the modification or degradation of the desired final product.

An important aspect is the flexibility of cells as biocatalysts. Different systems can be prepared from cells, depending on whether the transformation involves only one enzyme (in this case, dead cells can be used permeabilized as if they were enzymes, but without the need for purification) or multiple stages, for which it is necessary to use viable cells.

An intermediate situation between the two cases described, although much less frequent, is the immobilization of a multienzyme system, interesting in those cases in which a reduced number of enzymes are involved.

Regarding the main types of bioreactors in which the immobilized biocatalysts are used, which can be similar to heterogeneous catalysis reactors, the same type of reactors are used: fixed-bed or fluidized-bed reactors, reactors with membranes (which, as mentioned, constitute at the same time a form of immobilization), and, in a smaller proportion, the agitated reactors, which can be used in very specific cases, since agitation usually severely affects the physical integrity of most particles of immobilized biocatalysts.

10.4.1 Effect of Limitation by Internal Diffusion

The influence of the external mass transfer on the global reaction rate is classically expressed using the concept of efficiency factor, similar to that used for internal mass transport in catalytic reactors (see Chapter 9). The physical meaning of this factor is the quotient between the observed reaction rate (therefore, the one that takes place) and the reaction rate that would be had in the absence of limitations by external matter transfer, that is, if the concentration on the surface was the same as that of the fluid phase:

$$\eta_e = \frac{\text{observed reaction rate}}{\text{rate with no limitation by external mass tranfer}} \tag{10.52}$$

Figure 10.6 shows an scheme of the diffusion steps ocurring in an immobilized enzymatic system. The process will be produced at the rate, using the Michaelis–Menten equation:

$$r = k_C(S_f - S_s) = r_m \frac{S_s}{K_M + S_s} \tag{10.53}$$

The rates of diffusion and reaction should be the same. So, reordering the equations, we can do:

$$r = \frac{(S_f - S_s)}{\frac{1}{k_C}} = \frac{S_s}{\frac{(K_M+S_s)}{r_m}} = \frac{S_f}{\frac{1}{k_C} + \frac{(K_M+S_s)}{r_m}} \tag{10.54}$$

where the subscripts refer to the fluid (f) and the surface of the particle(s).

In Eq. (10.52), the rate with no limitation by external mass transfer equals $r_m \frac{S_f}{K_M+S_f}$, so the efficiency factor is

$$\eta_e = \frac{\frac{S_f}{\frac{1}{k_C} + \frac{(K_M+S_s)}{r_m}}}{r_m \frac{S_f}{K_M+S_f}} \tag{10.55}$$

Using the following definitions:

$$\xi = S_s/S_f \quad \text{and} \quad \kappa = S_s/K_M$$

and solving for the efficiency factor:

$$\eta_e = \frac{\left(\frac{\kappa\xi}{1+\kappa\xi}\right)}{\left(\frac{\kappa}{1+\kappa}\right)} \tag{10.56}$$

In the case of gel-containing biocatalysts or where the biocatalyst has been immobilized in the pores of a support, it is not expected that the reaction will only occur on the surface. Similarly, it is expected that substrate consumption by a microbial sheet or a floccule will occur at a certain depth in the microbial mass. This situation is more complicated than in the case of surface immobilization; in this case, the transport and the reaction occur in parallel. By analogy with heterogeneous catalysis, the substrate flow is related to the reaction rate with the use of an efficiency factor. In general, the substrate flow will be given by

$$r_X = \frac{dX}{dt} = \eta \cdot r_m \frac{S}{K_S + S} \tag{10.57}$$

where r_m is the maximum reaction rate per volume of immobilized catalyst.

The simplest case to consider is that of a layer of microbes or enzymes that are immobilized uniformly in a block of support material with infinite area but finite depth. Since it is assumed that the reaction rate is given by a Michaelis–Menten kinetics, a material balance for any point of the block gives

$$D_e \frac{d^2S}{dx^2} = r_m \frac{S}{K_M + S} \tag{10.58}$$

where D_e is the effective diffusion for the substrate in the block and "x" is the distance measured from the surface of the block. This equation can be made dimensionless by adding $\kappa = S/K_M$ and $Z = x/L$, where L is the thickness of the block. With this previous equation, the system is defined by

$$\frac{d^2\kappa}{dZ^2} = \Phi^2 \frac{\kappa}{1 + \kappa} \tag{10.59}$$

where Φ is the Thiele module defined by

$$\Phi = \mathrm{L}\sqrt{\frac{r_m}{D_e \cdot K_M}} \tag{10.60}$$

for a system of immobilized enzymes. For a microbial layer, the equation defining the system is similar to (10.58), but substituting r_m by μ_m and K_M by K_S. Finally, we can get

$$\Phi = \mathrm{L}\sqrt{\frac{\mu_m \cdot X}{Y_{X/S} \cdot D_e \cdot K_S}} \tag{10.61}$$

for a microbial layer.

Equation (10.59) can be solved by numerical integration using the usual boundary conditions:

$$\kappa = \kappa_s \quad \text{when } Z = 0$$

$$\frac{d\kappa}{dZ} = 0 \quad \text{when } Z = 1$$

to obtain the curves' rate position or concentrationposition. The aspect of these profiles is similar to those discussed in Chapter 9; for internal diffusion resistance, see Figure 9.5. The family of curves obtained shows that the global reaction rate decreases when the value of the Thiele module increases. The rate is controlled by the reaction kinetics at low values of the Thiele module.

10.5 Bioreactors

Bioreactors or fermenters are classified according to the terminology used in the design of chemical reactors in:

1. Stirred Tank: Can Be Batch, Fed-batch (with Discontinuous Feed), and Continuous (CSTR or CSTB).

Batch: it is the most traditional and the most widely used on an industrial scale. The necessary operating time can range from hours to several weeks depending on the conversion and the operating conditions. During the operation it is important to avoid contamination, maintain agitation, and control pH and temperature. They operate with low cell density, especially in the initial period, and the feeding of high concentrations of substrate must be avoided to prevent inhibitions by the substrate or by the product. It is used mostly in the food, pharmaceutical, and biotechnology industry, in general, since it is easy to reach and maintain aseptic conditions during the operation. This fact is important when trying to maintain processes that use very nutrient-rich media. Typical examples are the production of acetic acid, vitamin C and C_{12}, baking yeast, penicillin, and anaerobic sludge digestion (microorganisms produced in continuous wastewater treatments).

The disadvantages that batch reactors present are typical of this type of systems: loss of efficiency due to start-ups and shutdowns, lack of product homogeneity between loads, and difficulty in implementing energy integration schemes.

In general, the design variables are as follows:

1. Dimensions of the reactor,
2. Operation time,
3. Initial concentrations,
4. Volume of microbial mass per unit of reactor volume,
5. Power and aeration,
6. Heat transfer surface.

Fed-batch: the substrate is fed in successive loads and no product is removed, thus varying the volume of reaction. By varying the speed and concentration of the feed, the concentration of one or more nutrients or substrate in the culture medium is controlled or altered. This increases overall productivity. This strategy is applied in the production of bread yeast and antibiotics.

Continuous CSTR (or CSTB): this continuous stirred tank reactor (or bioreactor) is also called the chemostat. This reactor is similar to the batch reactor but allows the entry and exit of a flow. However, the chemostat mainly operates in a steady state, which implies uniform conditions not only from the geometric point of view but also with respect to time. The control of the variables of operation is much easier, but the feasible values of the variables are limited; if the flow is too high, it can happen that the production of microorganisms is lower than the number of those that are dragged by the output current and, consequently, if the situation persists it happens that no microorganisms remain inside the reactor. It is said that they have been washed (washout) from the fermenter. The situation can be lightened by adding microorganisms similar to those of the reactor in the feed. The natural source of these microorganisms is obviously the output current. Thus, the effluent is passed through a settler or a centrifuge, and a concentrated solution of microorganisms is separated, which is largely recycled to the reactor. This procedure not only prevents the washing of cells but also increases production as it increases the residence time of the microorganisms in the reactor.

2. Tubular Bioreactor (TB)

It should be noted for this bioreactor that the operation is not feasible when a sterile medium is fed, given that it is an autocatalytic process. This drawback is solved with a recirculation or with immobilized microorganisms. According to the disposition of the microorganism, they are divided into the following:

- *TB with flocs in suspension:* A floccule is an aggregate of microorganisms with a size an order of magnitude higher than that of the microorganism. The mechanism controlling this size is not very well known but can act on the agitation and flocculating agents such as aluminum chloride and calcium. A fixed-bed operation is not viable. The viable operation involves the entrainment of microorganisms and, consequently, a necessary feeding of them.
- *Tubular film bioreactor*: The microorganisms are arranged in the form of a film that grows on the surface of an inert filler. The thickness of the film is controlled by removing excess microbial production normally by mechanical methods. It does not present dragging problems. It is widely used in the treatment of wastewater and is called percolating or drip filter. It is an inert bed on which the microbial film is formed and in which the liquid circulates in a downward flow by gravity, while the oxygen usually flows upward. Its height varies between 2 and 15 m. The control of the thickness of the film is carried out by the organisms themselves, since by increasing oxygen diffusion problems are created, so that the microorganisms cannot metabolize the substrate, lose adherence, and eventually become detached.

10.5.1 Continuous Stirred Tank Bioreactor (CSTB)

Given the reactor scheme shown in Figure 10.7, the first step in the analysis or design, is to do the mass balances. It is first necessary to indicate that the cell balance is carried out on the so-called viable cells, which are those capable of dividing and forming a colony in the culture medium. So, you have

$$\text{Viable cells}: \frac{d(V \cdot X)}{dt} = Q_0 X_0 - QX + V(r_X - r_d) \tag{10.62}$$

r_d being given by Eq. (10.57). In this kind of reactors, usually $dV/dt = 0$ (as $Q = Q_0$), and in the steady state ($\mathrm{d}X/\mathrm{d}t = 0$) we have

$$0 = Q_0(X_0 - X) + V\mu X - Vk_d X \tag{10.63}$$

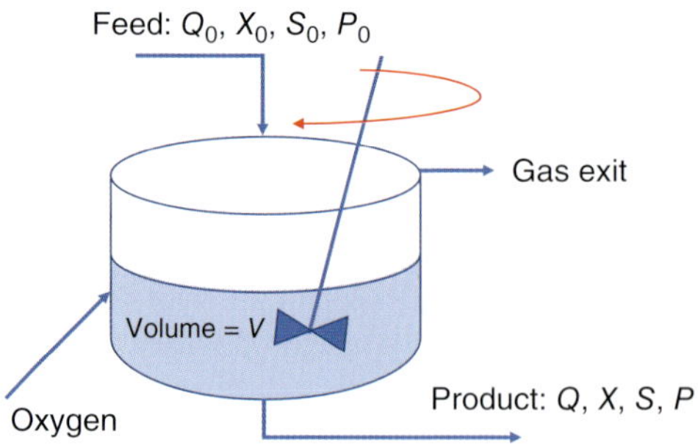

Figure 10.7 Scheme of a continuous stirred tank bioreactor (CSTB).

When working in the exponential growth ($k_d = 0$), and with sterile feed ($X_0 = 0$), we can write

$$Q_0 X = V\mu X = V\mu_m \frac{S}{K_S + S} X \tag{10.64}$$

Usually, the term Q_0/V is called "dilution rate" (D) and is equivalent to the inverse of the residence time. Using this definition:

$$D = \mu_m \frac{S}{K_S + S} = \mu = \frac{r_X}{X} \tag{10.65}$$

Therefore, $D = \mu$ in the steady state, so the specific rate of growth can be controlled by manipulating the dilution rate.

Introducing the Monod equation in the balance:

$$\begin{aligned} D &= \mu_m \frac{S}{K_S + S} \\ S &= D \cdot \frac{K_S}{\mu_m - D} \end{aligned} \tag{10.66}$$

For the substrate, we can say

$$\begin{aligned} Y_{X/S} &= \frac{X_0 - X}{S_0 - S} = \frac{-X}{S_0 - S} \\ S &= S_0 + \frac{X}{Y_{X/S}} \end{aligned} \tag{10.67}$$

and, equally, for the product:

$$Y_{P/S} = \frac{P_0 - P}{S_0 - S} \tag{10.68}$$

10.5.1.1 Influence of the Dilution Rate. Calculation of the Bioreactor Wash

For small flows and fixed volume, the value of D is also small and tends to zero when it does flow, $Q_0 \to 0$; $D \to 0$. In consequence of Eq. (10.65), it follows that S also tends to zero, $S \to 0$, which means that the substrate has been consumed practically since the microorganisms have had enough time to do it. When the flow increases and, therefore, D increases, it is observed that the substrate concentration increases first slightly linearly and then abruptly when D approaches μ_m.

From Eq. (10.66) it follows that when $D \to \mu_m$, $S \to \infty$, which does not make sense, since $S \leq S_0$. At most, the concentration of substrate can be equal to the initial input. In parallel, the concentration of biomass first decreases slowly and then abruptly tends to zero when $S \to S_0$.

That is, before $D = \mu_m$, $S = S_0$, and $X = 0$ have already been produced, which means that the cells have disappeared from the bioreactor, and the washing of the same has taken place. The value of the dilution rate for which such a limit situation is called "washout dilution rate" and is determined from Eq. (10.67) by doing $S = S_0$:

$$D_{\max} = D_{\text{washout}} = \mu_m \frac{S_0}{K_S + S_0} \tag{10.69}$$

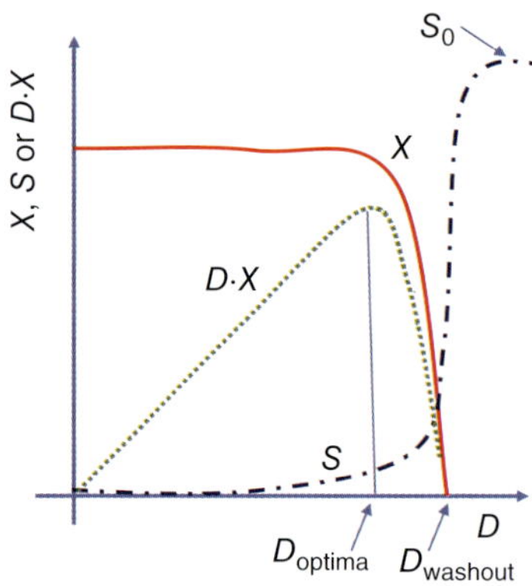

Figure 10.8 Variation of substrate, biomass, and cell production concentrations with the dilution rate.

Figure 10.8 reproduces the situation described where the sensitivity of the concentrations varies with the dilution rate. On the other hand, in the case of a non-sterile feed, $X_0 \neq 0$, no washing of the reactor takes place since when $D \to \infty$, $X = X_0$.

Example 10.4 ***Optima Microbial Growth with Monod Kinetics***
The microbial growth of *Escherichia coli* in glucose follows the kinetics of Monod, with $K_S = 4\,\text{g/m}^3$ and maximum velocity of $1.33\,\text{g/(m}^3\,\text{h)}$ cells. The value of $Y_{X/S}$ is 0.1 g/g. Determine the flow rate of glucose feed (concentration $S_0 = 60\,\text{g/m}^3$) to a continuous stirred tank fermenter of 1 m^3 that would give a maximum rate of glucose consumption and cell production. Determine these values.

Solution
In this case, we are looking for the maximum in Figure 10.8. So, in that point:

$$\frac{d(D \cdot X)}{dD} = 0$$

and we can write

$$S = D \cdot \frac{K_S}{\mu_m - D} \tag{10.66}$$

$$X = (S_0 - S) \cdot Y_{X/S}$$

and then:

$$\frac{d(D \cdot (S_0 - S) \cdot Y_{X/S})}{dD} = \frac{d\left(D \cdot \left(S_0 - D \cdot \frac{K_S}{\mu_m - D}\right) \cdot Y_{X/S}\right)}{dD} = 0$$

Deriving this expression:

$$\left(S_0 + K_S - D_{\text{opt}} \cdot \frac{\mu_m^2 \cdot K_S}{(\mu_m - D_{\text{opt}})^2}\right) \cdot Y_{X/S} = 0$$

Rearranging and clearing for the optimum dilution rate:

$$D_{\text{opt}} = \mu_m \cdot \left(1 - \sqrt{\frac{K_S}{K_S + S_0}}\right) = 1.3 \cdot \left(1 - \sqrt{\frac{4}{4 + 60}}\right) = 0.325\,\text{s}^{-1}$$

The rate of the microbial growth at this dilution rate is

$$r_{\text{opt}} = \mu_m \frac{S_{\text{opt}}}{K_S + S_{\text{opt}}} X_{\text{opt}}$$

$$S_{\text{opt}} = D_{\text{opt}} \cdot \frac{K_S}{\mu_m - D_{\text{opt}}} = \frac{0.325 \cdot 4}{1.3 - 0.325} = 1.33\ \text{g/m}^3$$

$$X_{\text{opt}} = (S_0 - S_{\text{opt}}) \cdot Y_{X/S} = (60 - 1.33) \cdot 0.1 = 5.86\ \text{g/m}^3$$

$$r_{\text{opt}} = 1.3 \frac{1.33}{4 + 1.33} 5.86 = 1.9\ \text{g/(h m}^3\text{)}$$

10.5.1.2 Cell Recirculation

A non-sterile feed is easily achieved by recirculating part of the previous product by separation and concentration of microorganisms. A schematic of this procedure is shown in Figure 10.9.

A balance of microorganisms to the bioreactor leads to

$$0 = Q_r X_0 - (Q_0 + Q_r) X_1 + r_X V = Q_r X_0 - (Q_0 + Q_r) X_1 + \mu X_1 V \qquad (10.70)$$

Using the definition of $D = Q_0/V$, and renaming $\alpha = Q_r/Q_0$ and $\beta = X_0/X_1$:

$$D = \frac{\mu}{1 - \alpha(\beta - 1)} \qquad (10.71)$$

whereas the microorganisms in the recirculation stream are more concentrated than in the output current of bioreactor results $\beta > 1$. Consequently, now the dilution rate is higher than the specific speed of growth and, therefore, with the same rate of growth, the use of cell recirculation allows to treat more food per unit of time and volume than when there is no recirculation. Comparing with Eq. (10.65), we have

$$D_{\text{with recirculation}} = \frac{D_{\text{without recirculation}}}{1 - \alpha(\beta - 1)} \qquad (10.72)$$

10.5.2 Tubular Fermenters with Flocs

As in the case of ordinary reactors, an overall balance for the entire reactor is not valid as the conditions there are not uniform with the position in the reactor. In

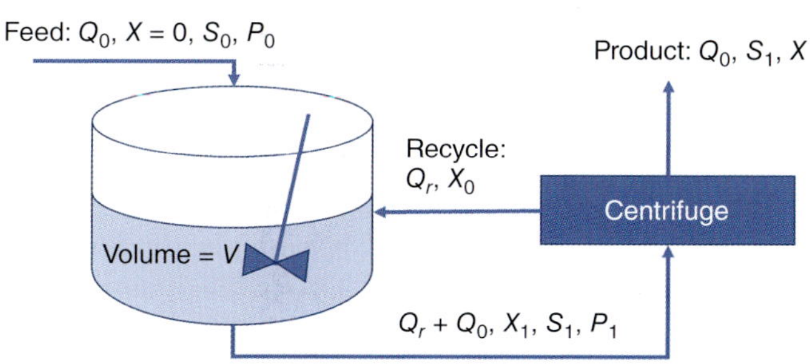

Figure 10.9 Continuous agitated tank fermenter with recirculation of microorganisms.

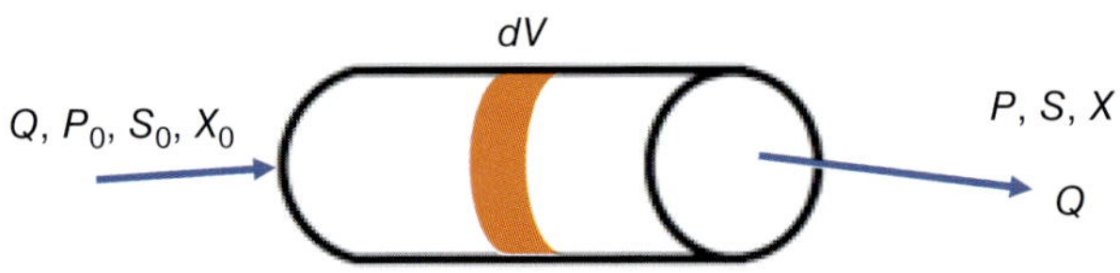

Figure 10.10 Scheme of a tubular bioreactor.

the ideal tubular flow-through fermenter (Figure 10.10), the substrate concentration decreases from entry to exit, and that of cells and products increase.

The most important factor to consider is the need for a non-sterile feed since, otherwise, microorganisms cannot exist inside. In a steady state, conditions vary with position but not with time. So, in an infinitesimal volume of bioreactor, the conditions and the rate are practically constant, and the balance for biomass in this volume is expressed as follows:

$$0 = Q \cdot X - Q(X + dX) + r_X \cdot dV$$

$$Q \cdot dX = r_X \cdot dV \tag{10.73}$$

Integrating between $V = 0$, $X = X_0$, $V = V$, $S = S$ and the design equation is obtained

$$\frac{1}{D} = \frac{V}{Q} = \int_{X_0}^{X} \frac{dX}{r_X} = \int_{S_0}^{S} \frac{dS}{Y_{S/X} \cdot r_X} \tag{10.74}$$

which will allow us to analyze and design the ideal tubular fermenter. It is required to integrate, of course, make explicit the reaction rate as a function of the corresponding concentration.

10.5.2.1 Tubular Fermenter with Recirculation and Monod Kinetics

Considering the substrate balance in the mixing point of the inlet of the reactor (see Figure 10.11):

$$Q \cdot S_0 + Q_r \cdot S = (Q + Q_r) \cdot S_{\text{in}} \tag{10.75}$$

S_{in} being the concentration of substrate entering the reactor. Expressing $(Q + Q_r) = Q{\cdot}(1 + R)$, we have, from Eq. (10.74):

$$\frac{1}{D} = \frac{V}{Q} = (1 + R) \int_{S_{\text{in}}}^{S} \frac{dS}{Y_{S/X} \cdot r_X} = (1 + R) \int_{S_{\text{in}}}^{S} \frac{(K_S + S)}{Y_{S/X} \cdot \mu_{\max} \cdot S \cdot X} dS \tag{10.76}$$

10.5.3 Fed-batch Bioreactor

Fed-batch culture consists of a biological process where one or more substrates are fed during operation, but the products remain in the reactor until the end of the run. From the reactor-type point of view, it is a semi-batch reactor system. In

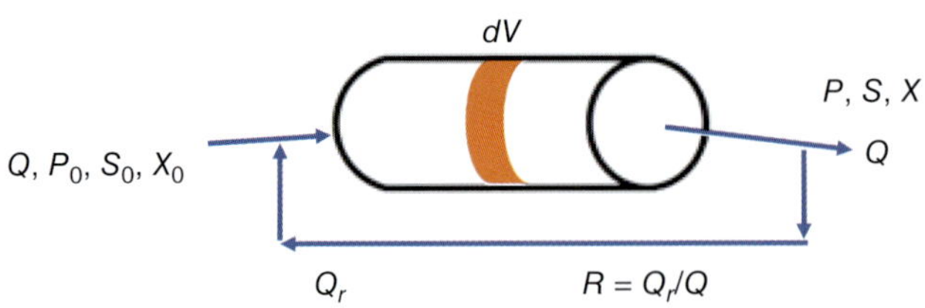

Figure 10.11 Scheme of a tubular bioreactor with recycle.

some cases, all the nutrients are fed into the bioreactor. An alternative description of the method is that of a culture in which a base medium supports initial cell culture and a feed medium is added to prevent nutrient depletion.

The advantage of the fed-batch culture is that one can control the concentration of the fed substrate in the culture liquid at arbitrarily desired levels (in many cases, at low levels).

Generally speaking, fed-batch culture is superior to conventional batch culture when controlling concentrations of a nutrient (or nutrients) affect the yield or productivity of the desired metabolite.

Assuming an input of Q_0 (flow rate), X_0 (concentration of cells) equal to zero, S_0 (concentration of substrate), and a V_0 (initial volume of the reactor), the mass balances are

$$\frac{d(XV)}{dt} = X\frac{dV}{dt} + V\frac{dX}{dt} = V \cdot r_X \quad \text{(Cells), (Eq. 10.77)}$$

$$\frac{d(PV)}{dt} = P\frac{dV}{dt} + V\frac{dP}{dt} = V \cdot r_P \quad \text{(Product), (Eq. 10.78)}$$

$$\frac{d(SV)}{dt} = S\frac{dV}{dt} + V\frac{dS}{dt} = Q_0S_0 - V \cdot r_S \quad \text{(Substrate), (Eq. 10.79)}$$

One aspect of the fed-batch model is that volume is not constant; therefore, the cell, product and substrate are subject to a dilution effect. Mathematically, the chain rule of differential calculus is applied to bear in mind the variation of the volume, $Q_0 = dV/dt$ and then, rearranging:

$$\frac{dX}{dt} = -\frac{Q_0 \cdot X}{V} + r_X \tag{10.77}$$

$$\frac{dP}{dt} = -\frac{Q_0 \cdot P}{V} + r_P \tag{10.78}$$

$$\frac{dS}{dt} = \frac{Q_0}{V}(S_0 - S) - r_S \tag{10.79}$$

Example 10.5 ***Fed-batch Reactor Simulation***

A fed-batch reactor is being set up in order to produce a product "*P*" in a cell culture. The following parameters were chosen: input cell concentration = 0.05 g/l, input substrate concentration = 10 g/l, initial volume = 1 l, and flow rate input 1 l/h. The feed has no any concentration of the desired product. Other data is available: $\mu_m = 0.20\ \text{h}^{-1}$; $K_S = 1.00$ g/l; $Y_{X/S} = 0.5$ g/g; and $Y_{P/X} = 0.2$. Simulate the behavior of this reactor calculating the evolution of all concentrations and the volume with time.

Solution

The present design is similar to that just mentioned, but the feed of the reactor is not sterile and has a known concentration of cells $X_0 = 0.05$ g/l. So, in this case, Eq. (10.77) is not valid, and the balance of cells is

$$\frac{d(XV)}{dt} = QX_0 + V \cdot r_X$$

Using the chain rule for derivation and bearing in mind that $dV/dt = Q$, we can see that:

$$V\frac{dX}{dt} = Q(X_0 - X) + V \cdot r_X$$

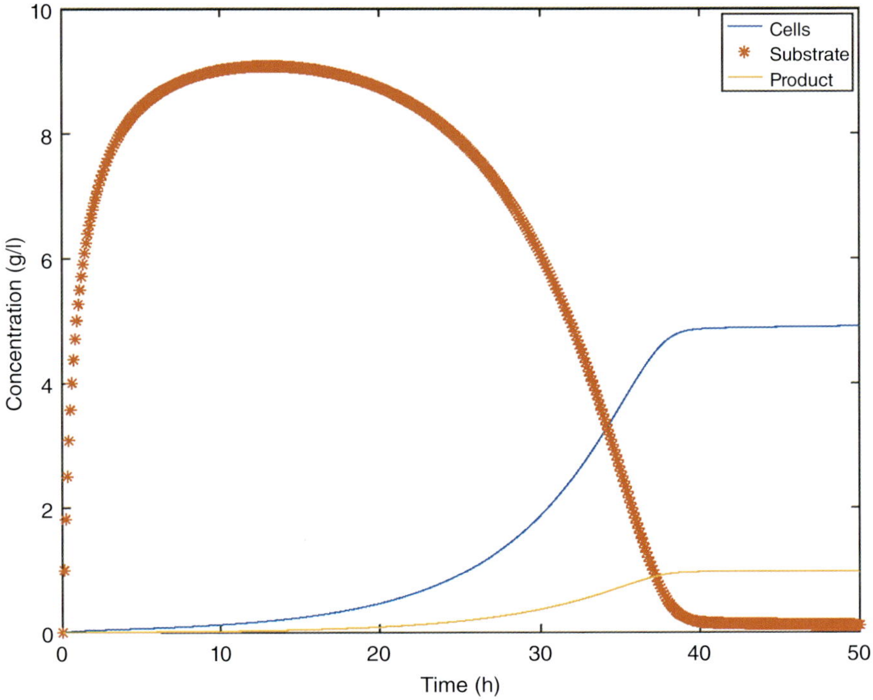

Figure 10.12 Solution of Example 10.5.

and then, using the previous equation, we can write

$$\frac{X^{t+1}-X^t}{\Delta t}=\frac{Q_0\cdot(X_0-X)}{V}+r_X \quad X^{t+1}=X^t+\Delta t\left(\frac{Q_0\cdot(X_0-X^t)}{V}+r_X\right)$$

In the same way:

$$P^{t+1}=P^t+\Delta t\left(-\frac{Q_0\cdot P}{V}+Y_{P/X}\,r_X\right)$$

$$S^{t+1}=S^t+\Delta t\left(\frac{Q_0}{V}(S_0-S)-\frac{r_X}{Y_{X/S}}\right)$$

Using an approach similar to previous programs, a valid Matlab calculation is given. The results are shown in Figure 10.12.

Program Listing 10.1

```
clear all
close all
mumax = 0.20; % h-1 Maximum Growth Rate
Ks = 1.00; % g/l Monod Constant
Yxs = 0.5; % g/g Cell yield
Ypx = 0.2; % g/g Product yield
S0 = 10.0; % g/l Feed Substrate concentration
Q=1; % L/h Input Flowrate

V0=1; % L Initial Volume
```

```
  V(1)=V0;
  X(1) = 0; % g/l Initial concent. of cells in the tank
  X0=0.05; % g/l concentration of cells in the input
current
  S(1) = 0; % g/l
  P(1) =0; % g/l
  t=0; % h
  tfinal=50; % h
  inct=0.1; % h

  i=1;
  tv=0:inct:tfinal; % Values of time used in the
calculation

while t<tfinal
   i=i+1;
   t=t+inct;
   rx=mumax*S(i-1)*X(i-1)/(Ks+S(i-1));
   rp=rx*Ypx;
   rs=rx/Yxs;
   X(i)=X(i-1)+inct*(Q*(X0-X(i-1))/V(i-1)+rx);
   P(i)=P(i-1)+inct*(-Q*P(i-1)/V(i-1)+rp);
   S(i)=S(i-1)+inct*(Q*(S0-S(i-1))/V(i-1)-rs);
   V(i)=V(i-1)+Q*inct;
end
plot(tv,X,tv,S,'*',tv,P)
legend('Cells','Substrate','Product')
xlabel('Time (h)')
ylabel('Concentration (g/l)')
```

As we can see in the figure, the consumption of substrate is too high when time approaches to 40 hours, so the best operation time is less than 40 hours, when the product concentration does not increase anymore and the concentration of cells behave in a similar way. In that case, the volume of the reactor needed will be approximately 40 l.

Example 10.6 ***CSTR Bioreactor with Recirculation of Cells***

The reaction of the previous example is done in a CSTR bioreactor with recirculation of cells. Determine the variation of the exit concentrations with the recycle relation, assuming that the concentration of cells in the recycling current is half of the concentration at the exit of the reactor. A volume of 10 l of reactor should be used. Repeat the simulation for a fixed value of recycle ratio and varying the concentration of cells at the exit of the reactor.

Solution

A scheme of such a system can be found in Figure 10.9. A balance of microorganisms to the bioreactor leads to

$$0 = Q_r X_0 - (Q_0 + Q_r)X_1 + r_X V = Q_r X_0 - (Q_0 + Q_r)X_1 + \mu X_1 V \tag{10.70}$$

Using the definition of $D = Q_0/V$, and renaming $\alpha = Q_r/Q_0$ and $\beta = X_0/X_1$:

$$D = \mu/(1 - \alpha(\beta - 1)) \tag{10.71}$$

In this example we have: $X_0 = 0.05$ g/l (note that in this case this concentration represents that of the recirculating current, Figure 10.9), $S_0 = 10$ g/l, $V = 10$ l (constant in this case), $Q_0 = 1$ l/h, $P_0 = 0$, $\mu_m = 0.20\ \text{h}^{-1}$; $K_S = 1.00$ g/l; $Y_{X/S} = 0.5$ g/g; $Y_{P/X} = 0.2$. With this data, we can calculate $D = (1\ \text{l/h})/10\ \text{l} = 0.1\ \text{h}^{-1}$.

The fraction of cells that are being recirculated can be calculated from

$$\text{Fraction of recirculated cells} = \frac{Q_r \cdot X_0}{(Q_r + Q_0) \cdot X_1} = \frac{\alpha \cdot X_0}{(\alpha + 1) \cdot X_1} = \frac{\alpha \cdot \beta}{(\alpha + 1)} \tag{10.80}$$

A mass balance in the centrifuge gives

$$(Q_r + Q_0) \cdot X_1 = Q_0 \cdot X + Q_r \cdot X_0 \tag{10.81}$$

and then:

$$X = X_1 \cdot [1 - \alpha(\beta - 1)] \tag{10.82}$$

From a global mass balance in the system:

$$0 = -Q_0 X + r_X V = -D \cdot X + \frac{\mu_m \cdot X_1 \cdot S_1}{K_S + S_1} \tag{10.83}$$

After substituting X by Eq. (10.82), we can get the relation giving the substrate concentration at the exit and in the recirculating current:

$$S_1 = \frac{D \cdot [1 - \alpha(\beta - 1)] \cdot K_S}{\mu_m - D \cdot [1 - \alpha(\beta - 1)]} \tag{10.84}$$

There are many ways to work in this type of system. One can change the value of the relation $X_1/X_0 = \beta$, and the system evolves accordingly. Let us take the value of $\beta = 0.5$ mentioned in the statement; in that case, we can use the previous equation to solve, and the following plot is obtained:

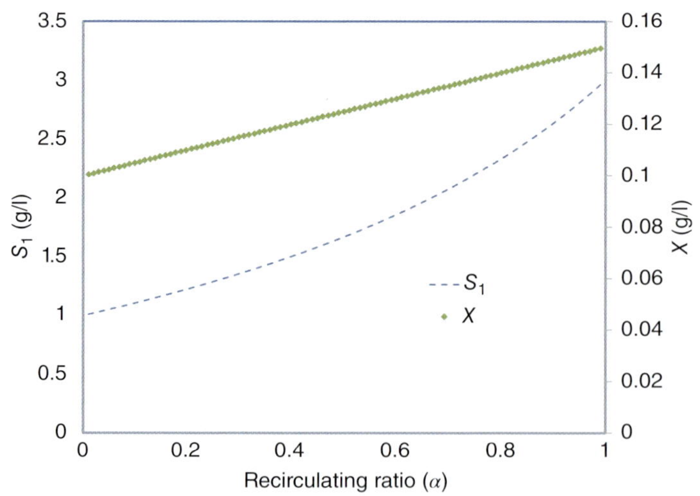

Note that it is possible to have a value of α higher than unity, as the Q_r can be higher than Q_0, but this usually is not useful and, furthermore, is expensive due to the high costs of impulsion of the currents. In the extreme case that α is infinity, the whole system will behave as a CSTR and it will be not possible to observe differences between X_1 and X_0.

If, on the other hand, the fixed parameter is $\alpha = 0.5$ (as an example), by varying the value of β we will get a different concentration X_1, if X_0 is fixed. Then, using the same equations, we can simulate the behavior of that system, and, in this way, we will find the next figure.

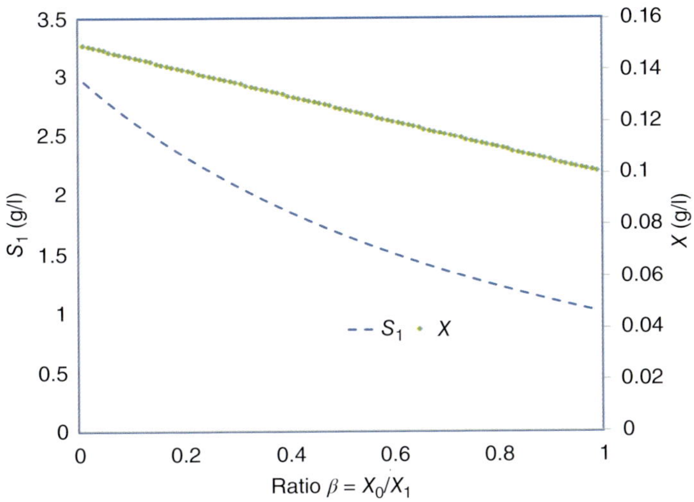

Bibliography

Doran, P.M. (2013). *Bioprocess Engineering Principles*. Elsevier/Academic Press.

Lee, K., Comolli, N., Punzi, V. et al. (2014). Survey: Matlab & Mathcad education in biochemical engineering. *Chemical Engineering Education* 48 (1): 59–61.

Levenspiel, O. (1999). *Chemical Reaction Engineering*, 3e. New York, NY: Wiley.

Index

Chemical Reactor Design: Mathematical Modeling and Applications, First Edition. Juan A. Conesa.